Lecture Notes in Geoinformation and Cartography

Publications of the International Cartographic Association (ICA)

More information about this series at http://www.springer.com/series/10036

Miljenko Lapaine · E. Lynn Usery
Editors

Choosing a Map Projection

Springer

Editors
Miljenko Lapaine
Faculty of Geodesy
University of Zagreb
Zagreb
Croatia

E. Lynn Usery
Center of Excellence for Geospatial Information Science
U.S. Geological Survey
Rolla, MO
USA

ISSN 1863-2246 ISSN 1863-2351 (electronic)
Lecture Notes in Geoinformation and Cartography
ISSN 2195-1705 ISSN 2195-1713 (electronic)
Publications of the International Cartographic Association (ICA)
ISBN 978-3-319-51834-3 ISBN 978-3-319-51835-0 (eBook)
DOI 10.1007/978-3-319-51835-0

Library of Congress Control Number: 2016963182

Printed on acid-free paper

This Springer imprint is published by Springer Nature
The registered company is Springer International Publishing AG
The registered company address is: Gewerbestrasse 11, 6330 Cham, Switzerland

Preface

The term "map projection" suggests the idea of a translucent globe with a light inside shining the map outlines onto a flat surface. As a matter of fact, two well-known and very useful projections for limited areas, the Gnomonic and Stereographic, are derived directly from this perspective model, but most projections are produced mathematically.

The great majority of world map projections may be grouped into classes; each class has a distinctive outline and distortion pattern. Except perspectives, they are all mathematical constructions that have a variety of attributes.

There are hundreds of ways to arrange the globe surface within an oval, rectangle, or circle; no one system is best for all maps. Choosing a projection from among the many options requires something like a cost-benefit analysis. For every desirable quality there is often some consequent drawback, and furthermore the distortions associated with any map projection are not uniformly arranged; some parts of a projection are always less distorted than other parts.

Anyone who looks at world maps, on the printed page or on TV, should learn to make allowances for the distortions of shapes, sizes, directions, and distances that are bound to occur. Similarly, a cartographer, designer, editor, or anyone who wants to make a map or have one made for some use should make the selection of a projection an important element in the creative process. Projections should be chosen to best serve the objectives, and just as important, to keep from making a serious error by selecting one that is inappropriate.

To display the whole Earth on a flat map is a complicated operation. Before anything else can be done one must select a map projection to serve as a base. There are many from which to choose in each of the basic classes, and each one can be constructed in any desired aspect. One must settle on the necessary attributes and then adopt an appropriate distribution of the inevitable distortion.

It is not easy to make a choice, but the advantages of providing the best possible portrait of the Earth are worth the effort.

One of the tasks of the Commission on Map Projections of the International Cartographic Association (ICA) for the period 2011–2015 was to promote proper use of map projections at all levels of education and for all map users. This book is the result of the four-year work.

In various publications on map projections, one can find a display of a great variety of map projections from which the choice of a particular map can be made. A well-chosen map projection is one which "extreme distortions are smaller than those in any other projection used to map the same area" and map properties match the purpose of the map. Yet, a map projection may have none of these general properties and still be satisfactory, or it can be widely used despite being considered a poor choice for a particular map.

Three small but valuable publications are:

(1) Arthur H. Robinson and the Committee on Map Projections (1986): Which Map is the Best? Projections for World Maps. Special Publication No. 1 of the American Cartographic Association, American Congress on Surveying and Mapping, Falls Church, Virginia.
(2) Arthur H. Robinson and the Committee on Map Projections (1988): Choosing a World Map: Attributes, Distortions, Classes, Aspects. Special Publication No. 2 of the American Cartographic Association, American Congress on Surveying and Mapping, Bethesda, Maryland.
(3) Arthur H. Robinson and the Committee on Map Projections (1991): Matching the Map Projection to the Need. Special Publication No. 3 of the American Cartographic Association, American Congress on Surveying and Mapping, Bethesda, Maryland. As e-publication of the Cartography and Geographic Information Society available since 1997 at https://courseware.e-education.psu.edu/projection/.

Although none of the three publications is new, they are very informative and educational. The first two publications were printed and sold out. The third one is also available in digital format. A proposal by M. Lapaine was for the ICA Commission on Map Projections to take these three publications as a model and a base, and after some consideration and modification, as necessary, prepare and publish a new book on map projection choice.

The book 'Choosing a Map Projection' has been prepared by experts, but with the intention to serve laymen. The authors of the book are prominent persons in the field of map projections, as are famous Waldo Tobler, Fritz C. Kessler, Sarah E. Battersby, Michael P. Finn, Keith C. Clarke, Vladimir S. Tikunov, Henrik Hargitai, Bernhard Jenny, and Nedjeljko Frančula, just to mention some of them.

The book editors were Miljenko Lapaine and E. Lynn Usery, Chair and Vice-Chair respectively, of the ICA Commission on Map Projections for the period 2011–2015.

It is our hope the reader is going to be stimulated to understand map projections and appreciate their versatility. It is aimed at both those who display maps and those who look at them. Computer-assisted cartographic techniques now make a variety of map projections (along with coastlines and boundaries) quickly and inexpensively available. What is still needed is a more critical concern for their intelligent use taking into account distortions that are immanent in any map projection.

Finally, we would first like to thank authors who contributed to this book by writing about its theme from various angles. We would also like to thank CaGIS for permitting us to publish the first three chapters. We thank Professors Emeriti Waldo Tobler and Nedjeljko Frančula who read and reviewed individual chapters. E. Lynn Usery was responsible for proofreading and Marina Viličić produced nearly 130 new illustrations for the first three chapters. Springer's team contributed to making this book a reality. We thank all of them very much.

Zagreb, Croatia — Miljenko Lapaine
Rolla, USA — E. Lynn Usery

Contents

Chapter 1
Which Map Is Best?

Projections for World Maps

Arthur H. Robinson and The Committee on Map Projections

1.1 Foreword

A map projection is what cartographers call the system by which the rounded surface of the Earth is transformed in order to display it on a flat surface. It is not easy. Numerous ways have been devised, but no matter how one does it, something will be wrong with the shapes and relative sizes of regions and the distances and directions among places.

We see a great many world maps. Many, probably most, of the projections used for them were chosen, not because they displayed geographical relationships well, but instead for "practical" reasons: already-made maps were available, a map could be easily copied, the map would fit conveniently a page format or a wall, or the map was a current fad, and so on. Equally disturbing is that most viewers seem not to care when blatantly distorted geographical shapes and sizes are displayed without reason.

This chapter is slightly modified version of the Special Publication No. 1 of the American Cartographic Association, a member organization of the American Congress on Surveying and Mapping, 210 Little Falls Street, Falls Church, VA 22046. The Special Publication No. 1, ISBN 0-9613459-1-8, has been published in 1986 by the American Congress on Surveying and Mapping. In that time, the Committee on Map Projections consisted of the following distinguished members: James R. Carter, Marshall B. Faintich, Patricia Caldwell Lindgren, Barbara B. Petchenik, Arthur H. Robinson and John P. Snyder, Chairman. Text and design was by Arthur H. Robinson, computer plotting of projections and coastlines by Waldo R. Tobler and preparation of graphics by the University of Wisconsin Cartographic Laboratory.

Permission granted by Eric Anderson, Executive Director for CaGIS, 2011–2016

A.H. Robinson (✉)
Madison, WI, USA

M. Lapaine and E.L. Usery (eds.), *Choosing a Map Projection*,
Lecture Notes in Geoinformation and Cartography,
DOI 10.1007/978-3-319-51835-0_1

1.2 Which Map Is Best?

For the least distortion the best map is one on a globe, since all the geographical relationships are correctly displayed. But globes are bulky, expensive, and do not permit the entire surface to be seen at once. Flat maps are better for most purposes.

Which projection is best for a world map? The flip answer is "none," but that is not a good question. A better question would be: Which map projection is best for what purpose? One can narrow the options by deciding whether one wants to show shapes of features as well as possible, or correct sizes, or a good compromise between the display of shapes and sizes, or to suggest the roundness of the Earth, or to show easily how far it is from here to there, or whatever.

A poorly chosen map projection can actually be harmful. We tend to believe what we see, and when fundamental geographical relationships, such as shapes, sizes, directions, and so on, are badly distorted, we are inclined to accept them as fact if we see them that way on maps. This can lead to very wrong impressions, for example, that the route from Chicago to Rome is due east or from San Francisco to Tokyo is due west, or that North America is larger than Africa. Our mental maps, the brain's geographical "data base," are generated only from what we look at. Since globes are uncommon, most mental images of the Earth come from flat maps. A badly distorted map seen regularly (such as one on a television news program backdrop) will look familiar after a while and thus "look right." This can cause one's mental map of the world to become permanently warped.

1.3 Scale, Globe Maps and Flat Maps

A map is not a simple drawing of the Earth's surface, but it is a model obtained on the basis of certain mathematical laws. First, we pass from the physical Earth's surface onto a mathematical surface—rotational ellipsoid or sphere. This transfer is realised by means of orthogonal projection of physical surface points onto the mathematical surface by using a geodetic control network enabling correct geographic location and orientation of map contents within the frame of some co-ordinate network. This is followed by the transfer from the rotational ellipsoid surface or sphere onto a plane. These projections are called map projections and are the subject of the theory of map projections.

Scale is the approximate size relationship between a map and the real world. It is often expressed by a ratio, such as 1:330,000,000, that tells how many units (inches, centimeters, etc.) on the Earth surface model (a sphere or an ellipsoid of revolution) are represented by one of the same units on the map. A globe of that scale would have a diameter of 39 mm (about 1 1/2 in.).

Conceptually the projection process begins by first reducing the Earth to a globe at a chosen scale and then flattening that globe to make a flat map. But—and it is an important but—whereas the distance scale is the same everywhere on the globe,

Fig. 1.1 Orthographic projection: It is one of the oldest and still a favorite. It "looks right" because the increase in distortion toward the edges matches the normal perspective foreshortening that occurs when one observes something spherical. In visual terms, it is a realistic view of a globe

it will be the same on the flat map only at a point or along certain lines in particular directions. Everywhere else it will be larger or smaller, which is why distortion occurs in all flat maps. One can say only that the nominal scale of the flat map is that of the globe.

The several views in Fig. 1.1 are of a globe with a scale of 1:330,000,000. All the other projections illustrated in this chapter are derived from this same globe; therefore, they all have the same nominal scale.

Actually the best way to appreciate the way a flat map distorts the surface of the globe is to compare it physically with a globe of the same scale. As an alternative to having one in hand, we show several views of a globe of the same scale alongside each projection so that the reader can make direct comparisons of what the map on the globe looks like and what the flattening process has done to it.

Although it can show only a hemisphere at a time, it can be considered a projection for world maps because a matched pair does display the whole Earth. Individual hemispheres are useful when interest is focused on one region. The projection can easily be made to place any region in the center where the shapes are not significantly distorted.

1.4 What Kinds of Projections for World Maps Are There?

There is a variety of map projections, and cartographers classify them in many ways. In this section we are concerned with kinds of projections distinguished by two visual characteristics: their overall shape and the arrangement on the map of the network of parallels and meridians (lines of latitude and longitude, or the *graticule*).

A globe's surface can be transformed to fit within any outline on a flat surface. As a matter of fact, such shapes as a valentine-like heart, a star, and even a stylistic butterfly have enclosed a map of the Earth. More traditionally, three simple shapes have been used: circles, ovals, and rectangles. Examples of each are included in this chapter.

Unless there is some compelling reason for not doing so, it is reasonable to assume that a general map of the whole world ought at least to suggest that we do live on a spherical Earth. Unfortunately, circular world maps have a great amount of distortion near their edges, and oval maps have been more widely used.

The meridians and parallels we use for the coordinate system define our cardinal directions, north-south and east-west. Except at the poles, they are everywhere at right angles to one another. Although this attribute of the globe can be duplicated on a flat map with curved lines, it is easier with straight lines. This forces the world into a rectangular shape, takes away the suggestion of sphericity, and no matter how it is done, much distortion is inevitable.

Because a sphere has no beginning or end it makes no difference how the transformation process is applied. But the Earth's coordinate system, anchored to the surface by the poles, will look quite different depending upon where the transformation originates. If it is centered at the equator, the graticule will appear orderly and symmetrical around two straight lines: the equator and meridian in the center of the map. Such projections are called *normal* or *conventional.*

The equator on a projection with the same nominal scale as the globe is often the same length. That is, if the globe were 10 in. around at the equator, the equator on the map would also be 10 in. long. In cartographic terminology, a line on a projection along which the scale is constant and that is the same length on the globe is sometimes called a *standard* line.

1.5 What Are the Kinds of Distortions?

All projections distort the globe surface in at least two and often all three of the following ways:

- Sizes of regions appear larger or smaller than on the globe.
- Distances between points are shown as longer or shorter than on the globe.
- Direct routes between points are not shown as straight lines.

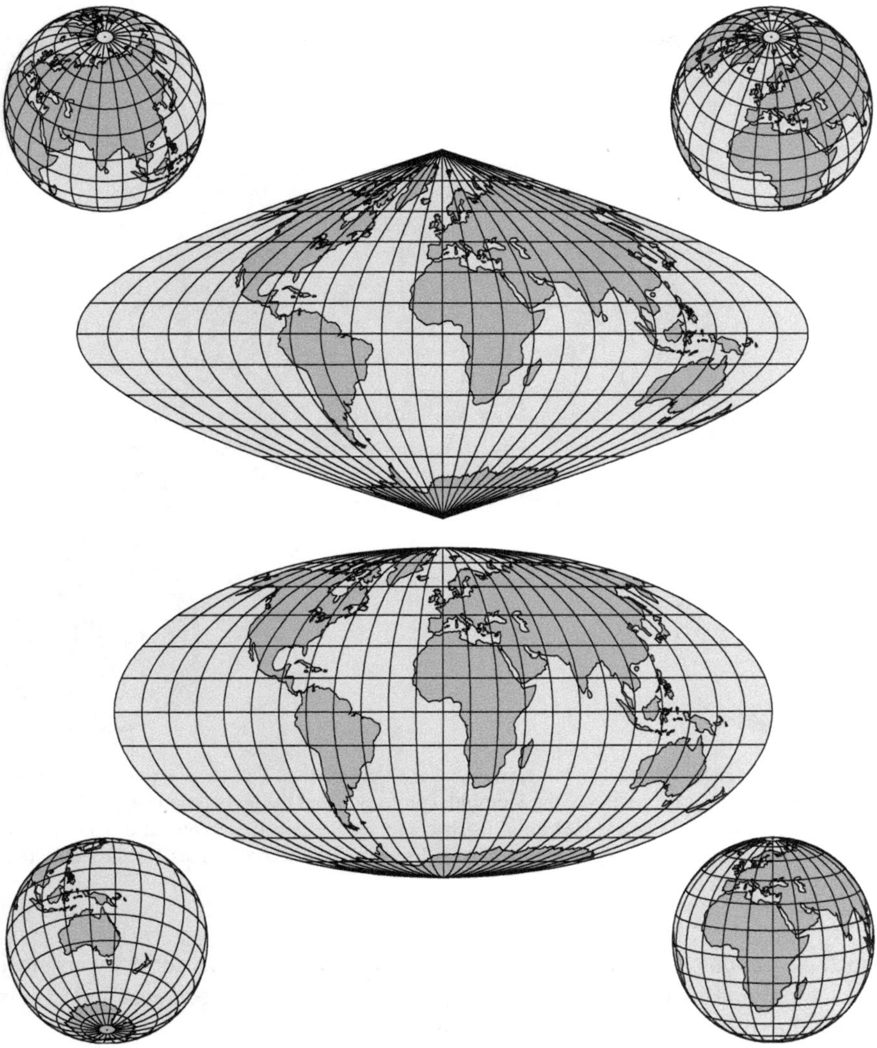

Fig. 1.2 Sinusoidal (*top*) and Mollweide (*bottom*) equal-area, oval projections. In the Sinusoidal all parallels and the central meridian are standard lines. In the Mollweide only one parallel, about 40°, is a standard line. Shapes are good in the central section of the Sinusoidal but are compressed in high latitudes. In the Mollweide, shapes are elongated in the lower latitudes but less crowded in the polar areas

The larger the area covered by the map the larger the amount of distortion; thus, it is greatest on world maps.

To a limited degree some of these distortions can be "controlled." For example, the correct relative sizes of regions can be maintained; such projections are called equal-area. Many other attributes of the globe surface can be retained, such as showing the direct routes from one place to all others as straight lines

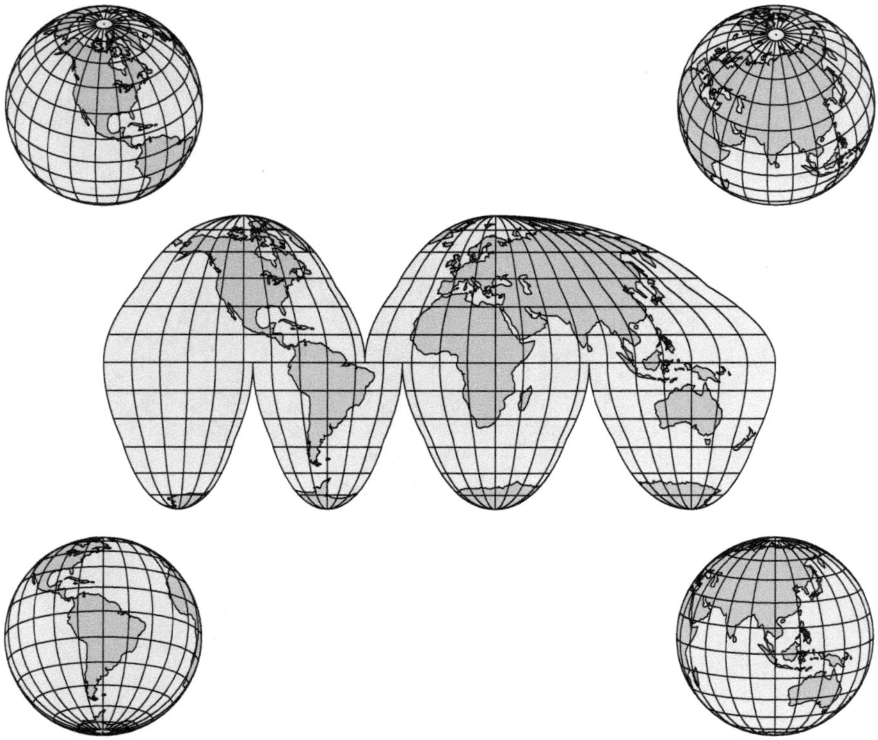

Fig. 1.3 Interrupted Goode Homolosine, equal-area projection. This is a combination of the better-shaped, equatorial section of the Sinusoidal projection and the less-compressed polar section of the Mollweide projection. In addition, each continental segment is centered around its own central meridian, thereby using the least distorted sections of the projections for the land areas. Ocean areas could be featured instead

(such projections are called *azimuthal*), or showing all directions correctly arranged at each point. This last property, called *conformality*, is quite technical and is very useful for specific purposes, such as navigation.

The representation of sizes can be controlled. Equal-area, oval projections, such as those in Fig. 1.2, greatly deform shapes. Among such projections, in the conventional aspect, less shape distortion in equatorial sections results in more in polar regions and vice versa.

There are two ways of dealing with the problem of shapes in oval, world map projections. One is to break up the smooth outline by separating it into lobes. Each lobe may be centered on a continent (or an ocean), thus lessening the crowding and shearing that occurs when all meridians converge to a single point at each pole. One example, an equal-area, *interrupted* projection, is shown in Fig. 1.3. Another way is to show the poles as lines instead of points, as in Fig. 1.4, thus decreasing the apparent distortion; it will still be there but in a less concentrated and noticeable form.

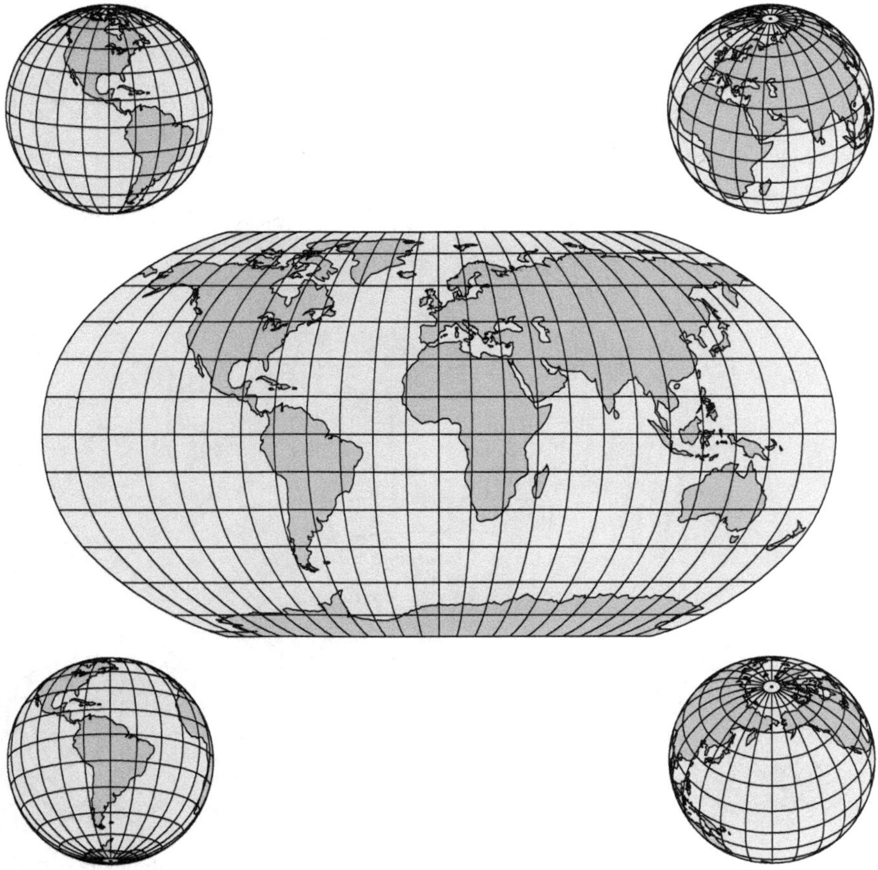

Fig. 1.4 Robinson projection. A compromise projection to show the whole Earth uninterrupted with reasonable shapes. It departs somewhat from being equal-area, especially in the high latitudes, in order to provide better shapes in the mid- and low-latitude regions

An alternative to the greatly distorted displays in uninterrupted, equal-area maps are *compromise* projections that allow some size distortion in order to improve the general representation of shapes. Three examples are shown in Figs. 1.4, 1.5 and 1.6.

1.6 Cylindrical Projections

To force the spherical globe into a rectangle produces extreme shape distortion, but surprisingly most people do not complain.

In the normal aspect of a cylindrical projection, the meridians are equally-spaced, parallel lines. They are crossed at right angles by the parallels of

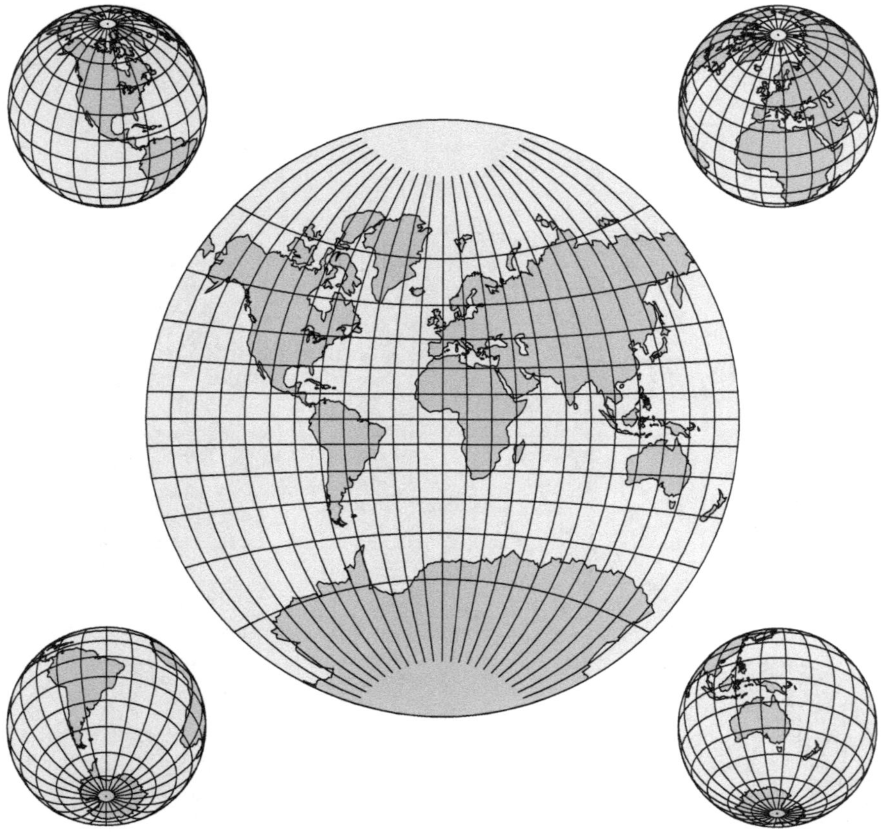

Fig. 1.5 Van der Grinten projection. A compromise projection showing the entire Earth within a circle. Shape and size distortions increase away from the central section, particularly in the high latitudes. Most of the greatly enlarged polar areas are left out when this projection is used, but then it is no more the map of the world. Four Van der Grinten projections were among the most popular new projections of the twentieth century. His first projection was adopted in 1922 for the National Geographic Society's official new world map. In 1988, the society chose the Robinson projection to replace the Van der Grinten, and in 1998 the Winkel Tripel was introduced and retained until today (Fig. 1.6)

latitude, just as they are on the globe. Although the orderly grid may seem right, there are some catches that make the graticule as bothersome as it is on oval maps. For example, on the globe, except at the equator, north-south lines are never parallel, and except on the equator, the shortest route to a place on the same parallel is not along that parallel. On a cylindrical projection the poles (mere points on the globe) are stretched to become lines as long as the equator, and all parallels of latitude are shown as being that same length.

Why are cylindrical projections rather common in spite of their extreme distortion of shapes, their lack of any suggestion of sphericity, and the unreality of

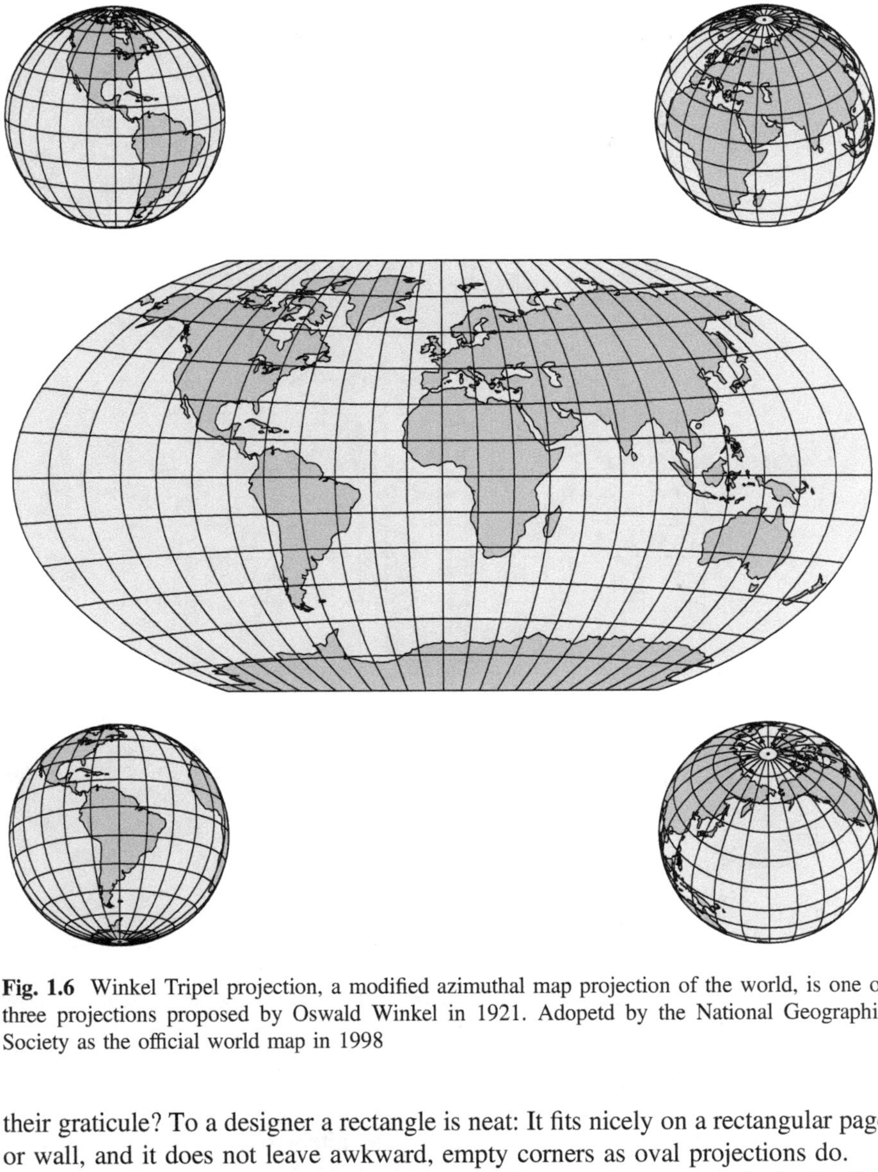

Fig. 1.6 Winkel Tripel projection, a modified azimuthal map projection of the world, is one of three projections proposed by Oswald Winkel in 1921. Adopetd by the National Geographic Society as the official world map in 1998

their graticule? To a designer a rectangle is neat: It fits nicely on a rectangular page or wall, and it does not leave awkward, empty corners as oval projections do.

Probably equally, if not more important, is that they became relatively familiar because a rectangular grid has been used as the base for all nautical charts since the 15th century. Sailors try to maintain a course with a constant direction, called a rhumb or rhumb line, and for them to know where it will lead, it should plot as a straight line on their chart. Only when the meridians and parallels are spaced a certain way on a cylindrical projection can this occur. Gerardus Mercator obtained the required spacing by increasing the north-south scale poleward at the same rate that the east-west scale was increased—a result of making the meridians parallel

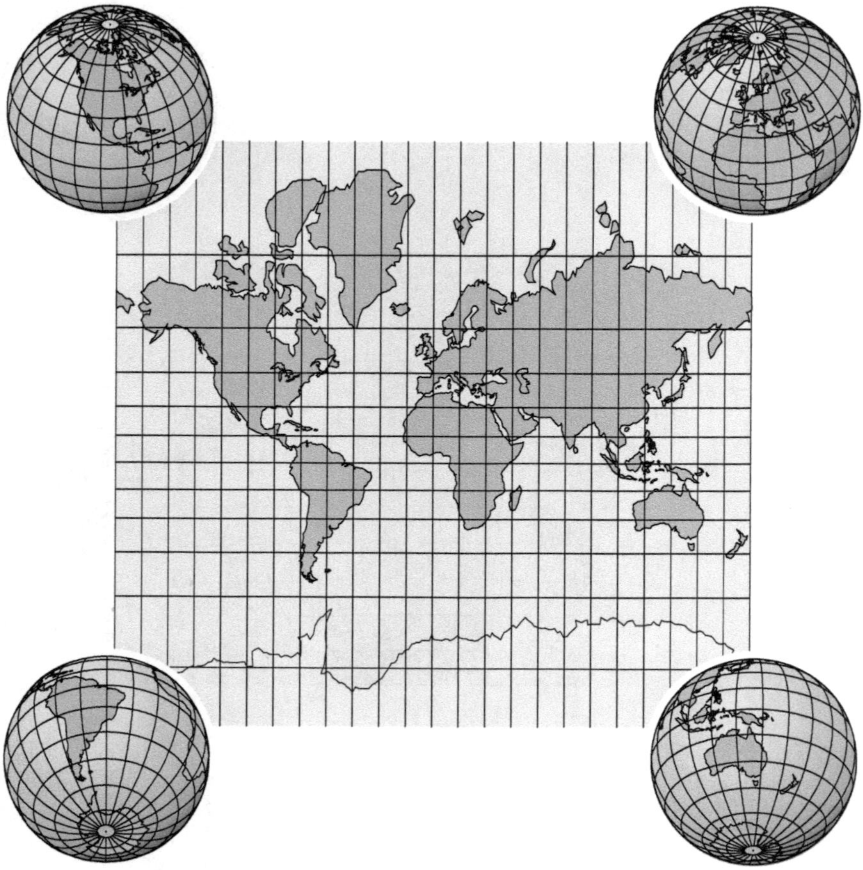

Fig. 1.7 Mercator projection. It is conformal and has the unique attribute that any rhumb is shown as a *straight line*. It greatly exaggerates sizes away from the equatorial areas, as a comparison with the accompanying globes shows. It does not show true direction since all great circles, other than the meridians and the equator, plot as complex curves. It should not be used as a base for a general purpose map of the entire world

straight lines instead of converging as they are on the globe (Fig. 1.7). Since the poles are only points, the projection extends indefinitely to the north and the south. Although he intended it to be used on sailing charts, Mercator first published the projection in 1569 as a world map to illustrate the concept.

The Mercator grid became very well known, and it began to be used for general world maps even though for such a purpose the fact that rhumbs are straight lines is not at all significant. Unfortunately, it is still often chosen for such maps. It is shown in Fig. 1.7. There have been several similar-looking substitutes devised to lessen the extreme distortion in the areas near the poles. None is conformal like the Mercator. One, the Miller Cylindrical, is shown in Fig. 1.8.

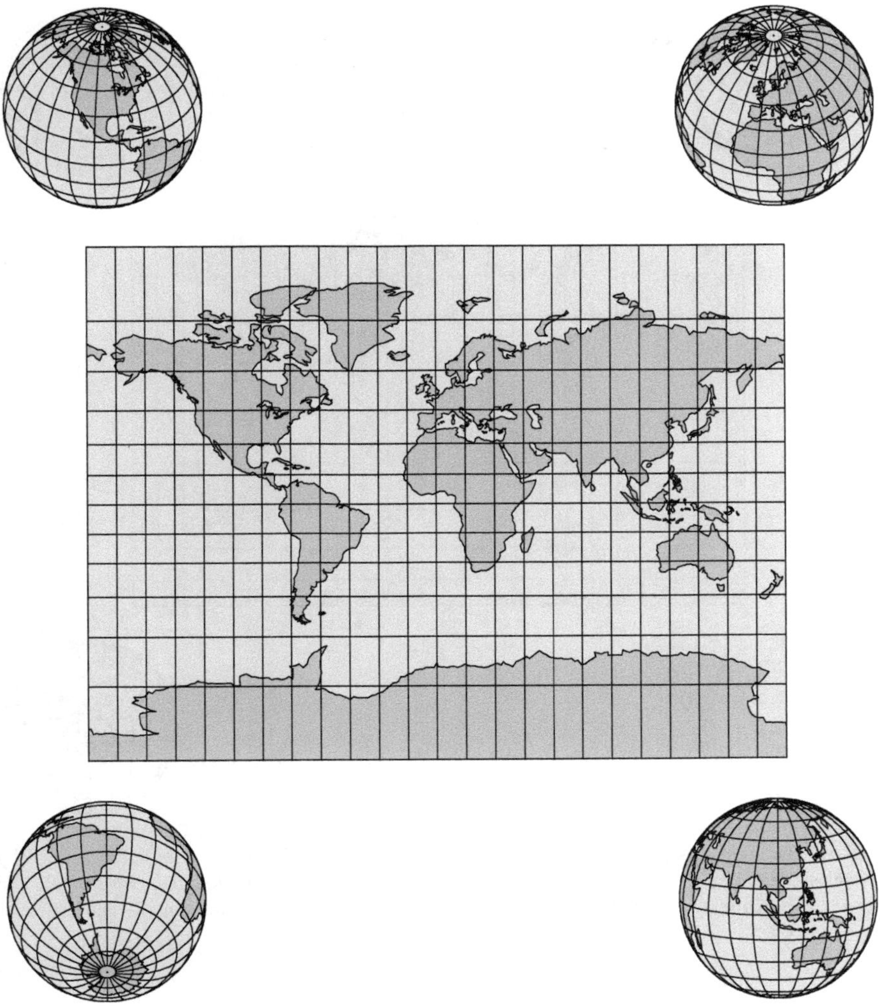

Fig. 1.8 Miller Cylindrical projection. A compromise projection showing the entire Earth in a rectangle without as much size exaggeration in the high latitudes as in the Mercator projection. It is not equal-area. In contrast to the Mercator, the poles can be shown

An infinite number of cylindrical projections can be made equal-area, simply by selecting a parallel to be standard and then spacing the others appropriately. Shapes are severely distorted in all of them. One such grid, first presented by Gall in 1855 and again by Peters in 1967, has been adopted by some non-cartographers who are not troubled by the unreality of the graticule and the extreme shape distortion. It is shown in Fig. 1.9.

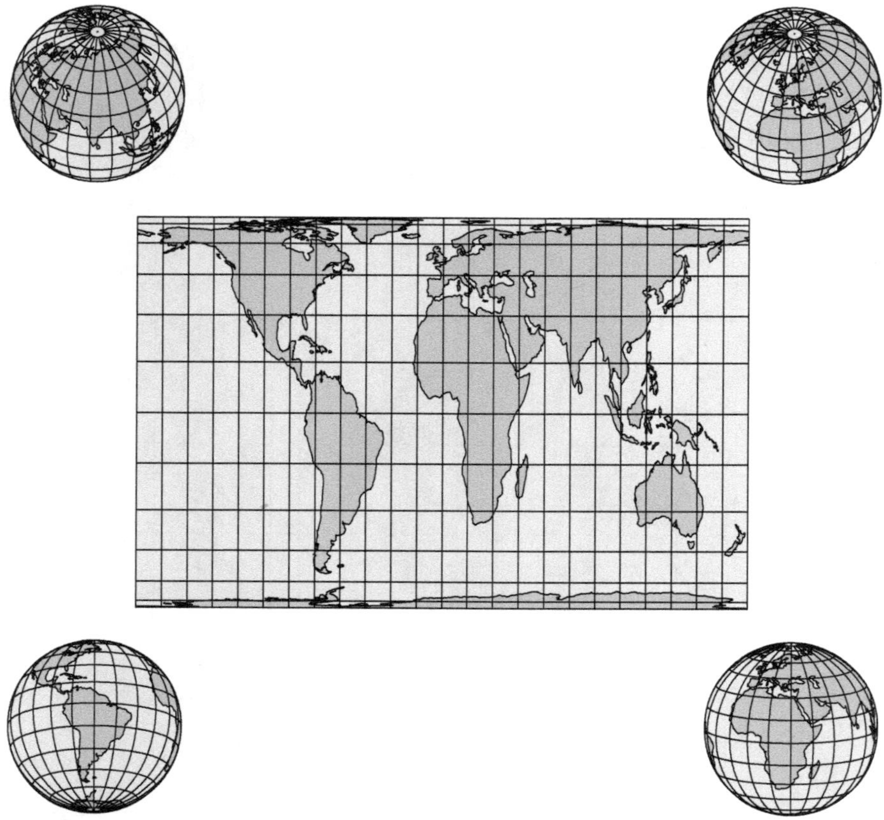

Fig. 1.9 Gall-Peters equal-area projection. To be equal-area within the rectangle determined by using standard parallels of 45° causes the equatorial regions to be greatly lengthened vertically and squeezed horizontally, while the higher latitudes are greatly lengthened horizontally and squeezed vertically

The name cylindrical projections comes from the fact that the image of the world in any of such projections is a rectangle, and a rectangle can be wrapped into cylindrical surface.

1.7 Direction and Distance on Map Projections

Direction on the globe is tricky. The spherical coordinate system that defines our directions is not internally consistent like the grid lines of a square, plane, Cartesian system. To appreciate how direction may or may not be shown on maps, we need to understand that the straight course, that is, the shortest route between points on a globe, is always along a line called a great circle. A complete great circle separates

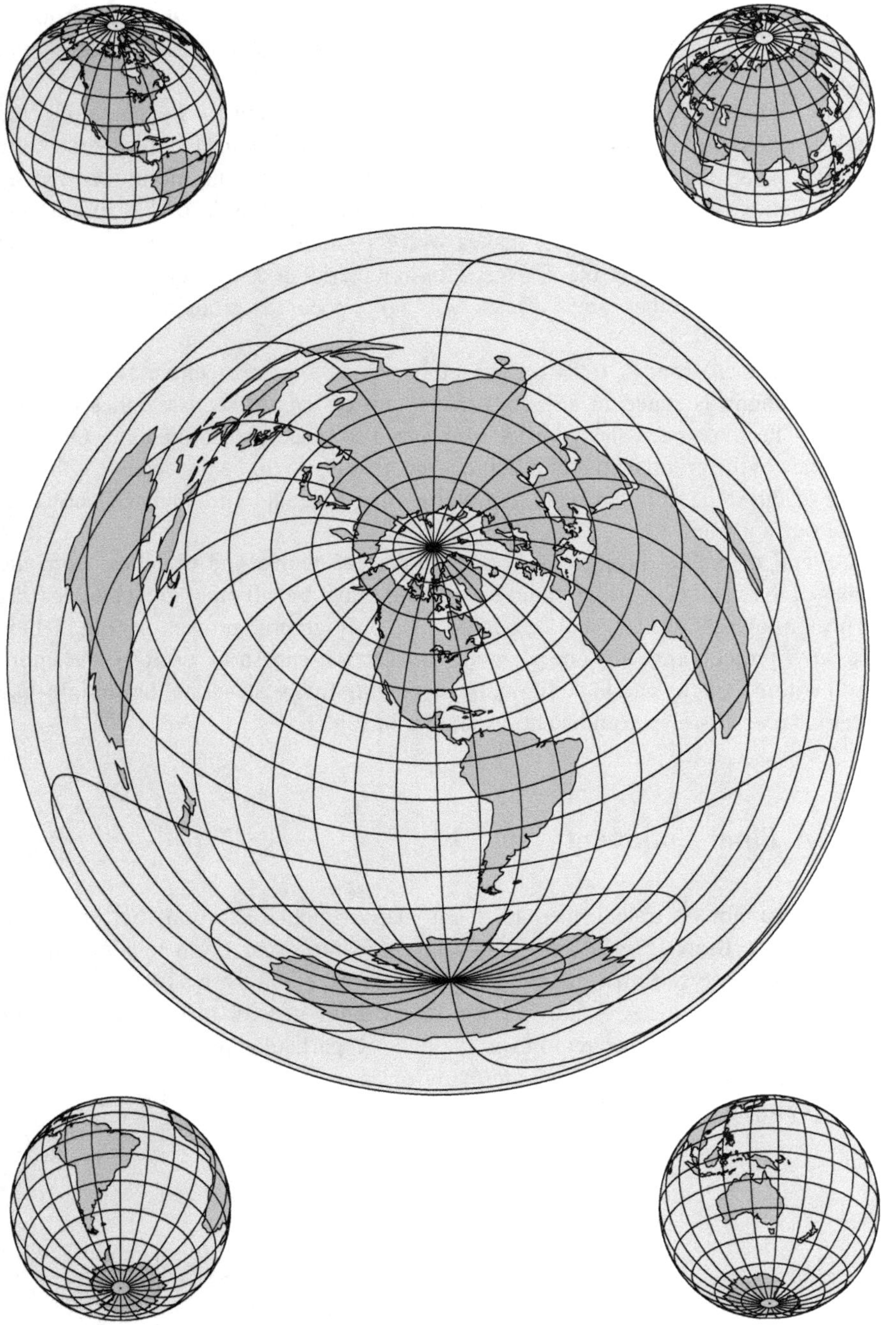

Fig. 1.10 Azimuthal-Equidistant projection centered near Chicago. Its purpose is to show all the great circle routes through the center as straight lines with correct azimuths at the center and to show the distances along the straight-line great circles with a uniform scale. It is not equal-area, and shapes in the outer half are greatly distorted. It is useful because it can be centered at any point, such as your home town

the globe into hemispheres. The equator is a great circle and each meridian is half a great circle. If you stretch a piece of string anywhere on a globe, it will lie on a great circle.

Any oblique great circle on the globe, that is, one that is neither along the equator nor along a meridian, that connects point A and point B will intersect the parallels and meridians at successively different angles all along its course. Consequently, the only way to designate the direction from A to B is to note the angle (direction) the great circle makes with the meridian at A. This is called the azimuth of B at A. Furthermore, except for any meridian or the equator which may be straight, most other great circles will not appear as straight lines on map projections.

The distance from A to B on the globe, along the great circle segment that connects them, is stated in any convenient unit of length, such as miles or kilometers. The scale necessarily varies from place to place on a projection. On standard lines of the graticule the unit will be true, that is, it will be the same length on a globe of the same nominal scale, but along other lines the units may be shown as equal but not true.

Several azimuthal projections show the correct course and direction from the center of the map to any other point by a straight line, but all other great circles (that is, not through the center) will be curved. The orthographic projection (Fig. 1.1) is one such projection. Only one azimuthal projection can show usefully the entire Earth within a circle, and in addition, by displaying true scale along the straight-line great circles, it is also equidistant, as shown in Fig. 1.10.

1.8 A Final Comment

A globe cannot be transformed to a flat map without distortion, but scores of systems have been devised to display various features and Earth relationships in advantageous ways. Some, like the Mercator, are excellent for a particular purpose but not for most others. A number can be used for general maps, and some for focusing attention on one region, or for making particular measurements, or for various other applications. There are many options.

Map projections should be selected and used with care.

Chapter 2
Choosing a World Map

Attributes, Distortions, Classes, Aspects

Arthur H. Robinson and The Committee on Map Projections

2.1 Foreword

For more than 2000 years mapmakers have been inventing ways to display the entire Earth on a flat surface—papyrus, vellum, metal, paper, or whatever—and now on a TV or computer monitor. The process requires a systematic conversion called a map projection. There is no single solution, and hundreds have been devised. The Earth is round like a ball, and no matter how one flattens the rounded surface the reality of the spherical surface will be variously deformed to a noticeable degree.

Ordinary vision modifies reality in several ways: diminution with distance, perspective convergence, foreshortening because of visual angle, and so on. Even as children we soon learn that the way things are and the way they look differ. Ideally, everyone would know enough about the Earth and the effects of map projection to see past the inevitable distortions in a world map to the Earth the way it really is.

This chapter is slightly modified version of the Special Publication No. 2 of the American Cartographic Association, a member organization of the American Congress on Surveying and Mapping, Suite 100, 5410 Grosvenor Lane, Bethesda, Maryland 20814. The Special Publication No. 2, ISBN 0-9613459-2-6, has been published in 1988 by the American Congress on Surveying and Mapping. In that time, the Committee on Map Projections consisted of the following distinguished members: James R. Carter, Marshall B. Faintich, Patricia Caldwell Lindgren, Barbara B. Petchenik, Arthur H. Robinson and John P. Snyder, Chairman. Text and design was by Arthur H. Robinson, computer plotting of projections, coastlines, and cartoons by John P. Snyder and Waldo R. Tobler, and preparation of graphics by the University of Wisconsin Cartographic Laboratory.

A.H. Robinson (✉)
Madison, WI, USA

M. Lapaine and E.L. Usery (eds.), *Choosing a Map Projection*,
Lecture Notes in Geoinformation and Cartography,
DOI 10.1007/978-3-319-51835-0_2

The ICA Commission on Map Projections hopes that this nontechnical chapter will make the subject of map projections for world maps more understandable. In it we describe the important geometric characteristics of the Earth that can be retained in a flat map of the whole world, as well as the kinds of distortions that may occur. We also show how various classes of projections affect the appearance and utility of world maps. In addition, we describe the patterns of distortion and how they can be displayed.

2.2 Flat World Maps

The Earth is complicated physically, and human life on it is so varied and intricate that we need a variety of world maps to try to understand it. To maintain reality we can make the map on a reduced model of the Earth, a globe; but only half the surface of a globe is visible at once, and the foreshortening toward the edges of the rounded surface greatly alters the appearance of shapes and sizes, Fig. 2.1.

A flat world map is usually much more useful because we can see it all. In addition, flat maps are convenient while globes are awkward.

To convert the entire rounded globe map to a flat map we must somehow "peel" the map from the continuous globe surface in order to lay it out, as in Fig. 2.2. As a matter of fact, the reverse of that process is how globe maps are made: a set of tapered sections, called "gores," is printed flat, as in Fig. 2.3, and then cut out and glued onto a ball.

In each gore there is not much stretching or shrinking, so the map in Fig. 2.3 is a comparatively true, flat, world map. But it is not useful. The multiple discontinuities ruin overall shapes, while important relationships, such as proximity, distance, and direction are essentially destroyed. Continuity in a map is extremely important, and it is far more useful to decrease the number of interruptions by arranging the map

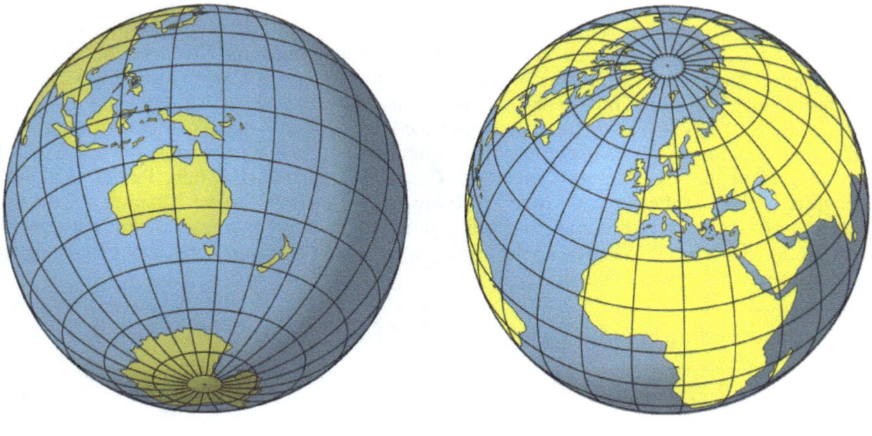

Fig. 2.1 A globe (in perspective projection)

Fig. 2.2 "Peeling" the globe

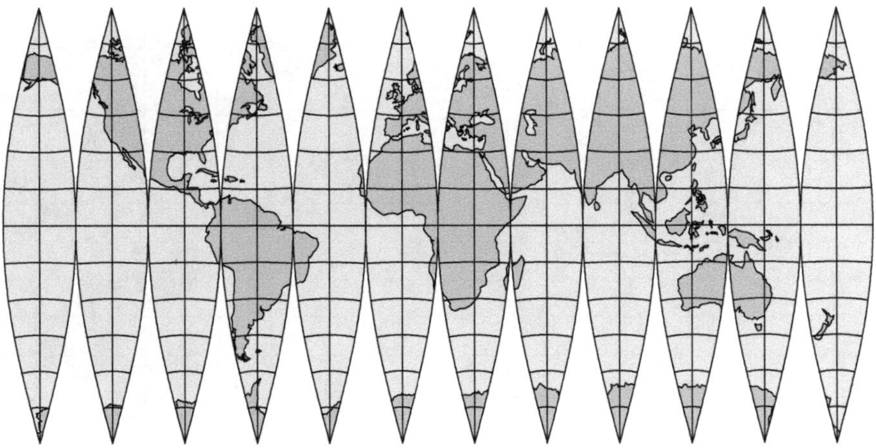

Fig. 2.3 Globe gores

within a single flat shape, such as an oval, rectangle, or circle. There is then only one interruption, the outside edge. Thus, the maximum continuity possible in a flat map is obtained, but only at the expense of altering shapes, distances, directions, and possibly areas.

Suppose we are looking at a globe straight on; the front is visible, the back is not. If we wish to make an oval world map we must first decide what system to use to flatten the rounded surface and apply that to the front half of the globe. Then, in effect, we cut the globe map down the middle of its back side and pull the two back

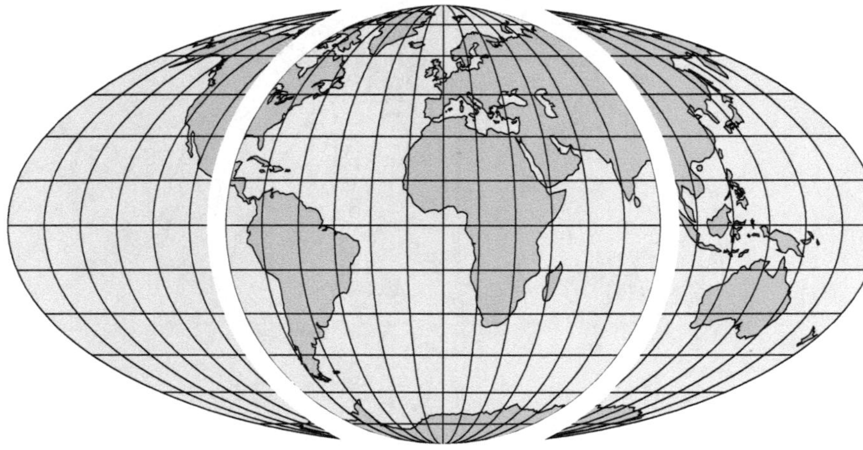

Fig. 2.4 Mollweide projection

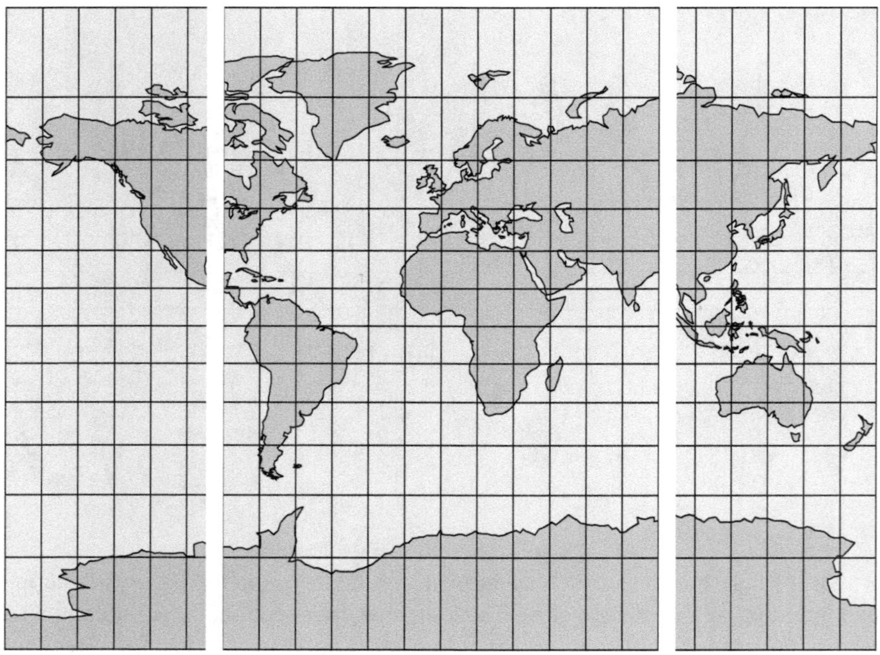

Fig. 2.5 Miller cylindrical projection

quarters around alongside the map of the front half as diagramed in Fig. 2.4. Obviously, the back quarters must be considerably altered in order to make them join the front half.

If we wished to make a rectangular world map we would proceed the same way, that is, by first flattening the front half of the globe map into a rectangular format, and then putting the two back quarters alongside, as illustrated in Fig. 2.5.

They, too, are greatly altered but in a different way as will be explained in a later section.

If we were to follow this same general procedure to make a circular world map we can imagine the back half of the globe being stretched almost beyond recognition and made to surround the map of the front half as in Fig. 2.6. It is important to keep in mind that these separation processes do not lose any part of the globe surface. Thus, the right and left edge of each map in Figs. 2.4 and 2.5 is the *very same* line on the globe, and the outside edge of Fig. 2.6 is the *point* on the globe opposite the center of the map of the front half.

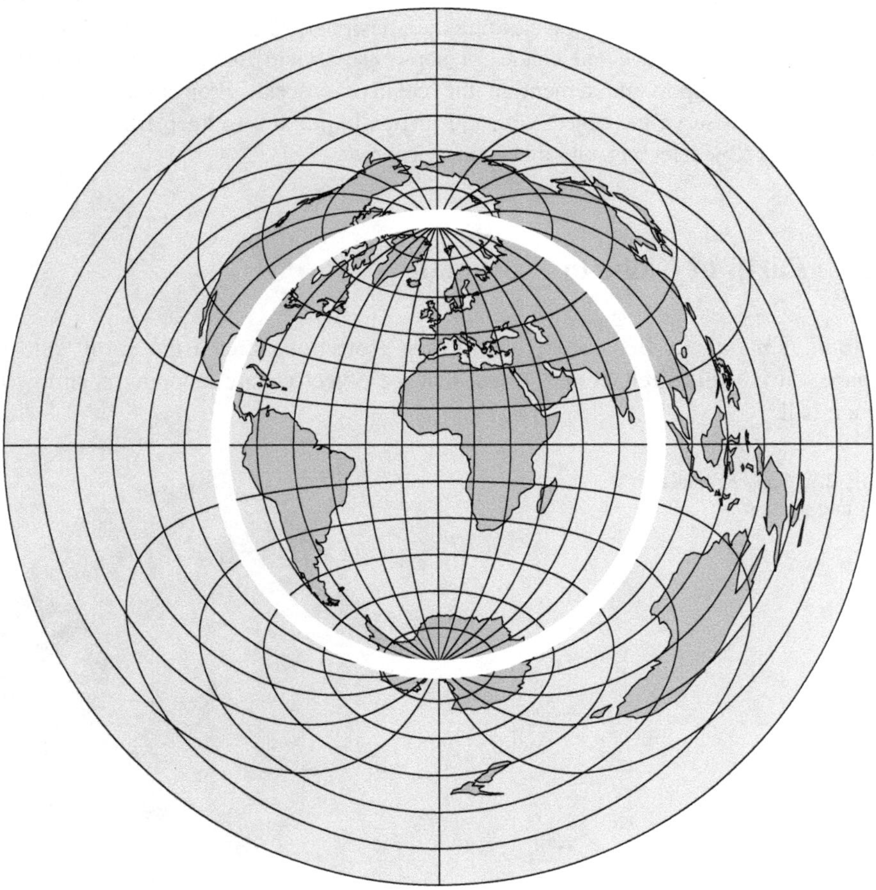

Fig. 2.6 Azimuthal equidistant projection

In the process of making the rounded surface lie flat it is evident that we must stretch or shrink it unevenly. This will alter some of the globe's geometrical qualities, and consequently shapes, sizes, distances, and directions may be changed. Technically, these alterations are called *distortions*. Unfortunately, the word "distortion" suggests something bad; that is an unhappy connotation. The stretching and shrinking can be manipulated so as to produce flat transformations that are more useful than a globe. For example, one can make the direct routes from one place to all others appear simply as straight lines radiating from a central point, as in Fig. 2.6.

There are hundreds of ways to arrange the globe surface within an oval, rectangle, or circle; no one system is best for all maps. Choosing a projection from among the many options requires something like a cost-benefit analysis. For every desirable quality there is often some consequent drawback, and furthermore the distortions associated with any map projection are not uniformly arranged; some parts of a projection are always less distorted than other parts.

Anyone who looks at world maps, on the printed page or on TV, should learn to make allowances for the distortions of shapes, sizes, directions, and distances that are bound to occur. Similarly, a cartographer, designer, editor, or anyone who wants to make a map or have one made for some use should make the selection of a projection an important element in the creative process. Projections should be chosen to best serve the objectives, and just as important, to keep from making a serious error by selecting one that is inappropriate.

2.3 Earth to Globe to Projection

A Ball Is a Ball … Because all world map projections stretch the Earth "out of shape," it is well to reflect on the fact that the object we are mapping is simply a huge ball.

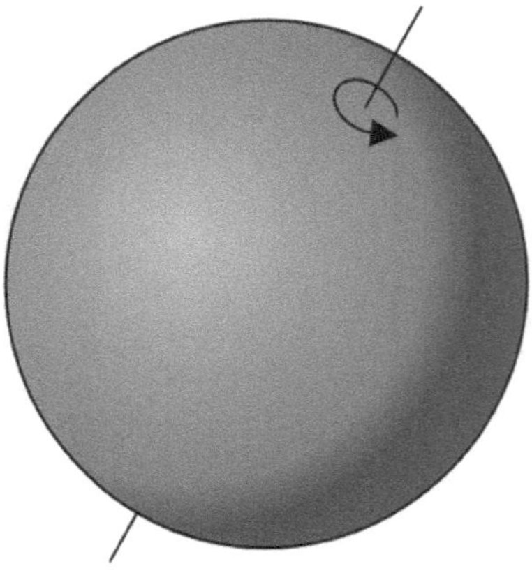

Fig. 2.7 Sphere rotates around an axis

The Earth's surface is almost a perfect sphere that rotates around an axis, Fig. 2.7. Compared to a perfect one of the same size it is only slightly "out of round" due to centrifugal force from rotation. A perfectly proportioned globe 12.0 in. in diameter through the equator would have a polar diameter only 0.04 in. shorter. The highest mountain would be only 0.0083 in. high above sea level (about the thickness of two sheets of this paper). These ratios are important for large-scale navigation and topographic mapping, but not for world maps.

Map Scale. An important key to the distortions in a map projection is an understanding of *map scale*. The scale of a map is not simply the relation between a given distance on the map and the Earth distance it represents. In fact, first of all, all measurements performed on the Earth surface or close to it should be in some way projected onto the Earth's model—a sphere. A map projection is a mapping from the Earth's sphere or a rotational ellipsoid into the plane of projection. It follows that the scale of a map is approximately the ratio between a given distance on the map and the Earth distance it represents. It may be expressed as a relationship, such as 1.0 in. represents 660 mi., or stated as a ratio or fraction with a numerator of one and a denominator designating how many of those same units on the Earth it represents. Suppose, for example, we have a globe with a diameter of 12 in.; its scale will be about 1:42,000,000 because at that size 1.0 in. on the globe represents 42,000,000 in. on the Earth, about 660 mi. In this chapter we will use the sphere as a model of the Earth, because the flattening of the Earth's ellipsoid is so small, that its influence could be neglected for world maps.

Because projection to a flat surface always stretches and shrinks the spherical surface unevenly, there will always be different scales in different parts of a map. That is particularly important on world maps where differences in scale may be extreme. Statements of linear scale for world maps apply only at particular places in given directions.

In describing the relation of the scale at one point on a map to the scale at another point, it is convenient simply to use the terms *larger* and *smaller* (scale) instead of trying to compare the actual numerical values. A "larger scale" means that the scale fraction is larger; for example, 1:30,000,000 is larger than 1:40,000,000, in the same way that 1/3 is larger than 1/4.

Reference Globe. An easy way to visualize the basic transformation procedure is to imagine the Earth first mapped on a sphere or a globe at some desired scale, and then the surface of that globe being laid out flat. The size of the globe that is the model for the flat map is called the *reference* or *generating* globe. The scale of the map will then be the same as that of the reference globe. However, while the scale on the globe is the same everywhere, the flattening process is uneven, and the *actual* scale on the map projection will be different from place to place. We can only say, then, that the scale of a world map is a *nominal* scale, that being the scale of the reference globe. Later in this chapter we will explain how simple inspection of a world map can reveal a great deal about the variations from its nominal scale.

An important note—All the world maps and views of a globe in this chapter on which the continents and seas are shown in colour are based upon a reference globe with a scale of about 1:500,000,000, two views of which *at that size* are shown on

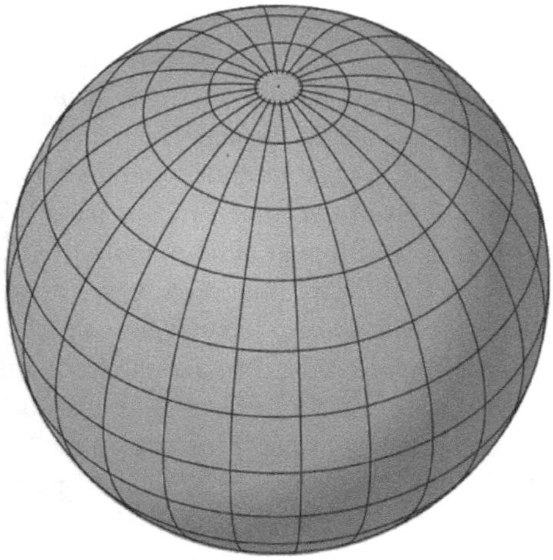

Fig. 2.8 Spherical coordinate system

Fig. 2.1. All the distortion cartoon maps at the end of the chapter also are that same nominal scale.

Coordinate Net. To the smooth, rotating ball the ancients fixed a spherical coordinate system, anchoring it at the two poles, Fig. 2.8. It is made up of (a) *parallels* of latitude, consisting of a set of *small circles* that denote distance north and south from the equator (0° Lat.) to the poles (90° north and south Lat.), and (b) *meridians* of longitude, composed of a set of *great circles* through the poles. Each meridian from pole to pole is half a great circle, and it designates distance from the Prime Meridian (0° Long.) east and west to 180° Long. (opposite the Prime Meridian). Great circles are the largest circles that fit a sphere. Any great circle divides the Earth in halves, so the equator and all paired meridians are great circles. The shortest route (the "straight line") between any two places on a sphere is along the arc of the great circle that joins them.

The parallels and meridians specify the system of geographical directions, east-west and north-south. It is important to keep in mind that the directions of the spherical coordinate system has rather different characteristics in comparison with the directions of the X and Y axes of a rectangular system. For example, except at the equator north-south directions are not parallel anywhere in a pole-centered hemisphere. When a selection of parallels and meridians appears on a map it is called a *graticule*. The arrangement of the graticule lines is important because directions are defined by them, and it is usually desirable to show directions consistently and as simply as possible throughout a map.

We often think of a particular arrangement of the graticule as the projection. However, that is not strictly true since it is the *surface* of the sphere that is being transformed, not the coordinate lines. A projection system can be oriented to the ball however we want, and it will not affect the resulting geometry of the projection,

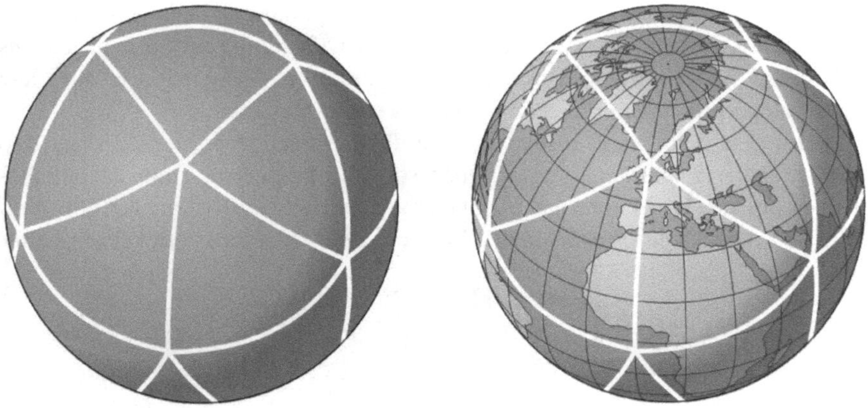

Fig. 2.9 Triangular net on the sphere

but it will greatly affect the appearance of the graticule. This is well illustrated by the series of orientations in Fig. 2.29.

Triangular net. A comparison of the graticule on a flat map with the parallels and meridians on the globe shows how the projection system has stretched, shrunk, and skewed the globe surface. It is, however, often difficult to make comparisons, even when a globe is at hand, due to perspective foreshortening and because the parallels and meridians on the globe are not evenly spaced except in a very restricted way.

More help in displaying comparative distortion is provided by placing a number of geometric figures on the globe and then showing how the projection system has changed them on the flat map. The most successful has been the employment of an icosahedron, a 20-sided geometric solid. When fitted to a sphere it provides a set of 20 exactly alike, equilateral, spherical triangles that cover the globe completely without overlap, and all sides of which are same-length arcs of great circles. The set is arranged on the reference globe for the colored maps in this chapter so that one triangle is centered on the North Pole with one vertex of that triangle located on the Prime Meridian, 0° Long., as illustrated in Fig. 2.9. That set of triangles is shown with white lines on the coloured maps.

An important note—Each world map (except Mercator) shows the set of 20 triangles in its entirety.

2.4 Attributes and Distortions

A globe map is "reduced reality" on which sizes and distances are made smaller according to scale, while other important attributes, namely angles, shapes, and direct routes remain the same as on the Earth itself. But when the globe map is transformed to a flat map some combination of these attributes always will be distorted in various ways. Only sizes and angles at points can be kept throughout a map, but not both on the same map and always with significant other distortions.

Because the several attributes are quite important for many map uses they have led to categories of map projections that have specific names.

Areas. It is desirable to display areas in their correct proportions if sheer territorial extent is important, such as when showing productive cropland or inhospitable climates. Projections that maintain the relative areas of all parts of the globe are called *equal-area* or *equivalent*. Unfortunately, when areas are kept in correct

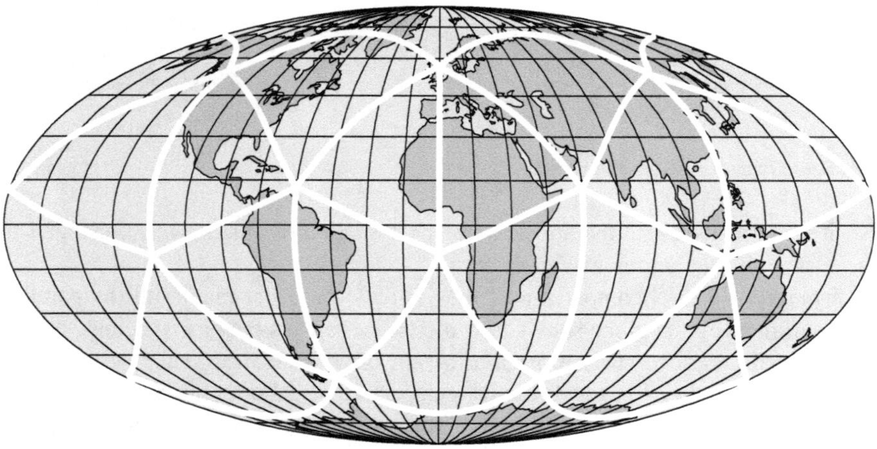

Fig. 2.10 Mollweide projection

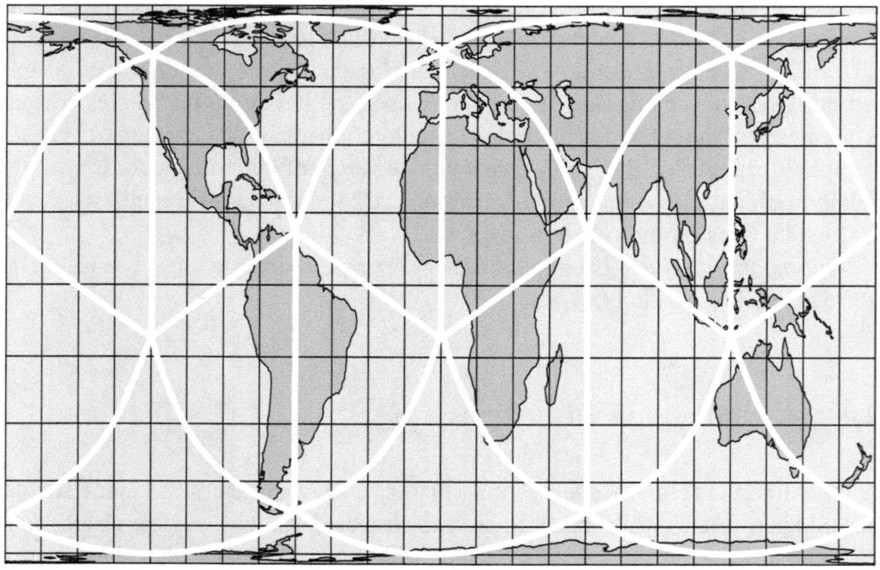

Fig. 2.11 Gall-Peters projection

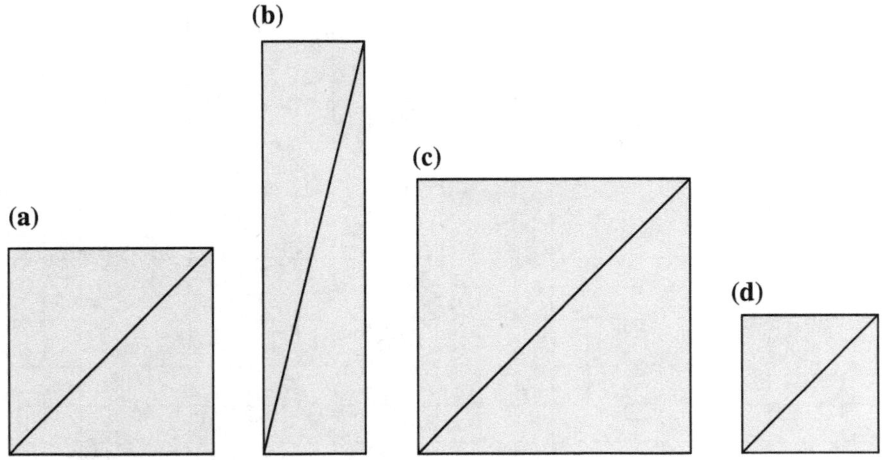

Fig. 2.12 Different transformations of a square

relation then some distances and directions will necessarily be distorted, as in Figs. 2.10 and 2.11.

Angles. On many projections some angles (and directions) at a point may be kept, while others at that same point are changed. For example, if, in Fig. 2.12, square a with the 45° diagonal from one corner is transformed to the rectangle b, the right angles of the sides have not been altered, but the angle of the diagonal has. It is possible to retain correct angles at all *points* only by maintaining the scale the same in all directions at each point, but in so doing the scale will usually differ from point to point. Compare c and d with a in Fig. 2.12.

Projections that maintain correct angles in every direction at each point are called *conformal* or *orthomorphic*. Both terms imply correct shape (and direction), but in practical application those qualities are limited to very small areas. Conformal maps are important tools for navigational, scientific, and technical purposes. Areas and sizes are greatly distorted on conformal projections of the whole world, and maps on those projections are not suitable for general use. Unfortunately, one well-known conformal projection, shown in Fig. 2.13, is too often employed for general world maps. Note the great exaggeration of sizes in the middle and high latitudes of Fig. 2.13. In reality South America is larger than Antarctica! The poles cannot be shown on the Mercator projection because it continues indefinitely to the north and south.

Shapes. The distinctive forms of all parts of the Earth, its continents, countries, seas, and so on, simply cannot be shown on a flat map without deformation due to the inevitable stretching and shrinking. Some representations are much better than others. Compare the nearly true shape of North America on the Orthographic projection (the same as looking directly at it on a globe) with its appearance at the same scale on the Gall-Peters and Mercator projections in Fig. 2.14. The shape is not easy to define. For example, for ordinary people a triangle and a square have completely different shape. In contrast, in the shape theory in mathematics, the two

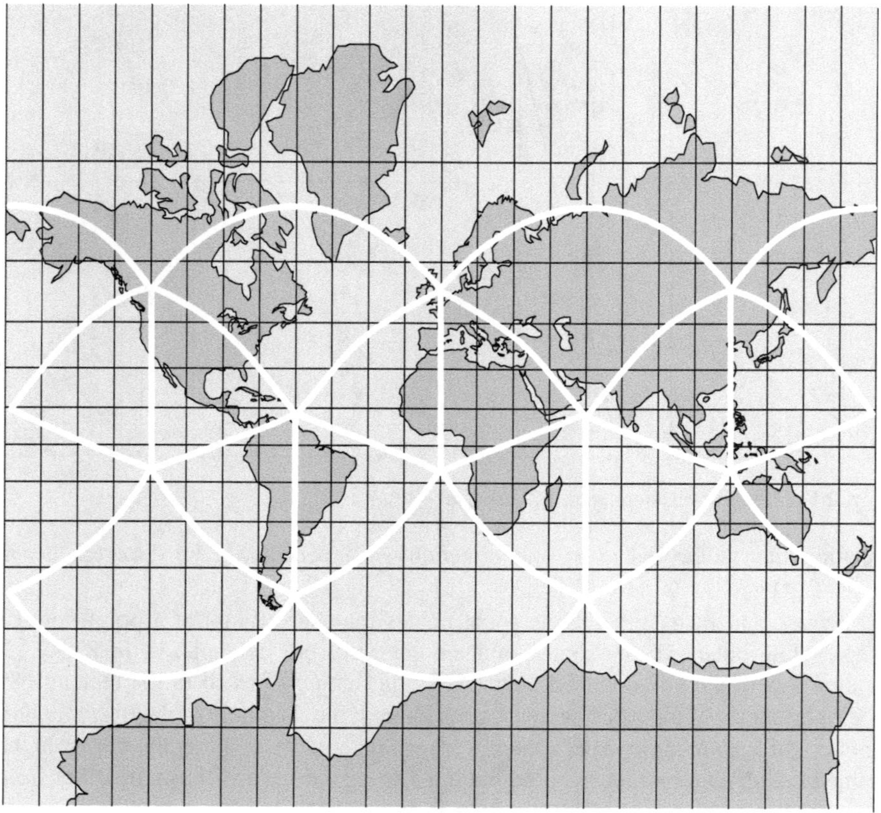

Fig. 2.13 Mercator projection

Fig. 2.14 North America in orthographic, Gall-Peters and Mercator projections

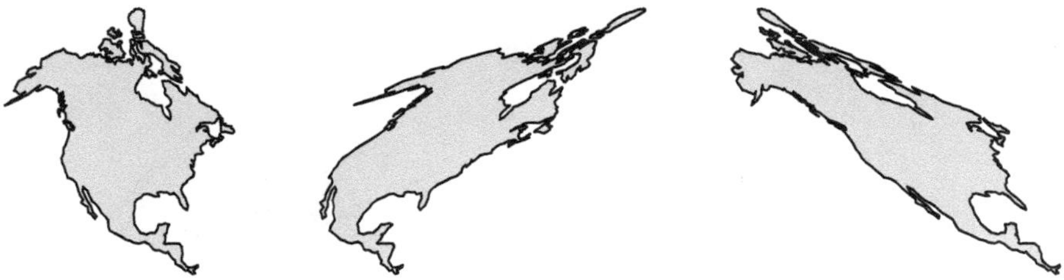

Fig. 2.15 North America in the *center*, on the *left* side, and on the *right* side of a Sinusoidal projection of the world

objects (triangle and square) have the same shape because there is a continuous mapping that transforms the triangle onto the square. On the other hand, if we accept that shape is something that remains unchanged in isometric mapping, then shape is always changed in map projections, because there is no map projection that is an isometry. Due to the fact that shape is not easy to define, this characteristic is generally not in use in the theory of map projections. We prefer areas, distances, and angles or directions as main attributes of any map projection because they are measurable and enable comparison between different mappings.

Unlike sizes or angles, which can be retained everywhere in a projection, shape representation can vary greatly from one part of a projection to another. For example, in Fig. 2.15 North America is shown the way it appears when it is in the center, on the left side, and on the right side of a Sinusoidal projection of the world (Fig. 2.19).

Direct Routes. The direct, or shortest, route from one place to another on the Earth's sphere or on the globe is along the arc of the great circle on which they both lie; that is the spherical equivalent of a straight line on a flat surface. All great circles are smooth curves on a globe, but in most parts of world map projections great-circle arcs become complex curves, except when both places are on the equator or the same meridian. Fortunately, some projection processes make it possible to "rectify" all the arcs from one place to all others so that they appear straight on the map. If the central point is made a pole, as in Fig. 2.16, then, since all meridians are great-circle arcs, all meridians will be shown as straight lines. The direction from one point to another is called an *azimuth* (the angle, in degrees from north, between the great circle arc and the meridian at the starting point). Because all directions from the center on these kinds of projections are correct they are called azimuthal; another term used is *zenithal*.

It is necessary to specify the direction (azimuth) of a great-circle route at the starting point because the coordinate directions in our spherical coordinate system has rather different characteristics in comparison with a rectangular system. Consequently, the relation between Earth's sphere or globe directions and any oblique great-circle arc differs from place to place along it as illustrated in Fig. 2.17 by the great-circle arc between two points on the same parallel.

A projection can be azimuthal and equal-area or azimuthal and conformal, but not both equal-area and conformal since those attributes are mutually exclusive.

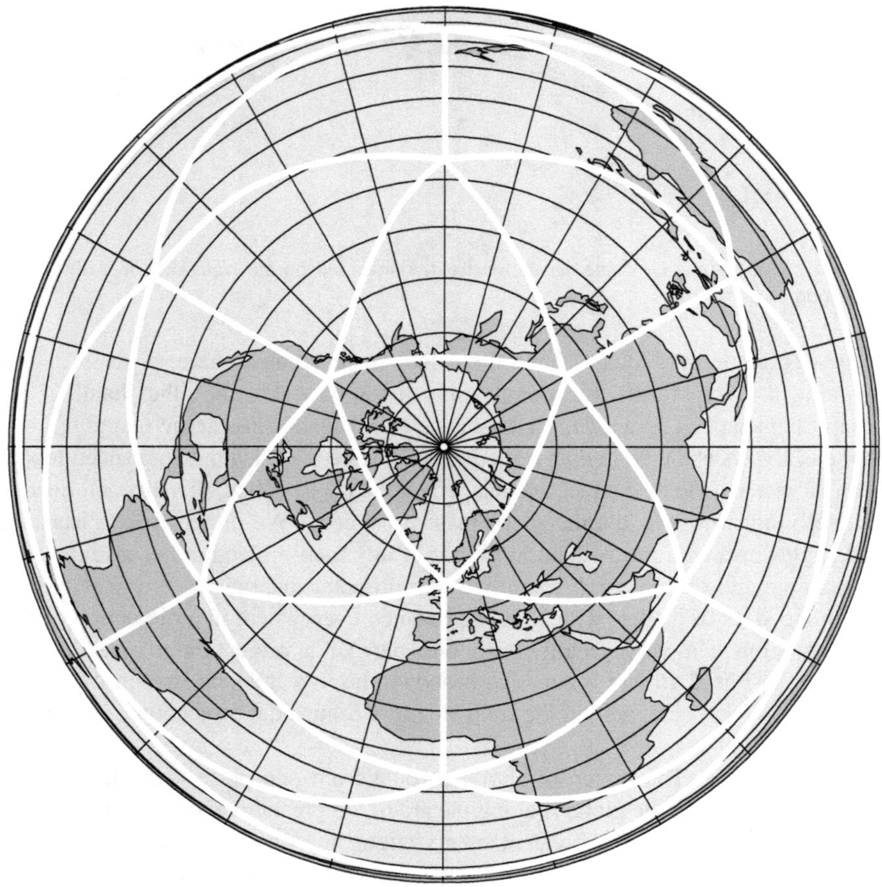

Fig. 2.16 Lambert Azimuthal equal-area projection

The azimuthal conformal projection, the Stereographic, can only show part of the Earth (in fact, theoretically the Stereographic shows the globe without only one point!). Even though the Lambert Azimuthal Equal-Area projection (Fig. 2.16) can portray the whole Earth it is not much used for world maps because shapes are severely distorted in the outer half, as in all azimuthal projections. Note that in Fig. 2.16 the bounding line of the map is the *point* opposite the center point of the map. Since the center is the North Pole, the entire outer edge of the map is the South Pole!

Distances. The topic of distance representation on world map projections is rather involved since the inevitable shrinking and stretching of the globe surface varies from place to place and at most points is even likely to be unequal in different directions. Nevertheless, some of the distance relationships on the globe can be retained in world map projections.

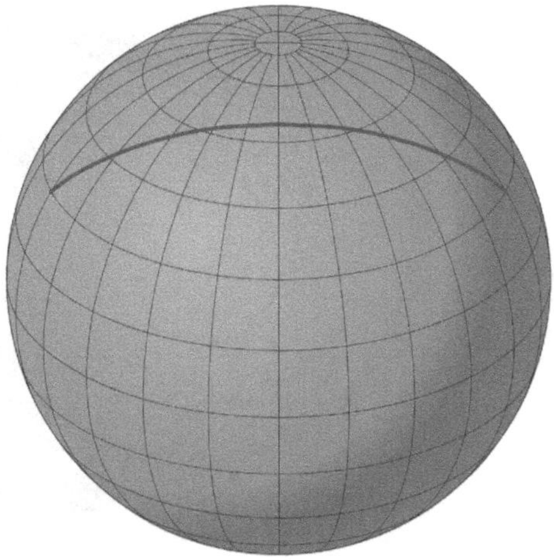

Fig. 2.17 Great-circle between two points on the same parallel

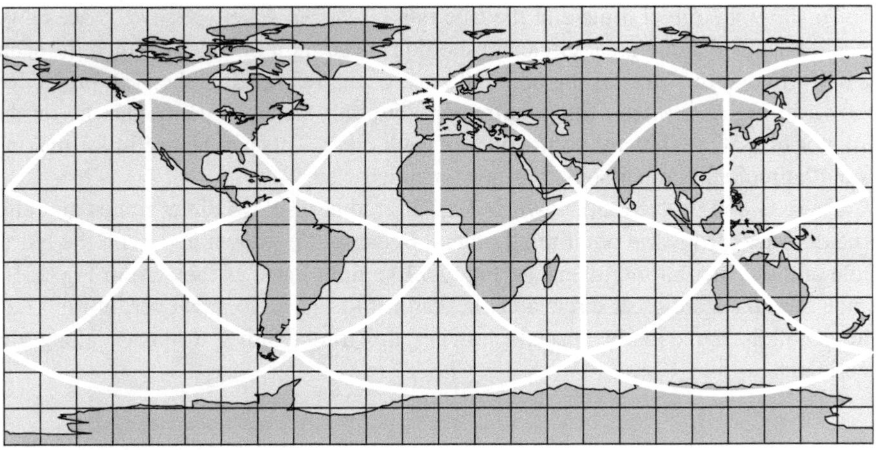

Fig. 2.18 Simple cylindrical projection (the equator and all meridians are length true)

A "true" distance on a map is a line the same length as it is on the reference globe, regardless of whether it is a great or small circle, straight or curved. In many world projections one or two lines may be their true lengths. In one projection (Fig. 2.18) the equator and all meridians are true, but no other line is. In another, shown in Fig. 2.19, the equator and all parallels are their true lengths. But remember that the shortest distance between two points on any parallel, except the equator, is not along the parallel.

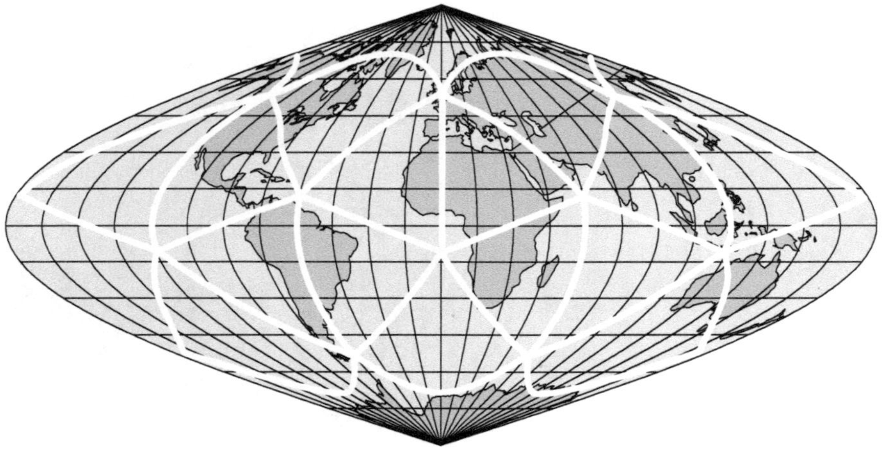

Fig. 2.19 Sinusoidal projection (the equator and all parallels are length true)

To show the direct route and the true relative distance between two places on a map demands that the great-circle arc should be a straight line along which the scale is uniform and the same as the nominal scale of the reference globe. The projection shown in Fig. 2.18 meets this requirement, but only between points that lie on the equator or that are directly north-south of each other. This can be useful for making some latitudinal comparisons, but not for much else.

There is one projection, shown in Fig. 2.20, that displays direct routes and also true distances from *one* point to all others. Because it is shown here with the North Pole at the center all meridians are Fig. 2.21 straight lines, as they are in Fig. 2.16, since meridians are great-circle arcs. In contrast to Fig. 2.16, however, in Fig. 2.20 the parallels are equally spaced showing the true relative distances along the meridians.

2.5 Classes of Projections for World Maps

The term "map projection" suggests the idea of a translucent globe with a light inside shining the map outlines onto a flat surface. As a matter of fact, two well-known and very useful projections for limited areas, the Gnomonic and Stereographic, are derived directly from this perspective model, but most projections are produced mathematically.

The great majority of world map projections may be grouped into classes; each class has a distinctive outline and distortion pattern. Except perspectives, they are all mathematical constructions that have a variety of attributes.

Azimuthal Projections. If we begin with the flat sheet to which the globe map is to be transferred, then one way to proceed is simply to place the translucent globe

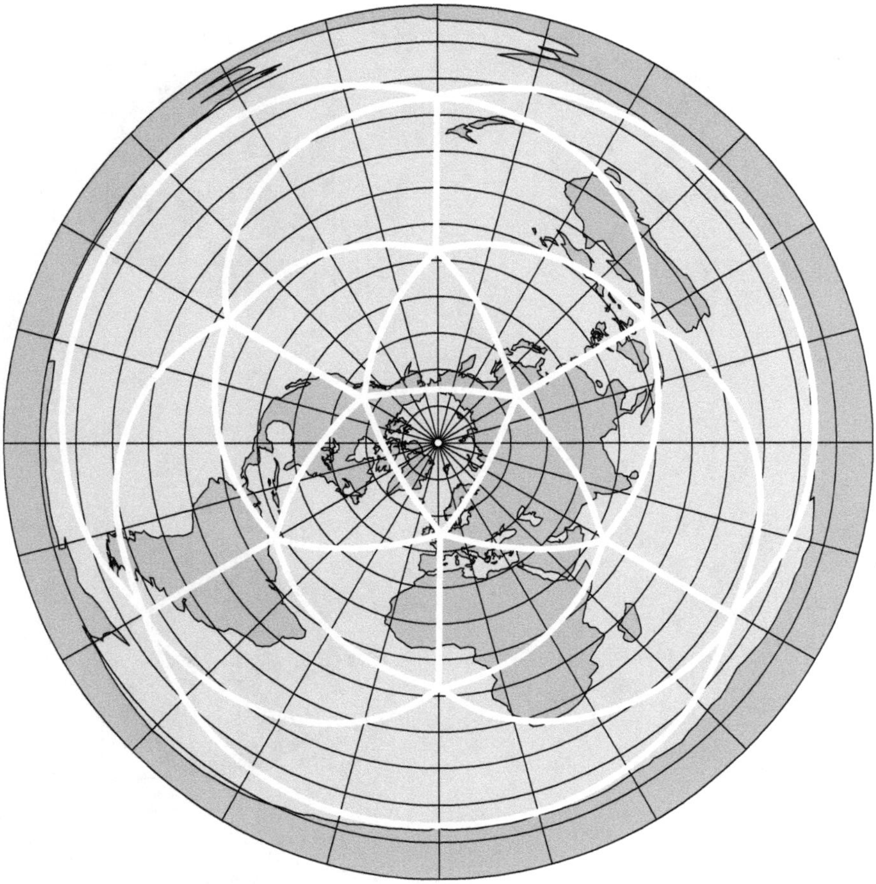

Fig. 2.20 Azimuthal equidistant projection (the true distances along all meridians)

on the sheet and "project" the map to the sheet by some system, such as placing a point of light some distance above the globe and located somewhere on the straight-line going through the center of the globe and that is perpendicular to the plane of projection. Any such perspective or perspective-like transfer will be symmetrical around a center point, resulting in a circular outline on the flat sheet. They are named "azimuthal" rather than "circular" because of their significant attribute that all directions from their centers are correct, which means that all great circles through the centers are straight lines on such maps. An azimuthal projection can be centered anywhere. Any complete azimuthal projection has a circular outline, but the central sections of these projections are often used in rectangular format. One azimuthal projection, the Azimuthal Equidistant (Fig. 2.20), is often used for world maps, but most azimuthal projections cannot be usefully extended to include the whole Earth.

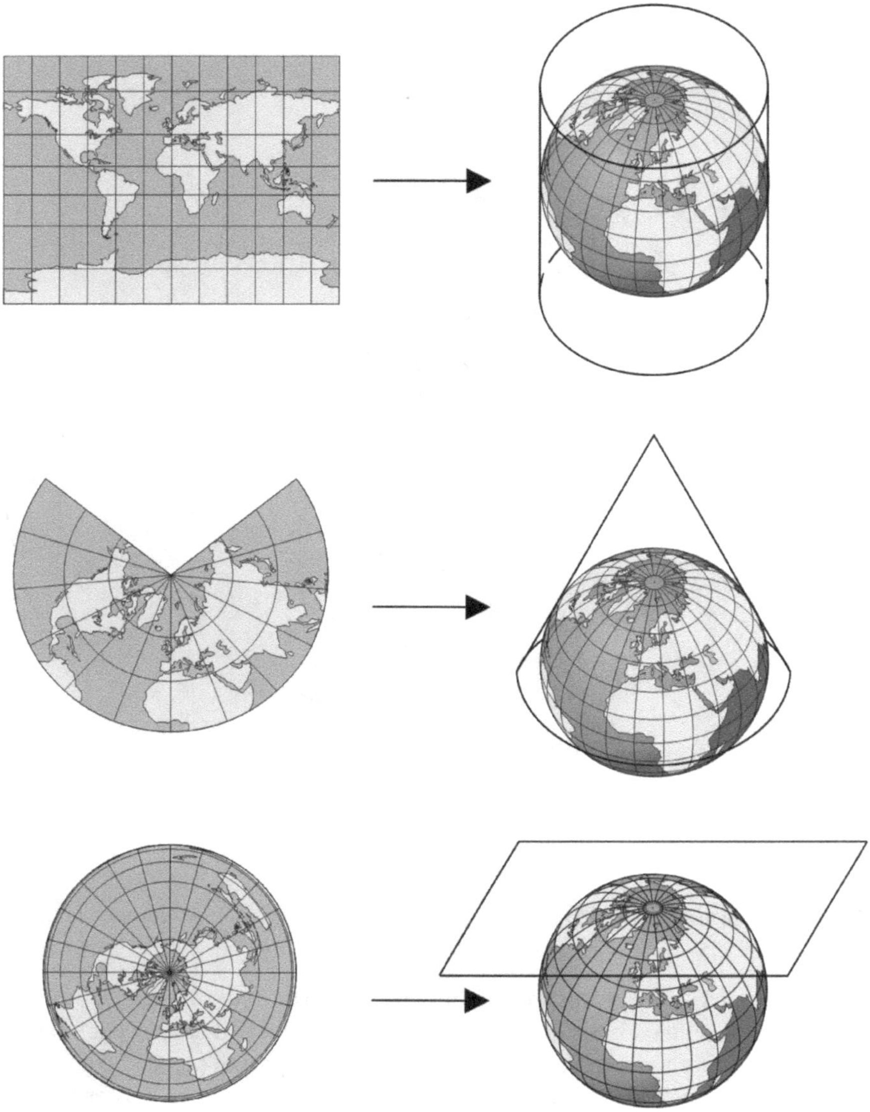

Fig. 2.21 Possible visualisation of map projections

Cylindrical and Conical Projections. Alternatively, the sheet may be formed into a cylinder and wrapped snugly around the globe, or the sheet can be rolled into a cone and perched on the globe. When the globe map has been somehow transferred to the surface of the cylinder or cone the form can be unrolled and laid out flat again. The described approach is rather old-fashioned and true for perspective projections only. The majority of map projections are not perspectives, and they are called cylindrical or conical because they are similar to cylindrical or conical perspectives by the appearance of graticule. A map designed in a cylindrical projection can be wrapped into cylinder, while one designed in a conical projection can be wrapped into cone.

All cylindrical projections show the Earth within a rectangle. Normally, these projections are arranged so that the parallels of latitude appear as a set of parallel, straight lines crossed at right angles by a set of parallel, straight meridians. Even though the overall shape is always a rectangle, they are called "rectangular" only when the graticule thus appears as a grid of perpendicular lines.

Various cylindrical projections have been devised as "compromise" alternatives to the conformal Mercator projection (Fig. 2.13) and the infinite equal-area series, to which the Gall-Peters projection (Fig. 2.11) belongs, such as the Miller Cylindrical, shown in Fig. 2.22.

Cylindrical projections are often used for world maps simply because a rectangular graticule seems to be psychologically comfortable and the format fits pages and walls. One should have better reasons.

The rectangle, circle, and especially the oval frame or border, include the great majority of projections used for world maps. Conical construction is widely used for maps of lesser scope, but it is rarely used for world maps.

Oval Projections. Although all these projections are basically oval, they can have noticeably different shapes. Compare the Sinusoidal projection (Fig. 2.19) with the Robinson projection shown in Fig. 2.23. Some, like these, are called

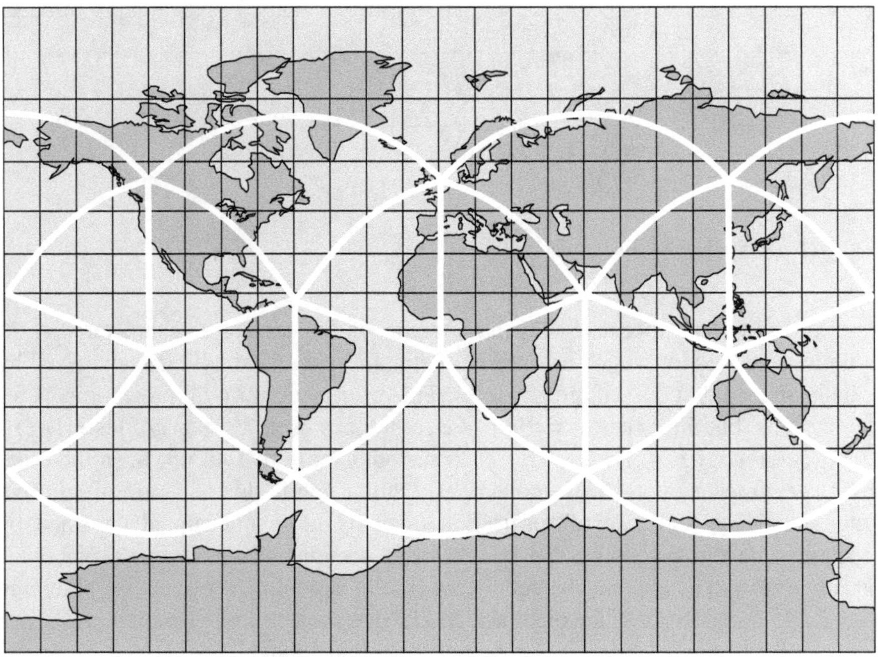

Fig. 2.22 Miller cylindrical projection

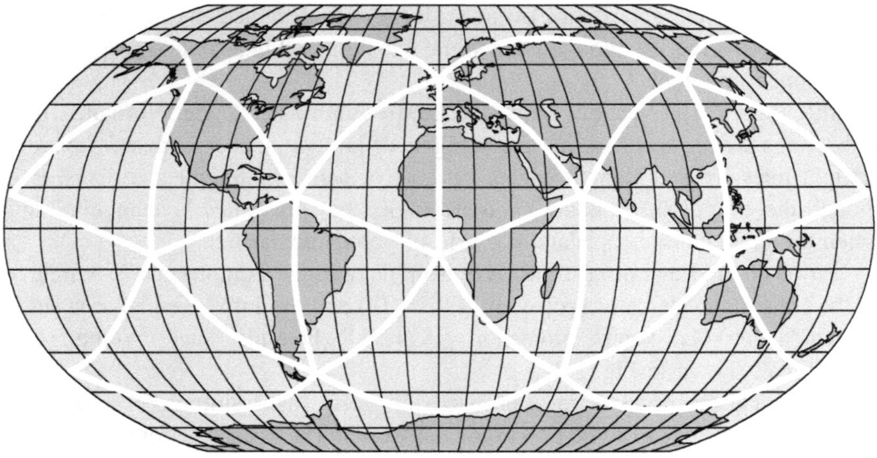

Fig. 2.23 Robinson projection

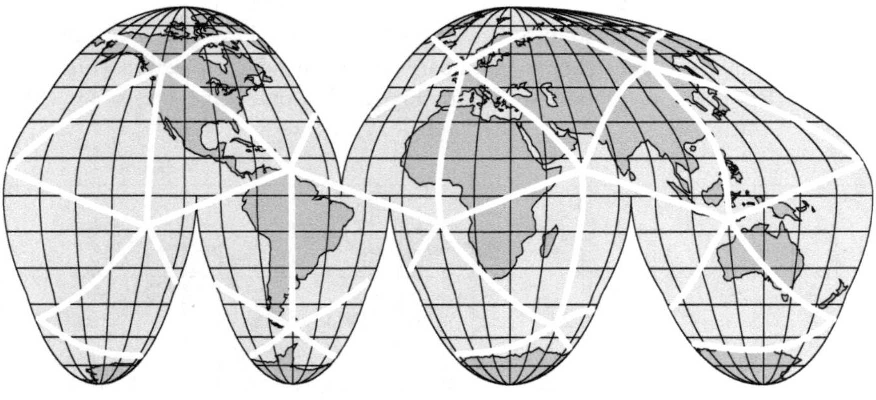

Fig. 2.24 Goode Homolosine projection

"pseudocylindrical" because they have some of the structural characteristics of cylindrical projections, such as one set of straight, parallel small circles crossed by equally-spaced, but curved, great circles. The primary use of oval projections is for world maps, but their central sections are sometimes used for maps of lesser areas.

Miscellaneous Projections. This class includes those that do not fit in the other three categories of world map projections. Only a few in this class are often used. One such is an "interrupted" projection (actually more interrupted!) formed by repeating and fitting together the less-distorted sections of one or two projections, usually oval. An example is the equal-area Goode Homolosine projection, shown in Fig. 2.24, that joins the Sinusoidal and Mollweide projections. The lands could just as well be interrupted to show the seas to better advantage.

Fig. 2.25 Van der Grinten projection

Another is the Van der Grinten, a circular but not azimuthal projection, shown in Fig. 2.25. Usually much of the polar sections are cut off because distortion is extreme there, as well as to give the map a rectangular format.

Map projections in the above classes may, of course, have some, none, or a combination of the various useful attributes.

2.5.1 Aspects of Projections

One part of a smooth sphere is curved exactly like any other part even though there are lines on it displaying a graticule and it shows geographical features. No matter how it may be turned with respect to a flat map surface a given projection system will produce maps that have exactly the same distortion pattern. At the same time, the way the globe is turned will have considerable effect on the looks of the graticule and the geographical features on the map because that shifts them in relation to the distortion pattern.

There are three classes of aspects. The three classes are generally called *direct*, *transverse*, and *oblique*. Sometimes a few other terms are used in place of these in particular cases: "equatorial", "conventional", "normal", and "polar." Their use will be identified below. Remember, no matter what the graticule looks like or how the geographical features are displayed on the map, the fundamental patterns of distortion are not affected. We illustrate the three aspects, with typical appearances of the graticule, by means of representatives of the three major classes of world map projections:

A—Simple *cylindrical* projection
B—Lambert *azimuthal* equal-area projection
C—Mollweide *oval* projection

When one is faced with deciding which world map to use, its aspect is an important consideration. The appearance of the graticule is particularly important. It is clearly an advantage to display the graticule in a way that allows easy recognition of basic directional relationships and so that its appearance does not perplex or perhaps even disconcert the viewer.

Normal Aspect. This is the name for the aspect of cylindrical and pseudo-cylindrical projections of the Earth's sphere when the projection of equator is a straight-line positioned in the middle of the map, while one pole is projected at the top and the other at the bottom of the map. This is also called the "equatorial", "direct" or "conventional" aspect for the cylinder or oval. In every case this results in a straight-line, central equator crossed at a right angle by a straight central meridian, as diagramed in Fig. 2.26.

Transverse Aspect. This is the name for the aspect that results when the projection of poles are points positioned on a meridian that appears horizontally (or vertically) in the middle of the map. In transverse cylindrical and oval projections one great circle (a pair of meridians) will form a central straight line through the long dimension of the projection outline, and a pole will be located somewhere along that line. If the pole is centered on an oval projection or, on a cylindrical projection, anywhere along the paired-meridian straight line, that will be crossed at a right angle by another pair of straight-line meridians. Except for the straight-line great circles and the bounding lines of cylindrical projections, most other elements of the graticules of these cylindrical and oval projections will be complex curves.

In all transverse azimuthal projections the meridians, being arcs of great circles, are straight lines radiating from the central pole at their correct angles. The parallels are circles concentric around the center pole. If the other pole can be shown it is the bounding circle. The only difference among the graticules of the polar aspects of the azimuthal projections is in the spacing of the parallels. The relationships of the transverse or polar aspects are diagramed in Fig. 2.27.

(a)

(b)

(c)

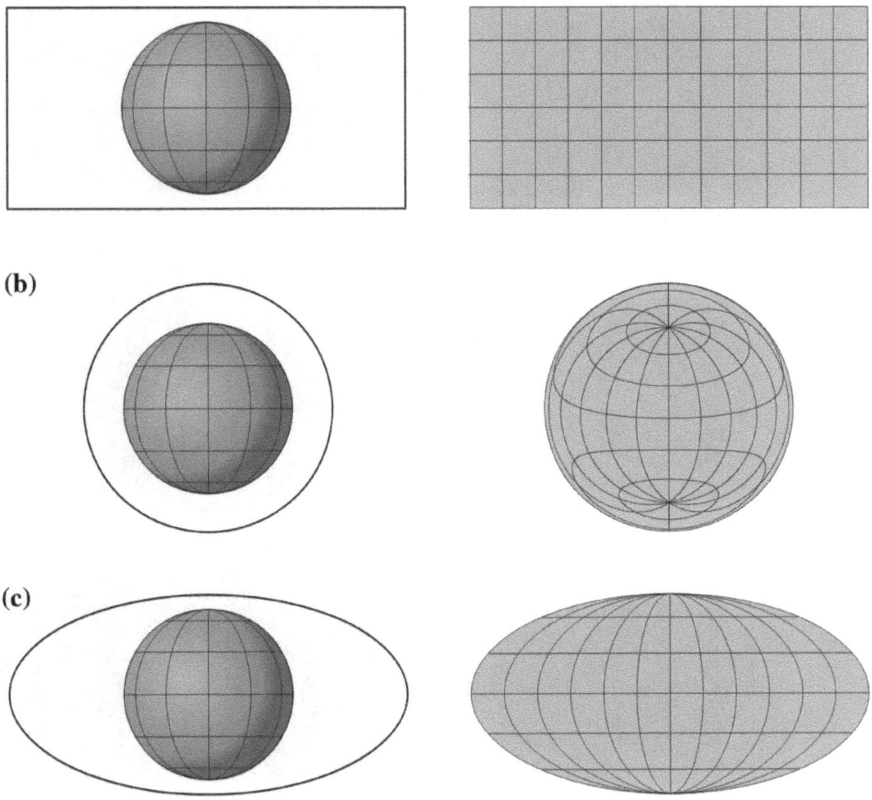

Fig. 2.26 Normal aspects

Oblique Aspect. This name applies to those aspects that are not direct nor transverse. In this aspect the graticules of cylindrical, azimuthal, and oval projections will usually appear as arrangements of complex lines, although the central sections of azimuthal projections may appear nearly "normal." These relationships are diagramed in Fig. 2.28.

Note that since the axis of the globe can make any angle with the plane between 0° and 90°, each world map projection system can have innumerable oblique aspects. For example, Fig. 2.29 shows how the appearance of the graticule changes on a Mollweide projection when the axis of the globe is tilted at progressive 10°-intervals from 0° to 90°. They are all the same projection system with the same pattern of distortion. Nonsymmetrical graticules for projection systems are produced if the axis of the globe is tilted in still other directions.

(a)

(b)

(c)

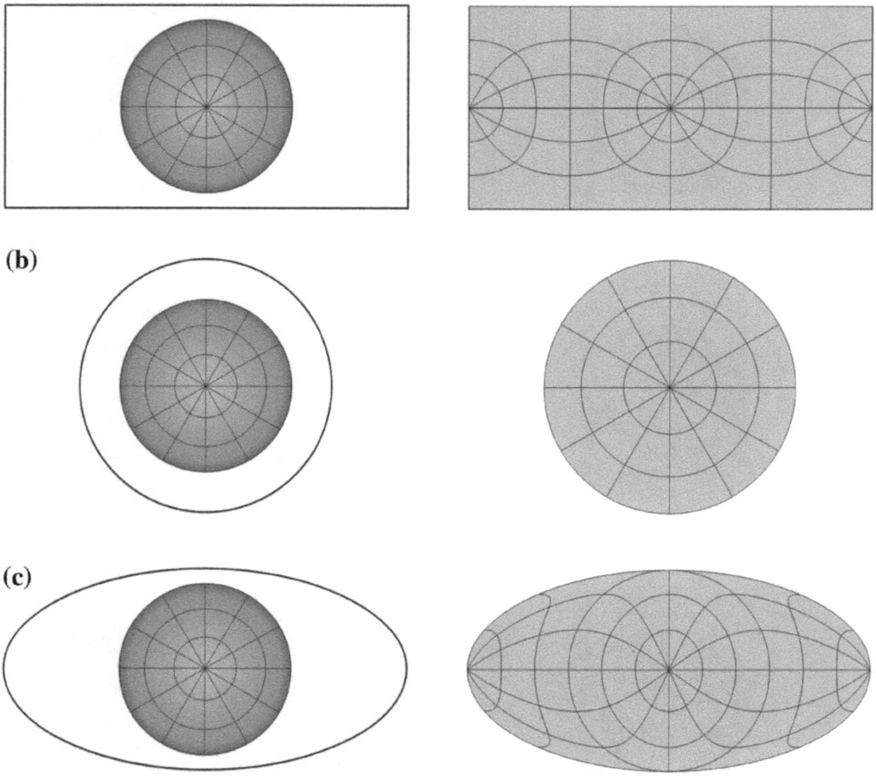

Fig. 2.27 Transverse aspects

2.6 Recognizing Distortion

Most people probably understand, perhaps only vaguely, that flat world maps must somehow distort the globe map. But just knowing that it occurs is not very satisfying; it would be much better to know how and where. Ideally there would be some kind of summary index, such as a letter grade of a, b, c, or a grade-point average, that could be used to rank world map projections. Unfortunately, that is not possible. As an alternative the average user of maps can easily deduce some aspects of distortion merely by observing the graticule. Another approach is to learn to recognize the general patterns of distortion associated with the overall shapes of world maps.

Analysis of Distortion. Only informational publications like this ever display, in addition to the graticule, any guide to the distortion, such as the set of equilateral spherical triangles. Unfortunately, the graticule is rather perplexing to many who are not very familiar with the looks of a globe, and this is especially so with regard

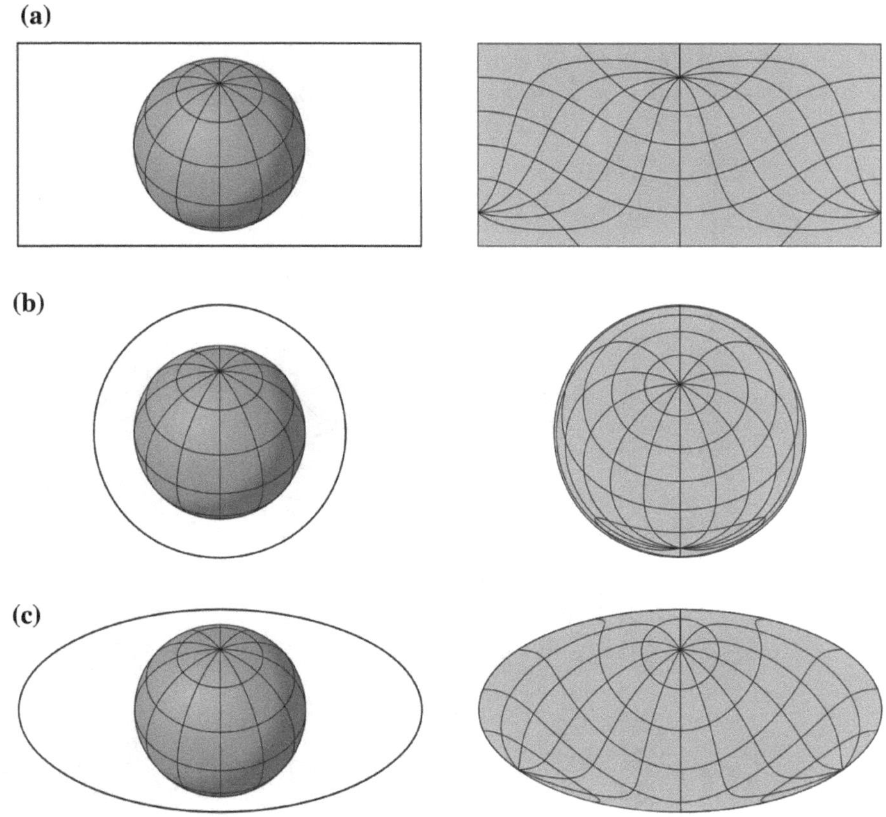

Fig. 2.28 Oblique aspects

to the complex of wavy lines on maps in oblique aspect. On the other hand, the graticule on maps in normal aspect can be a help in recognizing how the projection system has enlarged or diminished sizes, elongated or shortened distances, altered angles, and so on. One need only keep in mind a few basic facts about the spherical coordinate system.

Seven of the characteristics of the graticule on the globe are useful for this kind of reasoning. Assuming a perfect sphere, as we can for world maps, they are:

1. All meridians converge to a point at each pole.
2. All parallels are parallel.
3. All parallels and meridians intersect at right angles.
4. All meridians are spaced equally on each parallel.
5. All parallels are spaced equally on each meridian.
6. Parallels and meridians are spaced equally at the equator.
7. Meridians at 60° Lat. are half as far apart as are the parallels at that latitude.

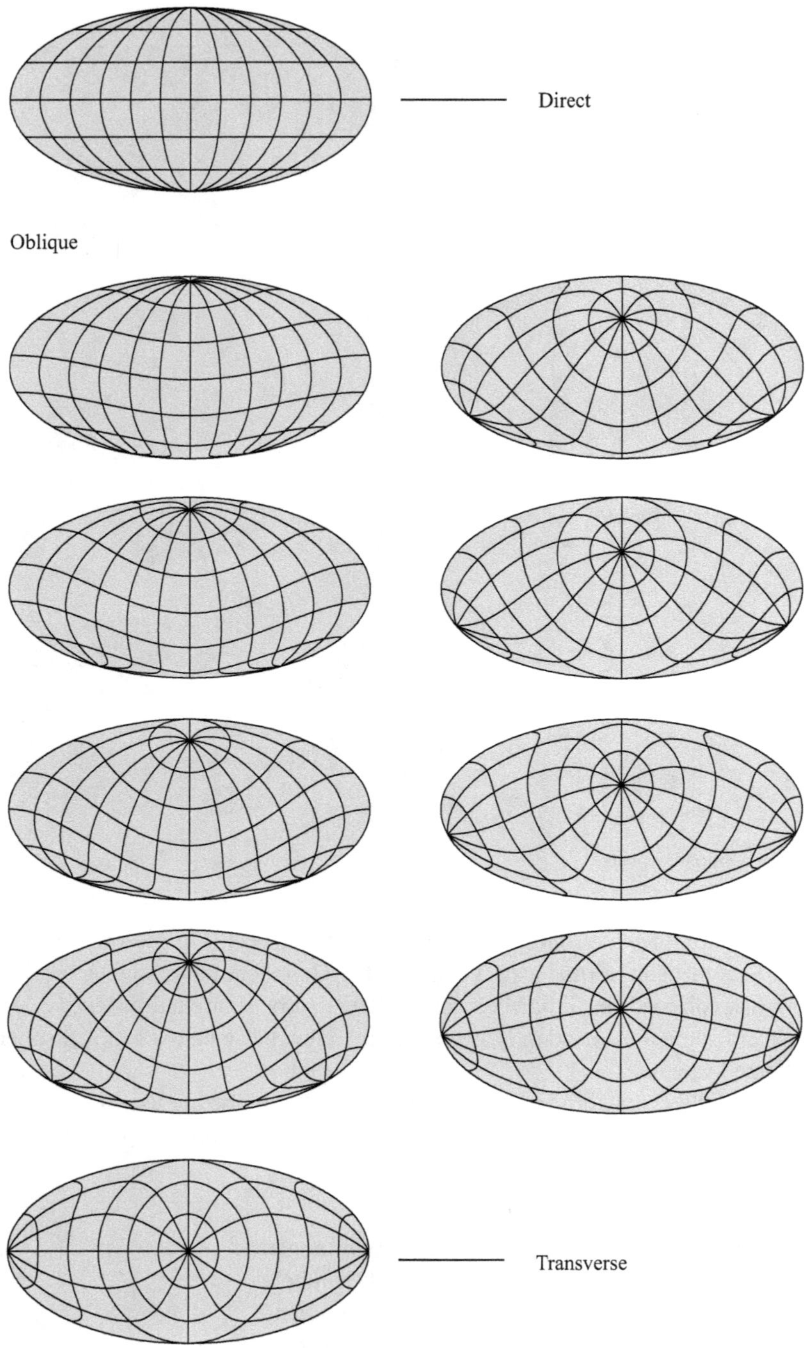

Fig. 2.29 Appearance of the graticule changes on a Mollweide projection

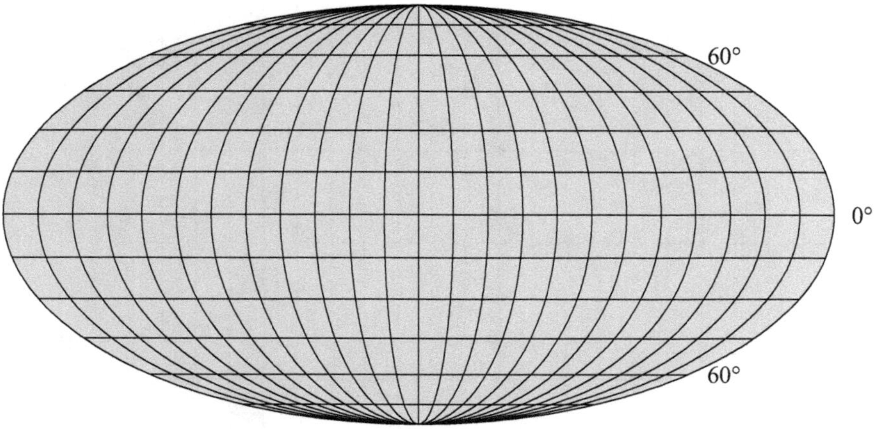

Fig. 2.30 Mollweide projection

When one compares these facts with the appearance of the graticule of a map on which both parallels and meridians are shown at equal-degree intervals, such as 10°, 15°, or 20°, some useful relationships may be inferred. Two projections shown in equatorial aspect, each with a 15°-graticule interval, provide examples.

An analysis of the graticule of the equal-area Mollweide projection, shown in Fig. 2.30, allows the following conclusions. The numbers in parentheses refer to the foregoing list.

Only characteristics (1), (2), and (4) are retained. Meridians and parallels do not intersect at right angles (3) anywhere, except along the equator and the central meridian, so angles elsewhere must be altered. Parallels are closer together toward the poles (5), so distances there must be somewhat shortened north-south. Meridians are nearly, but not quite, half as far apart as parallels at 60° Lat. (7).

Unfortunately, the conformal Mercator projection, shown in Fig. 2.31, is often used for general world maps. An analysis of its graticule provides abundant evidence of why it is an unsuitable choice. Characteristics (2), (3), (4), and (6) are retained, so one can assume that the angles and areas in the equatorial regions are quite well displayed. On the other hand, meridians obviously do not converge (1), so the higher latitudes must be greatly distorted. Parallels are not spaced equally on meridians (5), but become increasingly farther apart toward the higher latitudes. Since the parallel of 60° Lat., is shown the same length as the equator, it must be twice as long as it is on the globe (7), while the distance between parallels right at 60° seems about twice what it is at the equator. Accordingly, areas at 60° Lat. are four times as large as they are on the reference globe. Furthermore, because the expansion increases steadily poleward from the low latitudes, the middle and high latitudes must be much enlarged, and sizable areas in those regions must be greatly distorted.

Fig. 2.31 Mercator projection

Patterns of Distortion. The distortions of distances, sizes, and angles generally vary from place to place over a projection, and the overall outline of a projection is an important clue to its pattern of distortion, that is, to the way the distortion is arranged.

The cylindrical, azimuthal, and oval classes have distinctive patterns. In general, whatever the distortion of distances, sizes, and angles may be in a projection it will be a minimum along the true-scale lines or standard lines (the same length as those lines on the reference globe), and often the scales in other directions on those lines will not be greatly different. The distortion will increase perpendicularly away from the standard lines in cylindrical projections and radially from the true-scale point in an azimuthal projection, as diagramed in Fig. 2.32a, b, in which the darker the shading, the greater the distortion.

There could be two standard lines in cylindrical projections and the distortion will increase perpendicularly away from them in both directions as diagramed in Fig. 2.32c.

The oval projections have patterns of distortion quite different from the cylindrical and azimuthal patterns. In general, they have two "areas of strength" aligned with the perpendicular longer and shorter axes of the oval. Normally, these

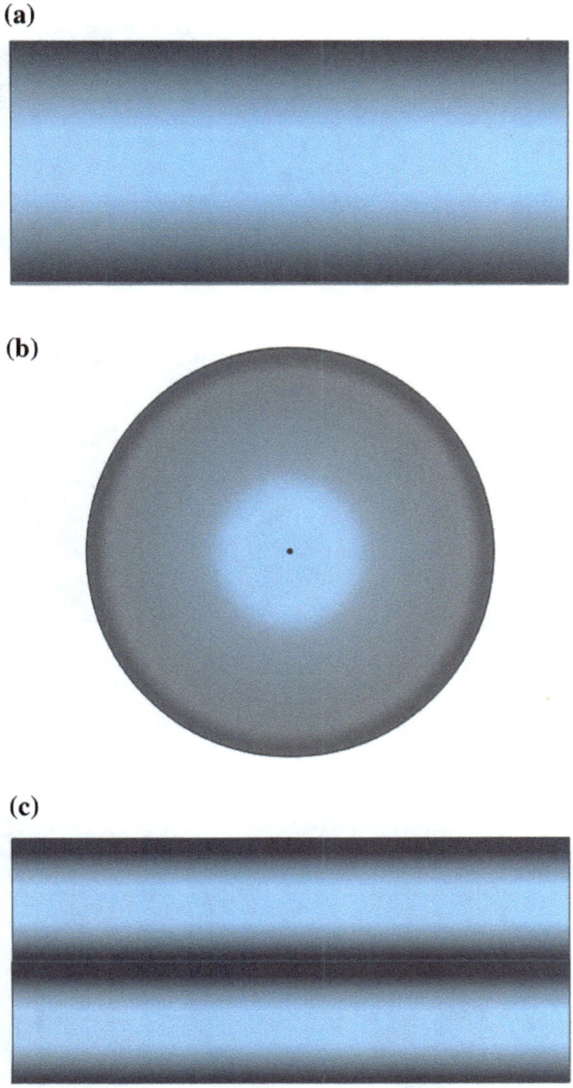

Fig. 2.32 Patterns of distortion in different map projections

projections are oriented with the long axis horizontal and the major difference among them is the way the distortion is arranged along the central, vertical axis. Two examples are shown in Fig. 2.33.

In almost all continuous oval projections there are four areas of maximum distortion symmetrically positioned in the outer sections of the four quadrants.

The advantage to be obtained by interrupting a projection so as to repeat the less distorted sections is illustrated in Fig. 2.34. In addition to being interrupted the projection is a combination of two equal-area projections, the Sinusoidal (better shapes in the equatorial areas) and the Mollweide (better shapes in the high

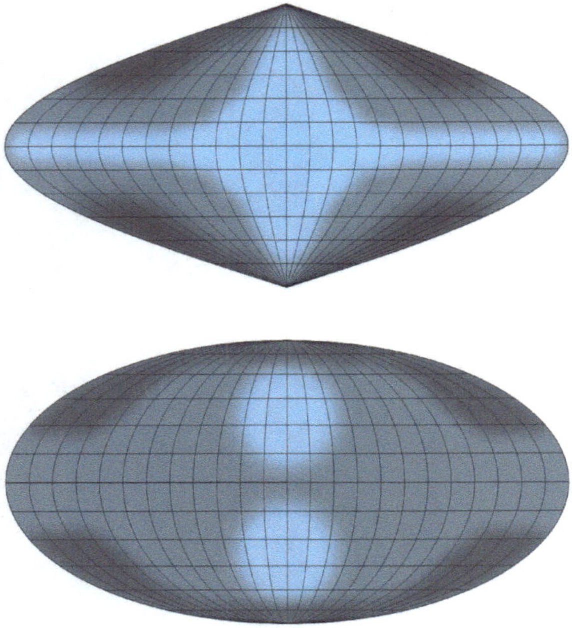

Fig. 2.33 Patterns of distortion in sinusoidal and Mollweide projections

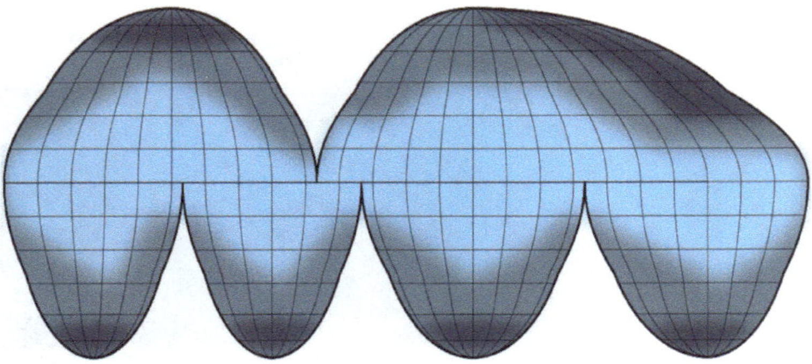

Fig. 2.34 Interrupted Goode Homolosine projection

latitudes), in order to minimize the total amount of inevitable distortion. The additional interruptions violate even further the inherent continuity of the surface of the globe, but as shown earlier, distortions may be considerably reduced by interrupting the surface of the globe map in many places. The question of whether the disadvantage of lack of continuity is worth the advantage of less distortions is one that must be decided in terms of the use to which the map will be put.

2.6.1 Distortion Cartoons

As described earlier, the graticule itself can serve as a "built-in" way of illustrating distortion, but it only functions easily when the projections are in normal, or sometimes transverse, aspects. A far better illustration is provided by the set of equilateral spherical triangles plotted on many of the colored world maps in this chapter. On the globe the triangles are alike in every way, lengths of sides, interior angles, areas, and so on, and because they neatly cover the entire globe surface one can compare the different parts of one projection as well as one projection with another. Other devices also have been used, such as spherical polygons or circles plotted in different parts of a projection, and arcs of great circles arranged strategically.

In an attempt to provide a more easily appreciated display of what a projection system does to the globe surface some authors have drawn a profile of a person's head on one projection and then mapped that same outline in the corresponding geographical positions on another projection. One can then compare the two. A major deficiency with this procedure is that the first projection always involves distortions. This flaws any comparison with reality since the profile will "look right" on whichever projection is used first.

There is considerable value in mapping a familiar figure, such as a human head, in the central sections of various world map projections. Everyone knows what a "face" should look like, and although the deformation resulting from the mapping is a complex mix of the various kinds of distortions, the overall modification of reality is readily apparent. Human faces or profiles tend to have such complex characteristics that many of the deformations are quite subtle. In order to employ a simpler and more geometric visage we use drawings of two masks, *Tragedy* and *Comedy*, derived from those worn by actors in ancient Greece, shown in Fig. 2.35. (In that period a few actors played all the parts and wore different masks to personify each role. *Tragedy* and *Comedy* were fundamental characters.)

Fig. 2.35 Comedy and tragedy masks

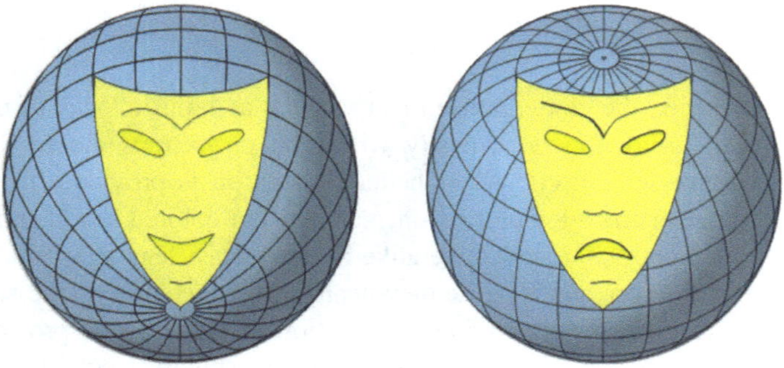

Fig. 2.36 Comedy and tragedy masks on a globe represented in orthographic projection

In order to be as realistic as possible we begin by plotting each mask on a globe (Orthographic projection), centering *Tragedy* at 45° Lat. in the northern hemisphere and *Comedy* at 45° Lat. in the southern. There is relatively little perspective distortion in the center of a globe view, and what there is is symmetrical about the center point. *Tragedy*'s chin is on the equator, and the upper tips of that mask are on the great circle formed by the meridians 90°E and 90°W of the central meridian. *Comedy*'s chin is at the South Pole, and the upper tips of that mask are on the equator.

In order that all the cartoons may be comparable they are shown at the same nominal scale. Two views of the masks on the reference globe at that scale are shown in Fig. 2.36. That scale is the same nominal scale as that of the colored world maps in earlier sections.

The plotting of the masks shown on Fig. 2.37 for the various projection systems is by computer. They are machine plotted just as would be the case were the masks continents with internal features. The masks are positioned in the central, longitudinal sections of the projections. Whatever the distortion, it is generally the least in these parts of the azimuthal and oval projections; mapping the masks farther to the right or left would add shearing. Distortion does not increase "sideways" in cylindrical projections.

2.6.2 *A Final Note*

To display the whole Earth on a flat map is a complicated operation. Before anything else can be done one must select a map projection to serve as a base. There are many from which to choose in each of the basic classes, and each one can be

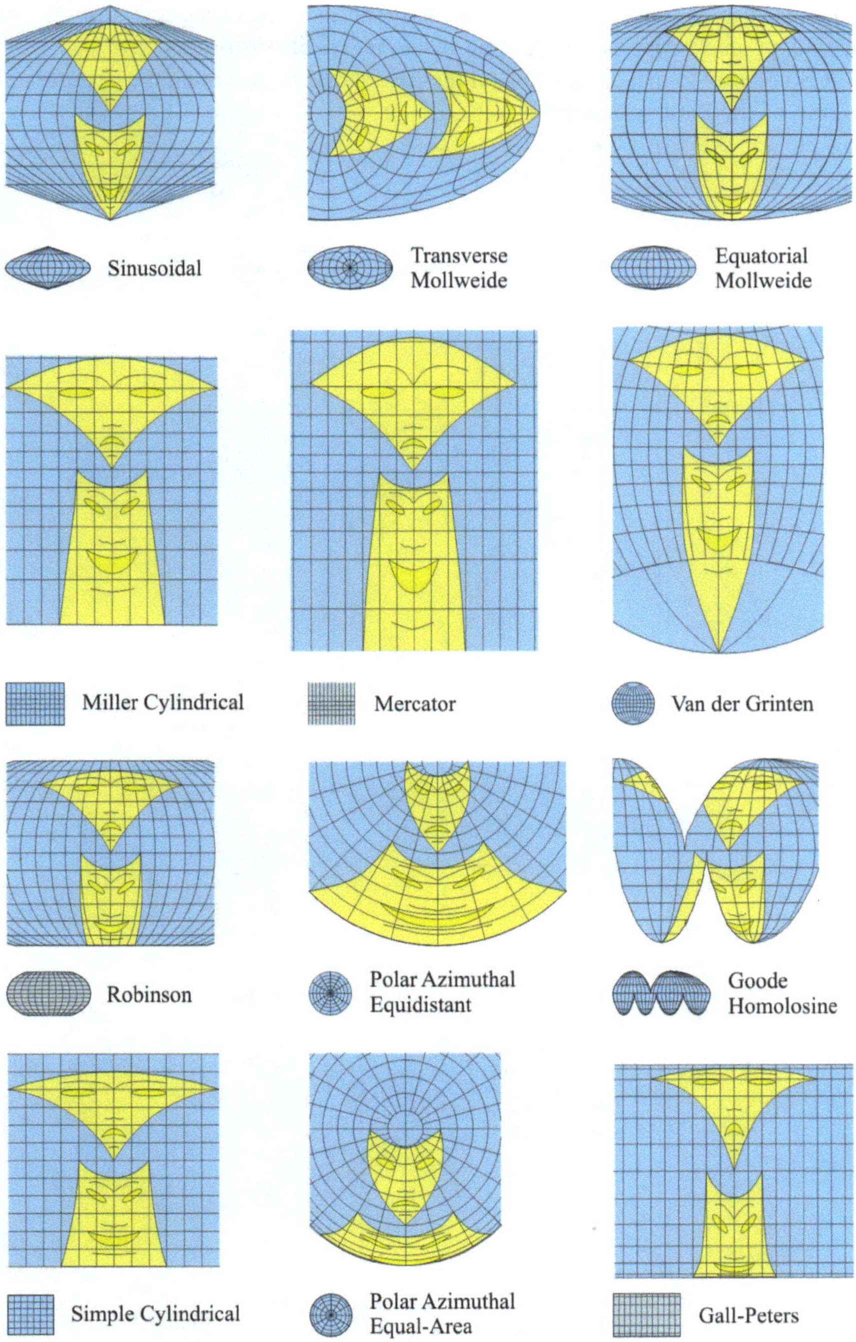

Fig. 2.37 Masks in various map projections

constructed in any desired aspect. One must settle on the necessary attributes and then adopt an appropriate distribution of the inevitable distortion.

It is not easy to make a choice, but the advantages of providing the best possible portrait of the Earth are worth the effort.

Chapter 3
Matching the Map Projection to the Need

Arthur H. Robinson and The Committee on Map Projections

3.1 Foreward

A globe is the only representation of the Earth that does not distort its geometry—except, of course, its size. Unfortunately, for many purposes a globe is an inconvenient way to display geographical relationships. Significant concepts, such as constant directions, shortest routes, and equal distances are difficult to measure on a curving surface. Often we want to see the whole world, but in looking at a globe we see only half. Finally, unless the globe can be turned in all directions, it is difficult to look directly at a region of interest.

The alternative, a map made by methodically flattening the globe surface, modifies its geometry but is easier to make, file, and (especially) use. A systematic transformation of the spherical Earth's surface to a flat map is called a *map projection.*

There are infinite ways to carry out this transformation. The changes that result from flattening the spherical surface are technically called "distortions." This is a

---

This chapter is slightly modified version of the Special Publication No. 3 of the American Cartographic Association, a member organization of the American Congress on Surveying and Mapping, 5410 Grosvenor Lane, Bethesda, Maryland 20814. The Special Publication No. 3, ISBN 0-9613459-5-0, has been published in 1991 by the American Congress on Surveying and Mapping. In that time, the Committee on Map Projections consisted of the following distinguished members: James R. Carter, Antony R. de Souza, Marshall B. Faintich, Patricia Caldwell Lindgren, Barbara B. Petchenik, Arthur H. Robinson and John P. Snyder, Chairman. Editors were Arthur H. Robinson and John P. Snyder, reviewer Barbara B. Petchenik, Copy Editor Gail Papa, and Design and Production by David DiBiase and Jeremy Crampton from Deasy GeoGraphics Laboratory, The Pennsylvania State University.

A.H. Robinson (✉)
Madison, WI, USA

M. Lapaine and E.L. Usery (eds.), *Choosing a Map Projection*,
Lecture Notes in Geoinformation and Cartography,
DOI 10.1007/978-3-319-51835-0_3

rather unfortunate term since it connotes something undesirable. Yet many projection systems can turn the inevitable distortions to our advantage. For example, direct routes can be made straight lines, regions of greater interest can be enlarged, ranges can be shown as circular, directions and distances from one point to all others (and vice versa) can be clearly displayed, areas of interest can be centered, and so on. The following chapters, prepared by experts, illustrate a number of ways to make map projections serve different purposes.

It is our hope that the reader will be stimulated to understand map projections and appreciate their versatility. They are not really mysterious.

3.2 As the Earth Turns

Arthur H. Robinson

The title of this chapter is similar to that of a well-known soap opera. It seems appropriate because the turning Earth is the stage on which the human drama is played. Like the ongoing soap-opera drama, the spherical Earth's surface is continuous, with no obvious beginning or end. How we view this stage and the relation of the turning globe to its display on world maps is the subject of this introduction to the more specific and detailed treatments in this book.

3.2.1 Meridians and Parallels

First, a few basics. The ball-shaped Earth spins around an axis that extends between the North and South poles. The poles are the anchor points for the geographical coordinate system of *latitude* and *longitude* (Fig. 3.1). Longitude designates

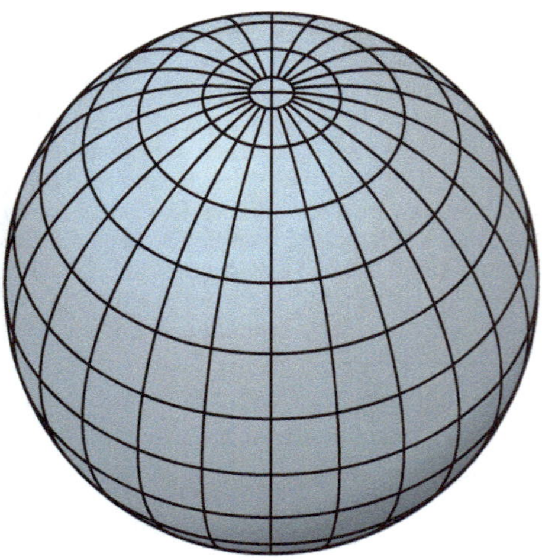

Fig. 3.1 The geographic coordinate system

distance E or W from some starting line. It is shown on world maps by a selection of great-circle-arc *meridians* that extend from pole to pole and converge to each. Latitude designates distance N or S from the equator, a line around the Earth halfway between the poles. It is shown by a series of smaller circles called *parallels*, that are concentric to the axis. On the round Earth all parallels and meridians are perpendicular to one another. The selection of parallels and meridians shown on maps is called the *graticule*.

The turning Earth is converted to something like a stage set on a flat world map that normally displays a graticule, coastlines, and a selection of boundaries. A spherical surface cannot be flattened without distorting it in several ways, so all maps must select and arrange the distortions by means of a systematic transformation called a *map projection*. There are scores of map projections, some more suitable than others for particular purposes. They display the Earth in a variety of shapes (and, of course, sizes), often as ovals and rectangles. Showing the entire Earth within a circle is not rare, but it is usually done only to show distances and directions from one point. A principle of stage design is simplicity, in the sense of not diverting attention from the drama being presented. It has its equivalent in the design of world maps.

Because the Earth is a ball, we can tilt its axis from the vertical in any way desired without affecting the inherent character or distortion pattern of the chosen map projection. But the way the axis is tilted will greatly affect the appearance of the graticule (Fig. 3.2). Since the parallels and meridians are simple shapes, it is usually not desirable to display them as complex curves. Accordingly, except when special needs require otherwise, the preferred design is as A in Fig. 3.2. The central meridian goes from top to bottom, and the equator becomes a horizontal straight line dividing the Earth neatly into northern and southern hemispheres. Eastern and western halves are separated by a straight central meridian.

3.2.2 *The Earth Viewed from Different Directions*

Some stage sets are constructed on a large circular platform that can be rotated to present different views to the audience. Similarly, we may rotate the Earth around its polar axis so as to bring any desired region "front and center." For example, in Fig. 3.3, on an Eckert IV *equal-area* projection, we turn the Earth as it might be desired by an American, an African, or an Asian.

The display of the Earth with north at the top of the page (or so that "north is up," as many people say) is a fairly recent convention. There is no "up" or "down" in space; those terms are earthbound, "down" being toward the center of the Earth and "up," the opposite. The convention stems partly from the beliefs that consistency is a virtue and that it is easier to recognize map shapes in familiar orientations, and from the fact that more mapmakers have lived in the northern hemisphere. Whatever the reasons, the convention rankles some who dwell in the southern hemisphere where some maps are made "upside down" as illustrated in Fig. 3.4.

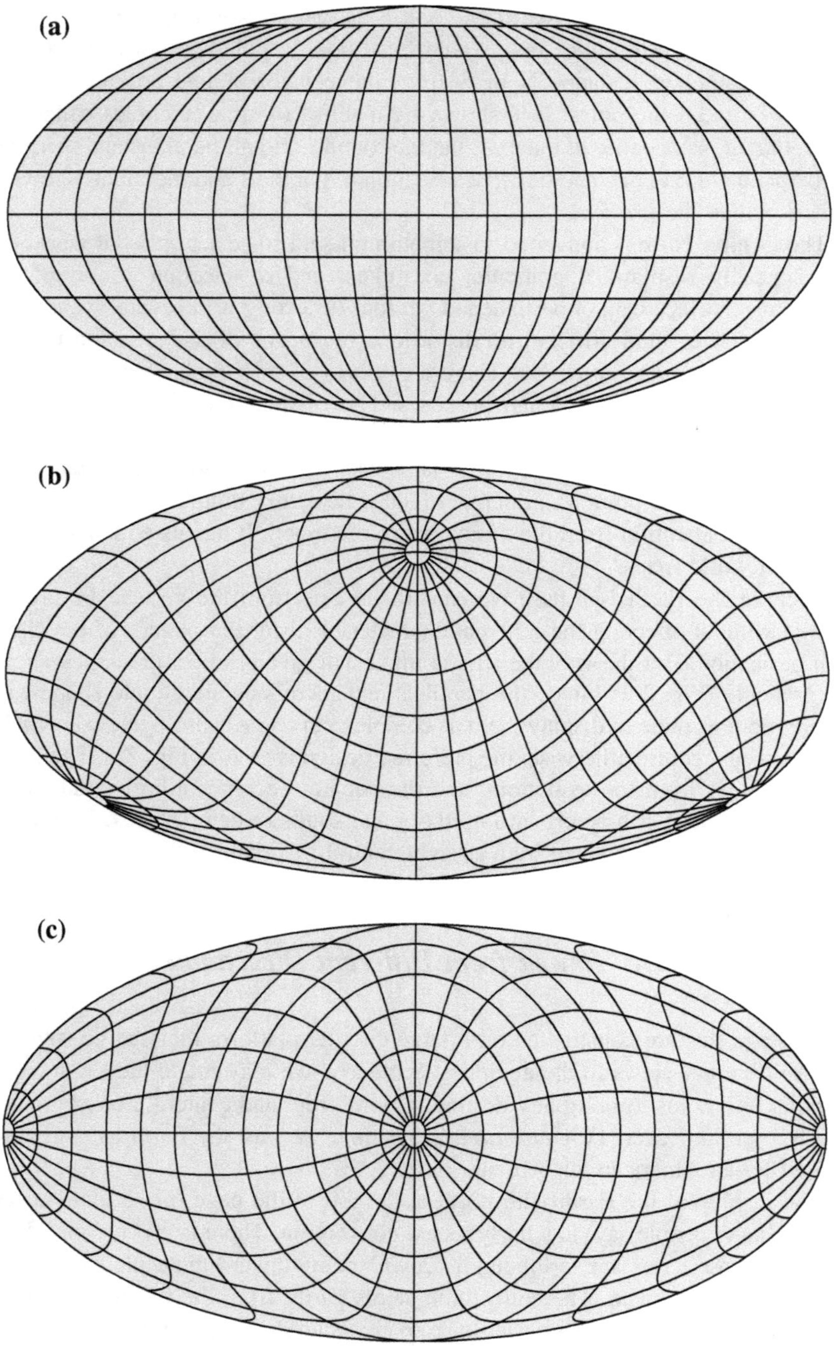

Fig. 3.2 The graticule on three Mollweide equal-area projections centered at **a** the equator, **b** 45°N latitude, and **c** a pole

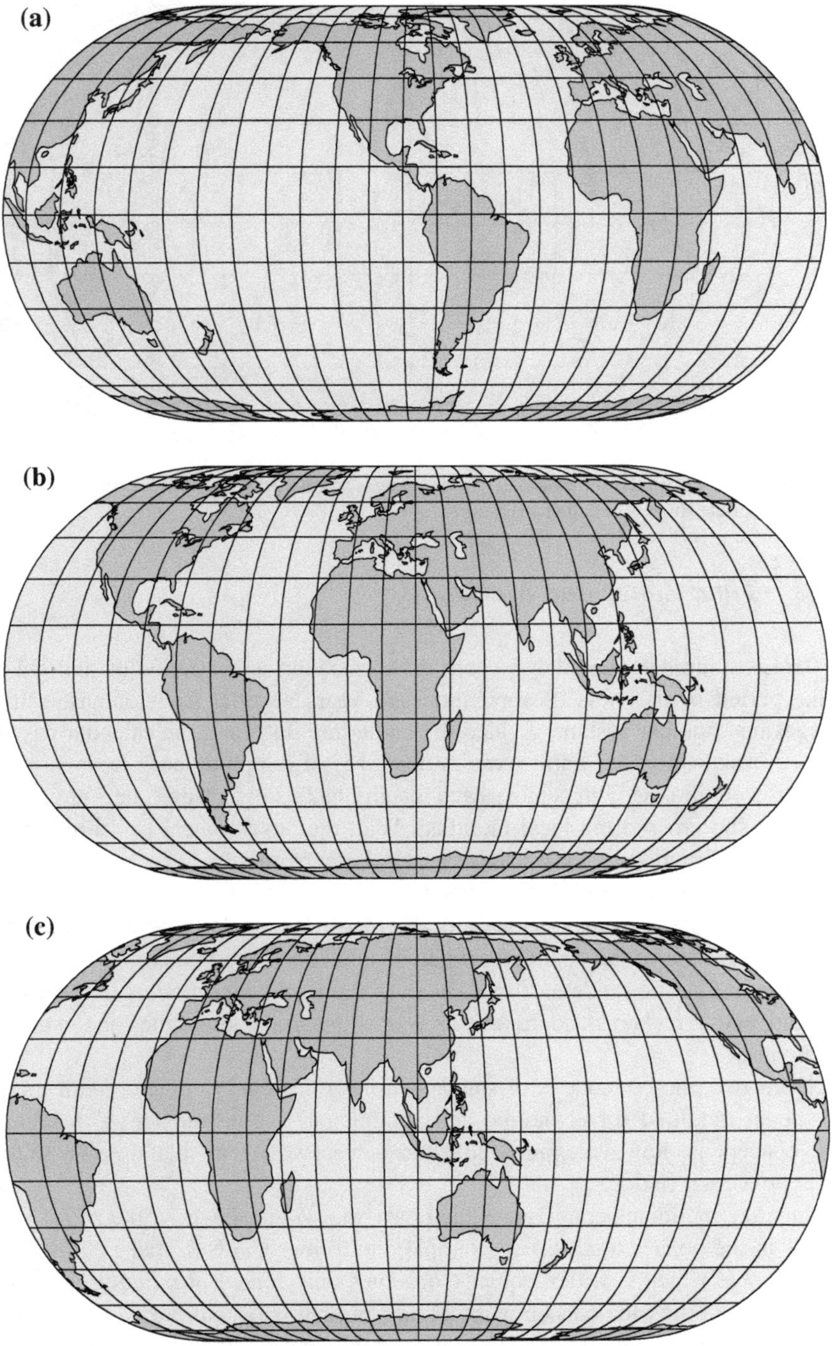

Fig. 3.3 The earth turned on an Eckert IV equal-area projection as might be desired by **a** an American, **b** an African, and **c** an Asian

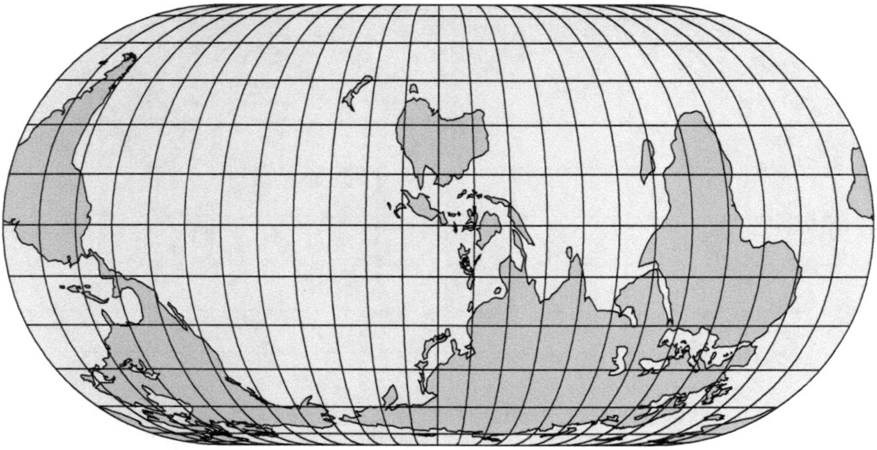

Fig. 3.4 The earth turned on an Eckert IV equal-area projection with south uppermost as might seem appropriate to an Australian mapmaker

3.2.3 Time Zones and Meridians

As the Earth turns it alternately sweeps the surface through a period of sunlight and a like period of darkness. Several thousand years ago the Babylonians, using a sexagesimal number system, divided the circle into 360° and the calendar day into 24 h. Consequently, the Earth turns through 15° of longitude each hour.

Until well into the 19th century each locality had its own "sun time," noon being when the sun crossed the local meridian. With the development of railroad transportation, the myriad of local times became bothersome. In 1884 an international conference in Washington, D.C., agreed on a system of "standard time" in which, ideally, everyone in each 15° longitude segment, from pole to pole, would set all clocks to the time of the central meridian of that zone. The conference also agreed to start counting longitude east or west from the *prime meridian* of Greenwich in England, and to reckon standard time as + or − the time in that time zone in units of one hour.

In practice, the boundaries of standard time zones rarely coincide with meridians, being modified for economic and administrative convenience. The standard time concept is, however, practiced almost everywhere, in that time is uniform throughout each region.

Since the meridians converge to the poles, the width of a 15° time zone is very narrow in the polar latitudes. For example, the former U.S.S.R. had 11 time zones while Africa, which is wider, actually observes only four. But an hour is an hour; therefore, to portray the concept with a uniform time scale and provide better detail in regions with crowded time zones, it is desirable to show the standard time system on a *cylindrical* map projection, as in Fig. 3.5, where the meridians are shown as parallel, thus having the same width (one hour) from pole to pole.

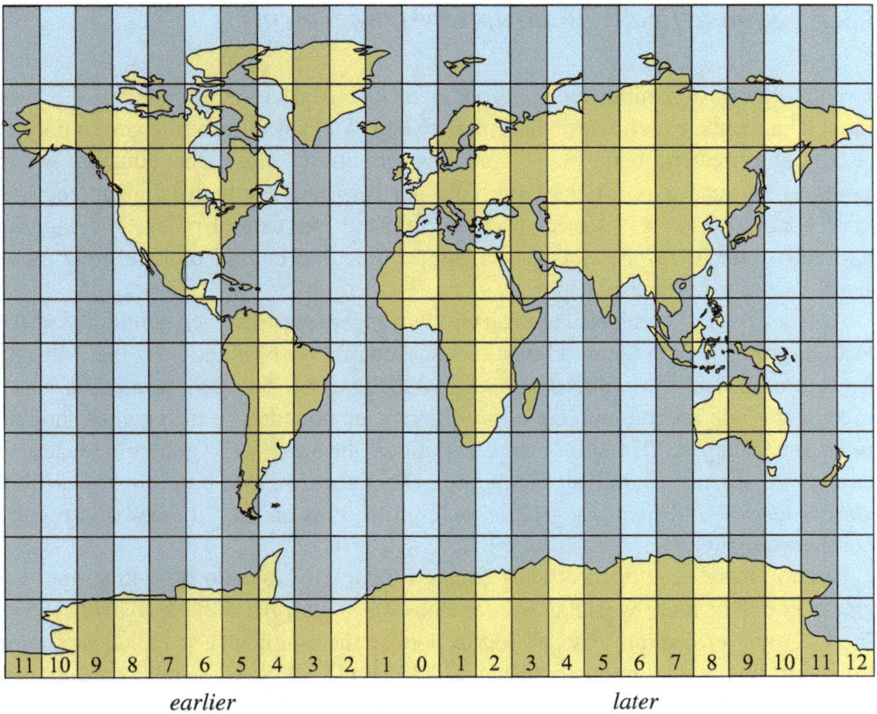

Fig. 3.5 The concept of standard time zones as displayed on a Gall stereographic projection. The actual boundaries of the time zones are extremely irregular

3.3 Interrupting the World Map

Richard E. Dahlberg

World maps constructed on *interrupted* forms of projections have appeared repeatedly during the last five centuries. In a sense, they represent a practical alternative to obtaining relatively undistorted views of parts of the Earth by rotating a globe.

Actually any flat map is interrupted, because a map always has an outer edge where the representation stops, even though the surface on the globe is continuous. In this chapter we refer to additional and internal interruptions. If we prepare individual maps of separate, smaller portions of the globe, using projections better suited to the particular area, we can join these maps into a single map, but it will have several discontinuities. The overall representation of regions can thus be improved, but at the expense of adding interruptions.

Earlier interrupted projections preserved strict symmetry by balancing the separate parts, but in the 20th century, inventors began to move away from strict symmetry, with notable improvements in the usefulness of a map for a given function.

3.3.1 Symmetrically Interrupted Arrangements

Perhaps the most familiar earlier types of symmetrical interruptions are the continuous hemispheres with various centers (Chap. 4). Projections arranged to portray eastern and western hemispheres preserve continuity along the equator, which serves as the map axis. Those portraying northern and southern hemispheres preserve continuity along a meridian connecting the two central poles. The familiar land and water hemispheres can be made to touch so that an oblique great circle serves as the axis for the map.

Another type of symmetrical projection arrangement preserves continuity of the portion of the globe in the middle latitudes. Probably the best known is the Butterfly projection of Cahill introduced around 1909 (Fig. 3.6). This arrangement is based on a regular polyhedron with eight faces, each face bounded by the equator and two meridians 90° apart. Of Cahill's three variations, the *equal-area* graticule illustrated here is perhaps most familiar. Using another polyhedron as a base, in 1943 Fisher constructed a world map using Gnomonic projections on the 20 faces of a regular icosahedron (Fig. 3.7).

Bartholomew developed at least four symmetrical projections of this type that appear in one or more world atlases, such as *The Times Atlas of the World* (1958). The Regional projection was devised to portray the continents. It includes a conic projection that is continuous between 22° 30′N and 80°N. Of similar concept and design, the Lotus projection was used for a world bathymetric chart, and avoids interruption between 22° 30′S and 80°S.

Continuity in a polar region is preserved by centering the projection on a pole, with numerous radial extensions. The best-known star projection of this class was

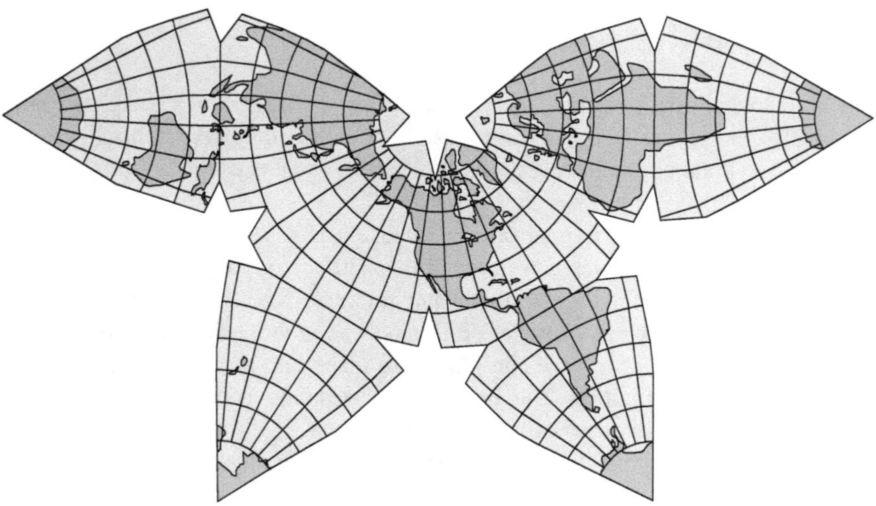

Fig. 3.6 The Cahill butterfly projection (equal-area variant)

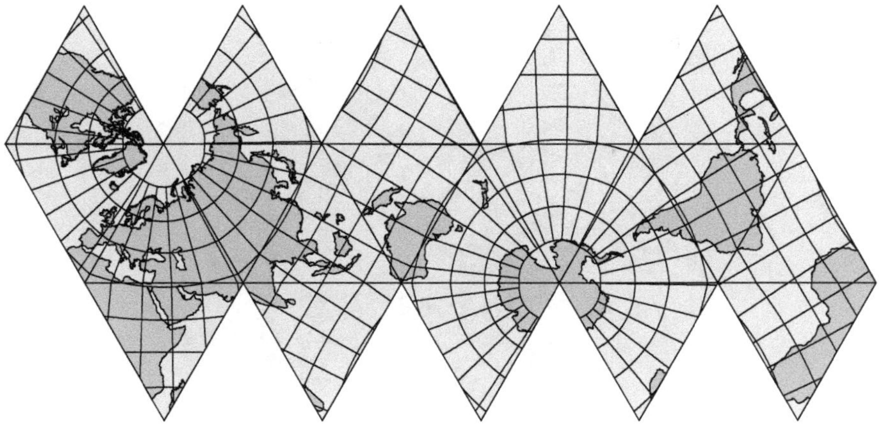

Fig. 3.7 A plane arrangement of Fisher's icosahedron on Gnomonic projections

introduced by Berghaus in 1879 (Fig. 3.8). It was used widely in textbooks and atlases in the late 19th century and has had high visibility ever since its adoption in 1911 for the logo of the Association of American Geographers.

3.3.2 Non-symmetrically Interrupted Arrangements

Non-symmetrical graticules differ in at least two ways from those of the symmetrical group: (1) the projection axes are chosen so as to pass through the middle of the main areas of interest; and (2) the widths of the map components are adjusted so as to provide continuity to these regions. Both adjustments are based on the purpose of the map and result in a loss of symmetry.

The most widely used projections in this group have straight parallels, equally spaced curved meridians, and oval outlines. When interrupted, continuity is preserved along the equator with the northern and southern hemispheres separated into lobes. The mid-meridian of each lobe is selected to fit the distribution of the items of interest (such as landforms or oceans); this often results in mid-meridians in opposite hemispheres being offset along the equator.

The Goode interrupted Homolosine was introduced in 1923 and quickly gained popularity. This equal-area projection combines high-latitude sections of the Mollweide projection with a low-latitude (central) Sinusoidal section. Kinks in the meridians result where the elliptical curves of the Mollweide join the sine curves of the Sinusoidal near the 40th parallels. The standard arrangement of the Homolosine for continental areas is shown in Fig. 3.9). That for the oceans is shown in Fig. 3.10. In each of these examples, the projection comprises central portions of six separate developments of the Sinusoidal and Mollweide projections.

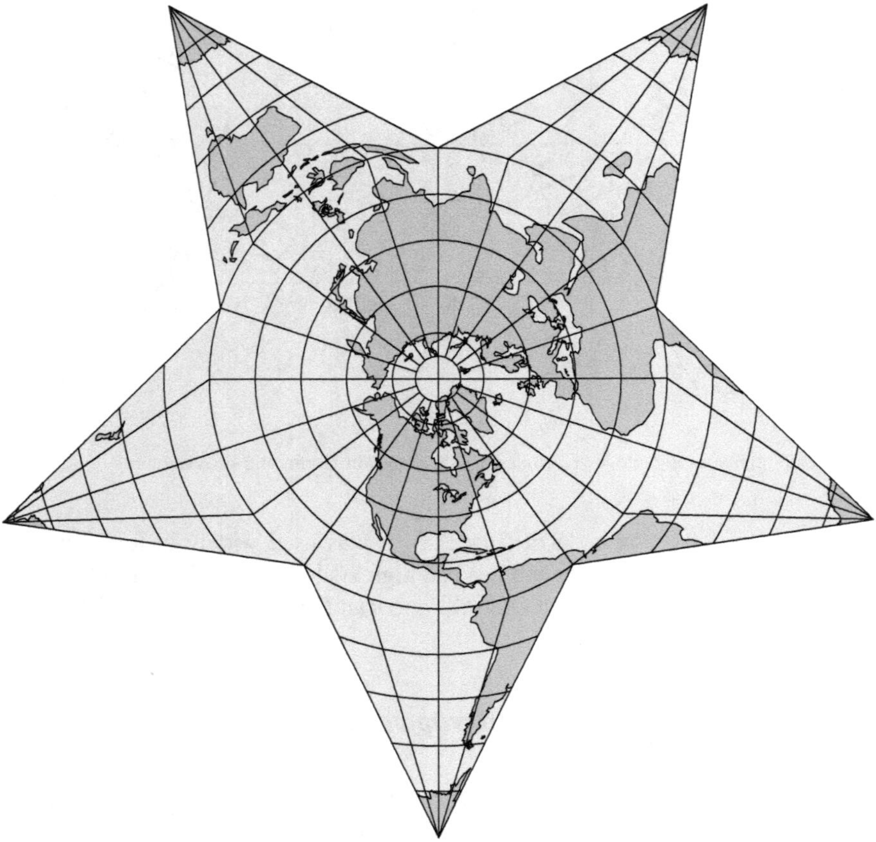

Fig. 3.8 The Berghaus star projection

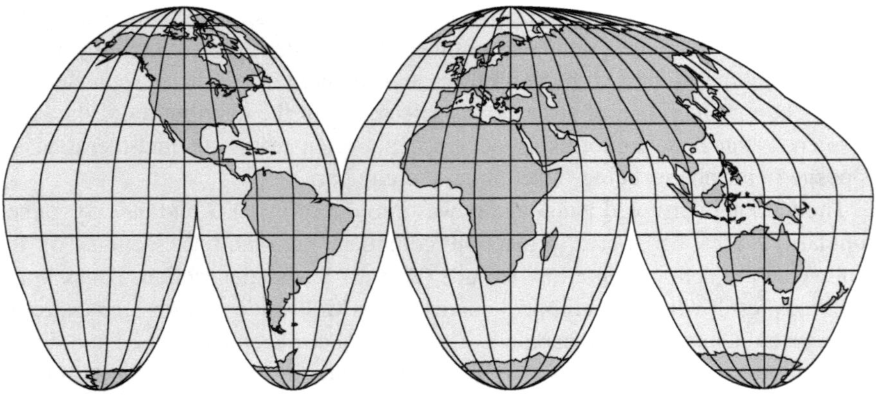

Fig. 3.9 The Goode interrupted Homolosine equal-area projection (continental version)

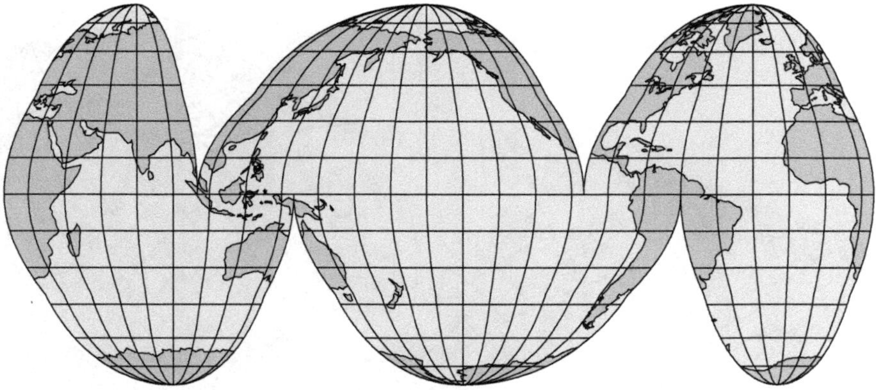

Fig. 3.10 The Goode interrupted Homolosine equal-area projection (oceanic version)

A different type of interrupted projection, a hemisphere with extensions introduced by Frye in 1895, consisted of the land hemisphere on an oblique Azimuthal Equidistant projection with extended lobes of different construction for Australia and South America (see Fig. 3.13). This was followed by many similar arrangements in the first half of the 20th century.

Some interrupted projections have lobes attached at either the North Pole to portray continental distributions or at the South Pole to portray the ocean basins. In 1928 Goode introduced a complex equal-area arrangement, with nine views of the Werner projection, centered at the North Pole, to depict the continents. In 1904 Schjerning introduced an equal-area projection, shown in Fig. 3.11, to depict the major oceans favorably. It comprises the central segments of three cases or centerings of the heart-shaped Werner projection joined at the South Pole. In recent years, Spilhaus proposed various interruptions and combinations of projections to show the continuity of the world's oceans (see Sect. 3.5).

In 1943, Fuller launched a non-symmetrical interrupted world map called the Dymaxion Globe in *Life* magazine. It first consisted of six squares and eight triangles that could be arranged on the plane in a variety of patterns. Fuller later changed to the icosahedron with its 20 triangles.

3.3.3 Summary

A wide variety of interrupted arrangements of projections have been proposed or devised by inventors from many disciplines and nations over the past five centuries. These arrangements are often controversial; few have made their way into the mainstream of map projection use. However, relaxation of the earlier symmetrical constraints stimulated a wealth of experimentation for non-symmetrical functional arrangements.

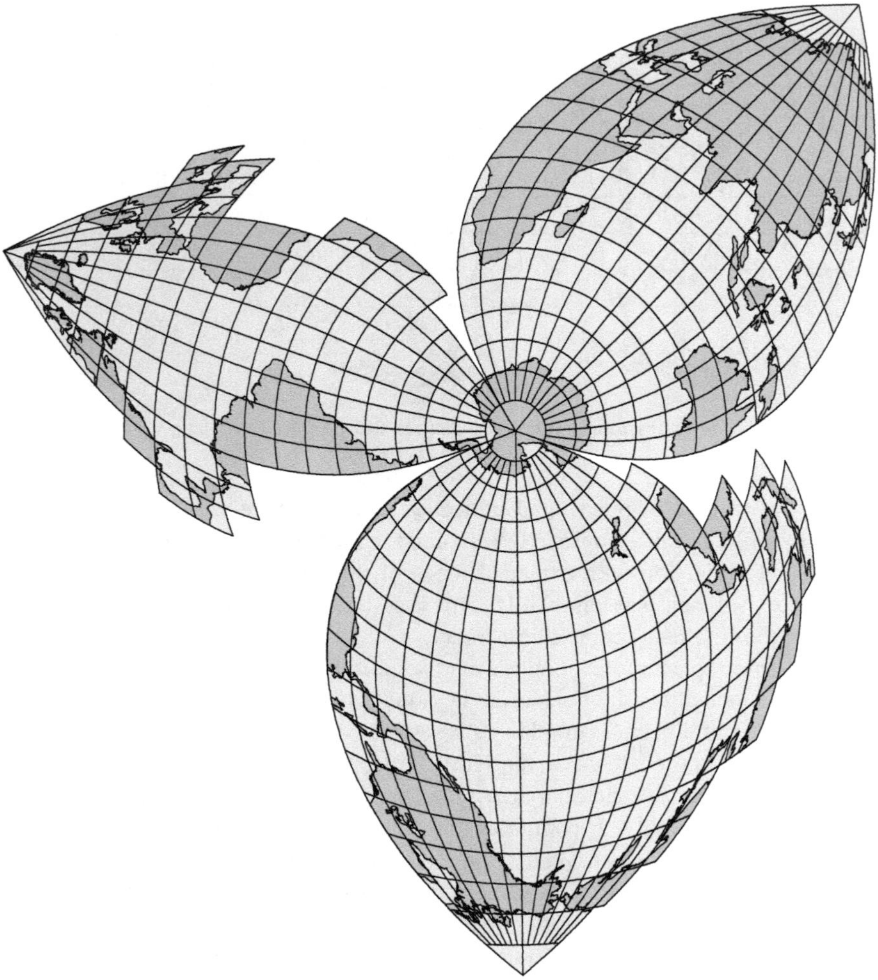

Fig. 3.11 The Schjerning equal-area projection for the oceans

3.4 Shaping the World Map

Richard E. Dahlberg

The outline shapes and general appearances of world maps are chosen to meet design requirements, improve feature displays, or create novel designs. To the map user the outline shape of a map is significant for several reasons. First and foremost, the shape of the frame often affects the internal properties of the map projection and thus the patterns of scale variation and distortion. The frame is also an edge along which the continuous surface of the Earth is interrupted.

Shapes of maps vary from simple to complex. Most conventional projections have simple, compact, and familiar outer boundaries such as circles, ovals, and rectangles. Circular and oval outlines evoke the fundamental roundness of the Earth. The rectangle, with its square corners, is unrelated to the smooth roundness of a sphere but is often seen, perhaps because its shape fits nicely in the common format of a printed page or poster. The focus of this chapter is on less familiar outlines.

Among the less conventional forms are perspective-like world map projections introduced by Raisz in 1943. One, his Armadillo projection, frequently used on reference and distribution maps (Fig. 3.12), resembles that mammal when curled up for protection. That the scale decreases away from the center is evident from the graticule, and this concept is reinforced by the curled shape of the external frame.

The discontinuities of such features as continents or oceans that result from the use of a frame bounded by meridians and parallels have led to a variety of enhancements. The simplest is to extend the map on both sides beyond its 360° width to an extent that avoids feature discontinuity through feature repetition. Minor marginal extensions to preserve continuity of continental edges are common.

A succession of views, each relatively undistorted, can be portrayed with interrupted projections as described in Sect. 3.4. The more interruptions, the more complex the frame.

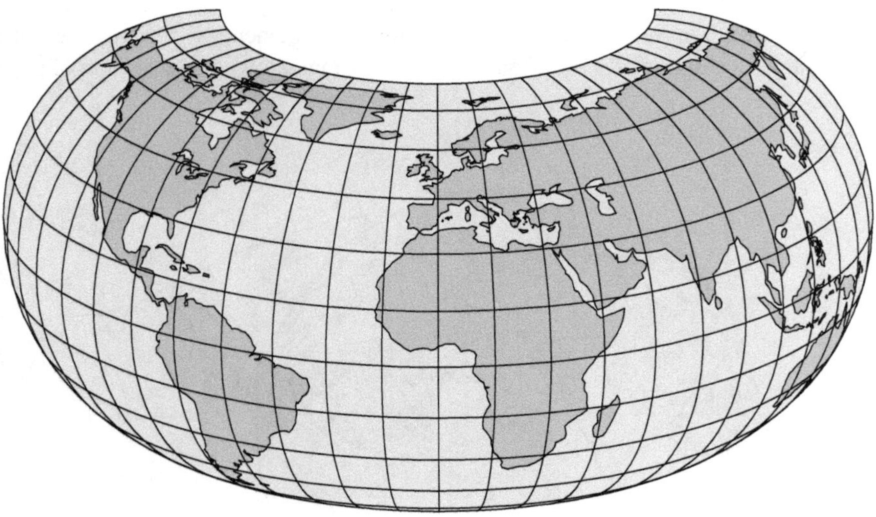

Fig. 3.12 The Raisz Armadillo projection

3.4.1 Concept-Linked Outlines

The diverse and ingenious world maps of this group result from the eternal struggle to comprehend and communicate spherical concepts via plane representations.

Continental Distributions. The optimum "land hemisphere" (see Fig. 3.20a), centered in France, excludes southern South America, Antarctica, New Zealand, Australia, and southern Asia. This creates major problems in fully representing the continents since scale variations and distortions increase markedly as coverage extends into the second hemisphere. One approach was introduced by Frye in 1895 to illustrate land-based features with the Atlantic and Arctic basins on one side and much of the Pacific Ocean omitted (Fig. 3.13).

Oceanic Distribution. The vast extent and complex outline of the oceans present an especially challenging representation task. As with continental distributions, this task also has led to the use of interrupted projections. One of Spilhaus's maps of the world ocean is an interrupted (nearly) equal-area arrangement (Fig. 3.14), which he says resembles a "crazy tulip". Preserving ocean continuity adds to the difficulty. This problem has been addressed recently with three lobes based on transverse Hammer equal-area projections by Spilhaus and Snyder (Fig. 3.15). This arrangement is centered at the South Pole and uses shorelines to replace the curved graticule-like lines normally bounding the projection.

Tectonic Plate Distributions. Spilhaus contributed numerous maps in a study of tectonic or crustal plate arrangements, some including polyhedral arrangements. On an interrupted transverse Sinusoidal equal-area arrangement with four lobes, the seven major tectonic plates were mapped in a way that allowed comparison of plate

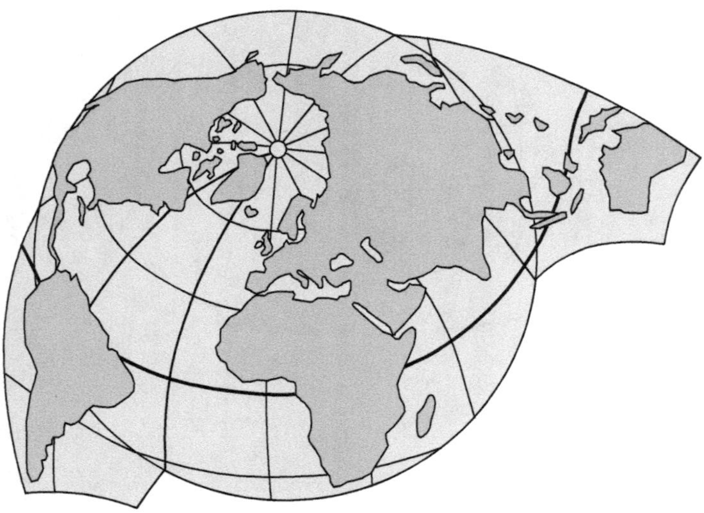

Fig. 3.13 Frye's extension of an oblique azimuthal equidistant hemisphere

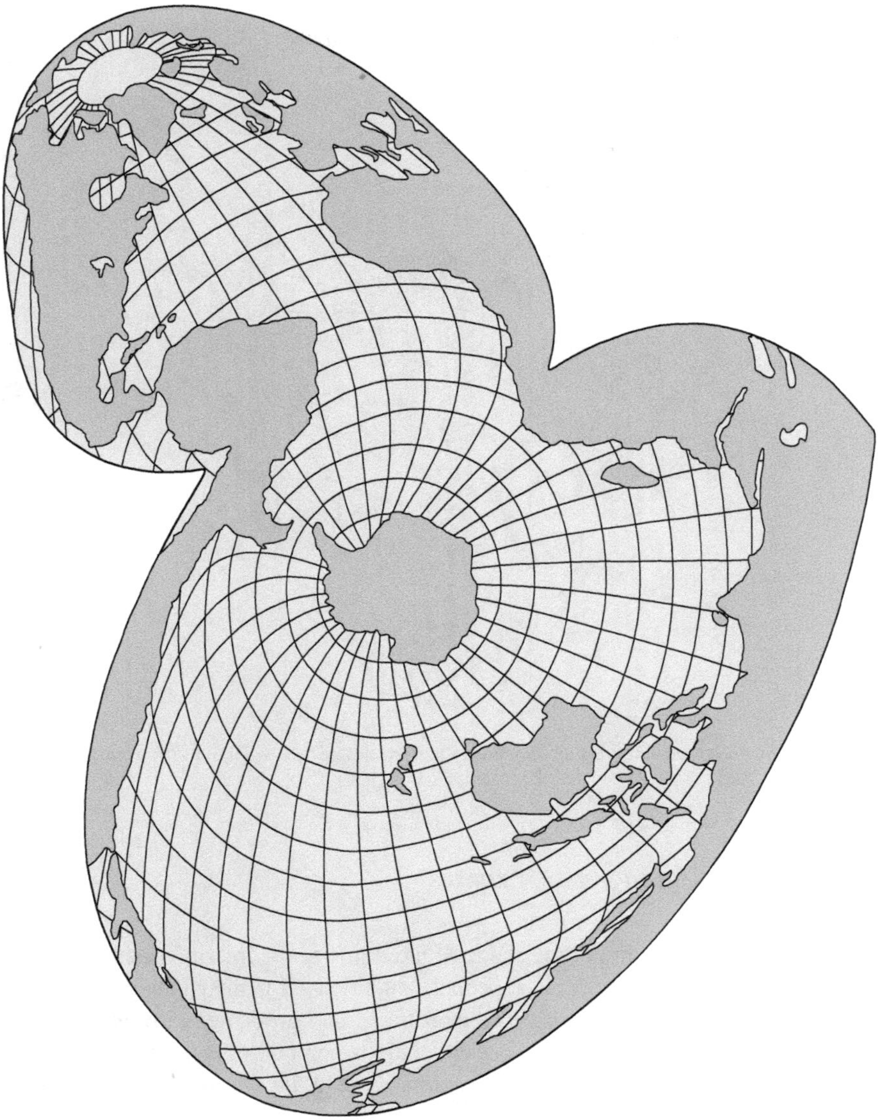

Fig. 3.14 An interrupted south-pole centered nearly equal-area projection (Spillhaus 1979)

areas and preserved plate shapes reasonably well (Fig. 3.16). Spilhaus also mapped the major plates on the seven faces of an irregular polyhedron.

Artistic Views. In addition to their informational function, the varied shapes of world map projections have led to their use as emblems on atlas and book covers, on letterheads, and in advertising. Selected examples are provided in Fig. 3.17.

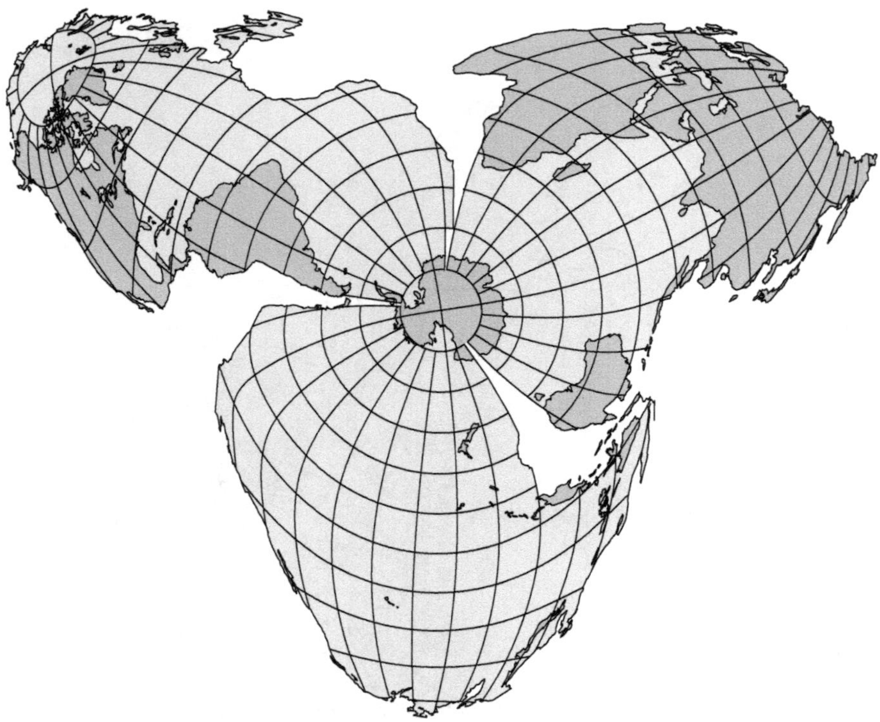

Fig. 3.15 An interrupted south-pole-centered transverse Hammer equal-area projection (Spilhaus and Snyder 1991)

3.4.2 Missing and False Frames

The important relationship of the shape of the frame to the internal design of a map projection, mentioned earlier, is especially significant for interrupted projections. Misleading maps can result when the map frame and graticule are omitted from a map projection interrupted in the oceans. This is sometimes done for design reasons, but on black-and-white maps oceanic areas and the non-map background are indistinguishable. This misleading effect is visually enhanced on multicolored maps where the blue "ocean" extends over non-map areas.

On some maps the actual outline of a world map has been replaced by an arbitrary enclosing line, which can be called a false frame because it is intended to create a different visual impression from the one provided by the true frame. This technique is sometimes used in advertising literature. One finds oval outlines enclosing areas mapped on a rectangular projection, and vice versa.

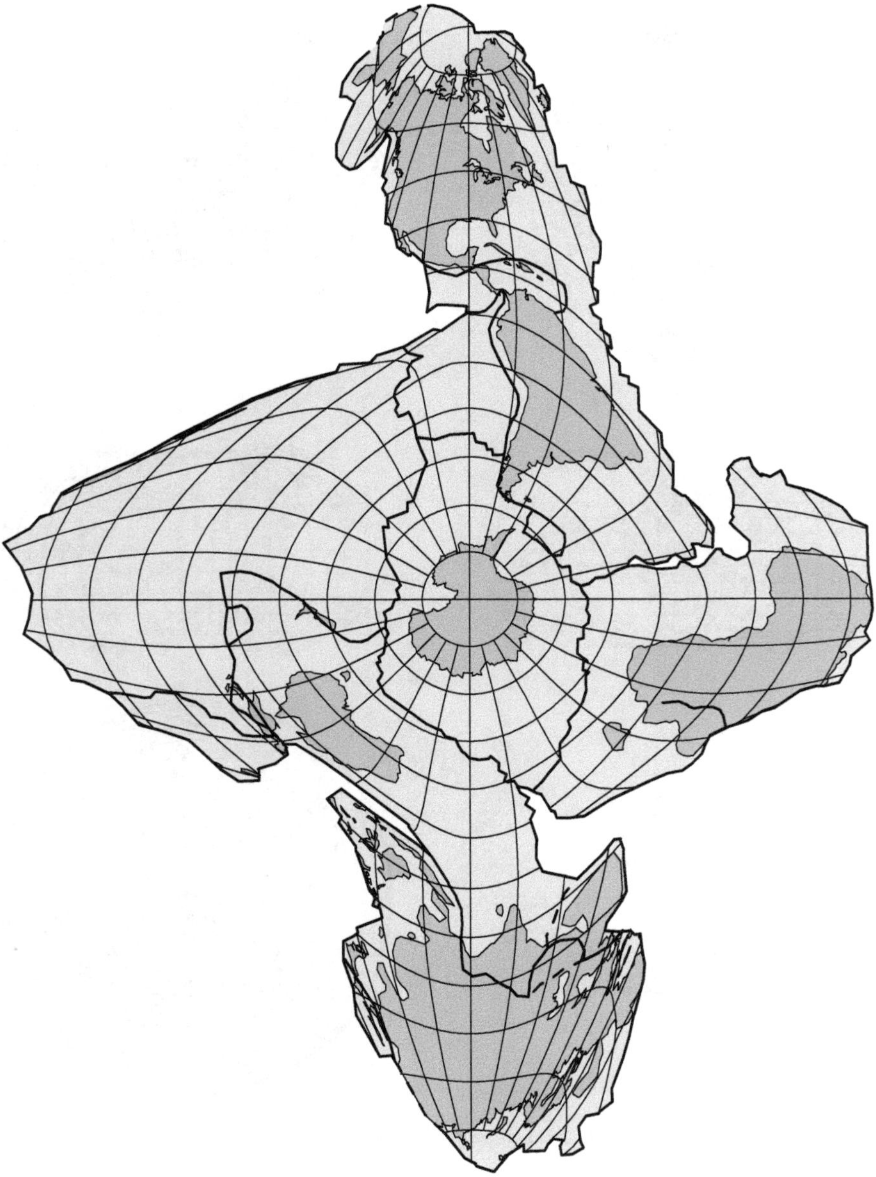

Fig. 3.16 An interrupted south-pole-centered pair transverse sinusoidal equal-area projections (Spilhaus 1990 ms.)

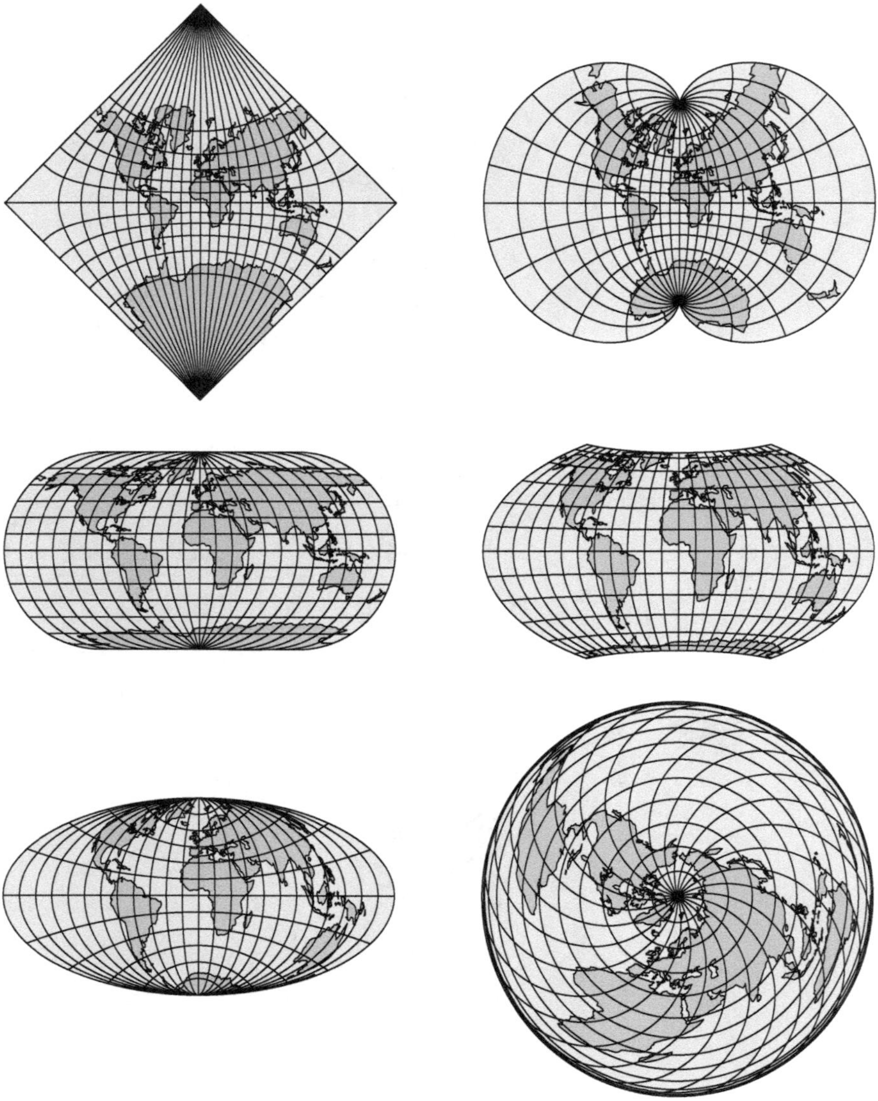

Fig. 3.17 Examples of less conventional projection shapes

3.4.3 Summary

Numerous modifications and combinations of standard projections have resulted in the creation of world maps with a wide variety of shapes. They may seem eccentric, but the group represents serious efforts to improve the portrayal of particular concepts or distributions. Attention to the shapes of these maps is important because they affect the internal properties of projections, internal grid discontinuities, and the spherical concepts and patterns being represented.

3.5 Projecting the Hemisphere

Judy M. Olson

The fact that people often create hemisphere maps is undoubtedly related to a common perceptual phenomenon: Looking at a spherical object such as a globe, we can see only half. Actually we see slightly less than half, but at any reasonable viewing distance "seeing the globe" inevitably suggests seeing half of it.

The term *hemisphere* derives from French, Latin, and Greek words of the same meaning, suggesting that human beings have long recognized the need for a term to designate half a sphere. There are no terms in the dictionary for a quarter of a sphere or for an eighth of a sphere, even though they would be just as easily derived etymologically. Clearly they lack perceptual justification.

3.5.1 Hemispheres by the Dozen

The Earth can be divided into an infinite number of hemisphere pairs. Any plane passed through the center of the Earth will intersect the surface along a *great circle* (a line of maximum circumference) that bounds two equal portions of the Earth.

Only a few of the hemisphere pairs are of sufficiently wide interest or utility to have been given names. The northern and southern hemispheres are centered on the North and South poles and are bounded by the equator (Fig. 3.18). They are the most "natural" pair of hemispheres and we encounter them primarily in connection with such natural phenomena as star patterns and atmospheric and oceanic circulation systems.

The western and eastern hemispheres are bounded far more arbitrarily and are not well-named. One might think they would be bounded by the prime meridian and the International Dateline, thus separating the portions of the Earth that are labeled with east and west longitude. They generally are not (Fig. 3.19). They are usually bounded by approximately 20°W and 160°E longitude, more conveniently separating North and South America from Europe, Africa, Asia, and Australia.

The lack of connection to west longitude and the close identification of the term western hemisphere with North and South America has led some to refer to the American hemisphere, rather than western hemisphere. Physically, the eastern and western hemispheres have little, if any, significance, but politically and historically they are the Old World and the New World to Asian and European-based culture. The western hemisphere is a particularly common reference in political rhetoric.

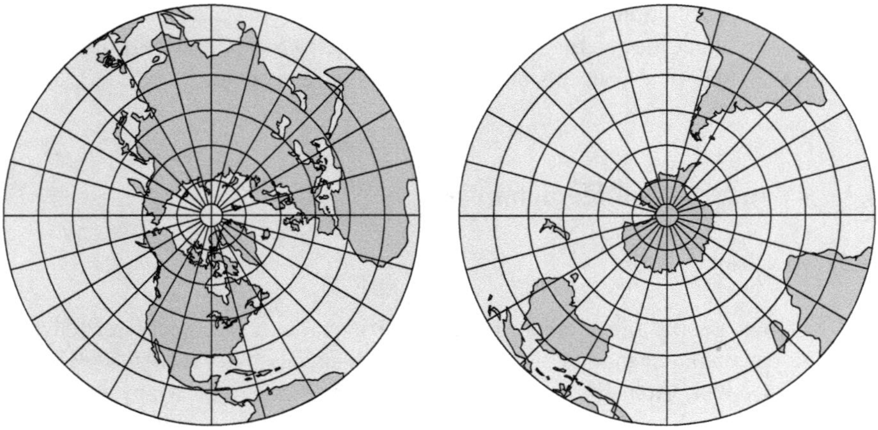

Fig. 3.18 The northern hemisphere (*left*) and the southern hemisphere (*right*) on the azimuthal equidistant projection

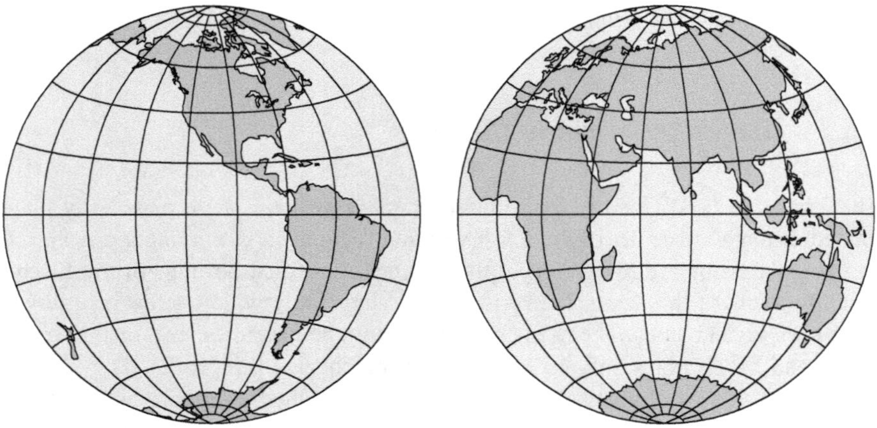

Fig. 3.19 The western hemisphere (*left*) and the eastern hemisphere (*right*) on the Lambert azimuthal equal-area projection

Among other halves of the globe that have interested us are the so-called land and water hemispheres (Fig. 3.20). The Earth is divided such that maximum land area is contained within one of the divisions and maximum water area in the other.

Another named half of the Earth is the so-called Pacific hemisphere. This term is used in referring to the economic and diplomatic matters of the Pacific Rim and the intervening islands, many being of strategic interest in the modern world.

3.5.2 Centering a Hemisphere

When we pick up a globe and look at it, seldom will any of the variously named hemispheres be the ones at which we are looking. We generally turn the globe until the particular point or area of interest is right in the middle. The location in the world of that place is best seen in that position, i.e., we can see it within a hemisphere whose edge is equidistant from it (Fig. 3.21). Positioning the globe in that way is so fundamental that we would find it incredible if someone tried to use a

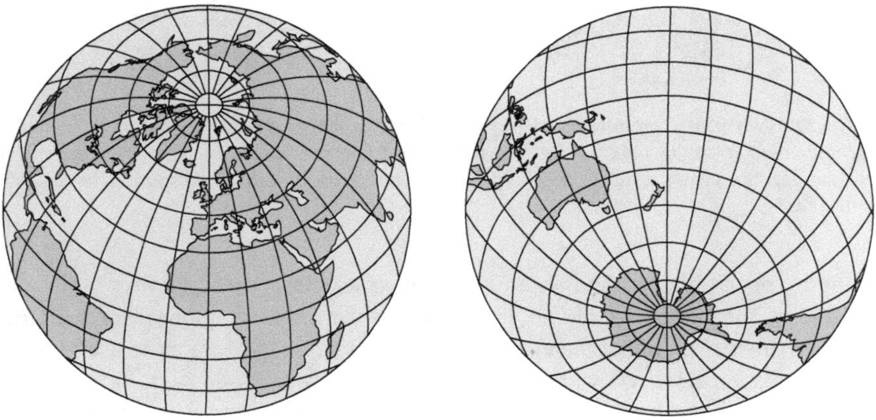

Fig. 3.20 The land hemisphere (*left*) and the water hemisphere (*right*) on the Lambert azimuthal equal-area projection, centered on 47° 13′N, 1° 32′W, and 47° 13′S, 178° 28′E, respectively (After Boggs 1945, p. 345)

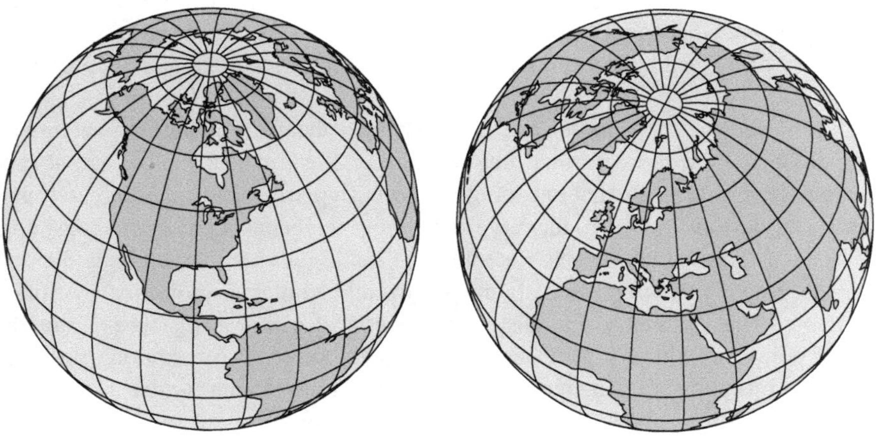

Fig. 3.21 The orthographic projection centered on East Lansing, Michigan (*left*), and centered on the Baltic countries (*right*). The orthographic projection simulates what we see when we center a globe on these points

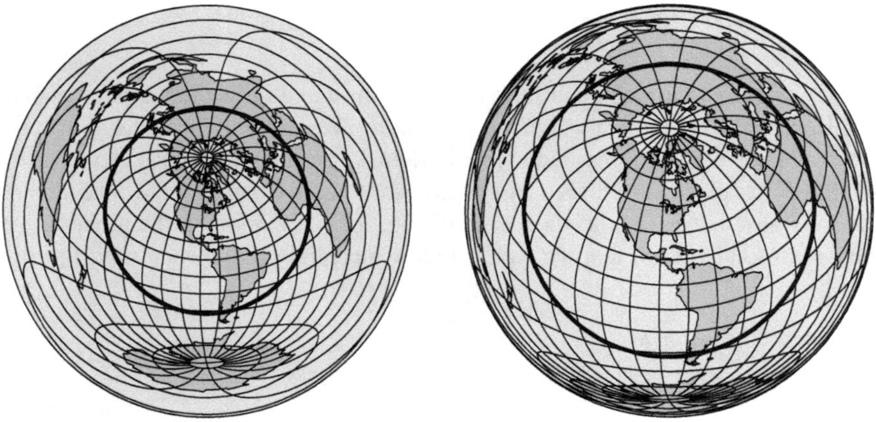

Fig. 3.22 Two planar projections, the azimuthal equidistant (*left*) and the Lambert azimuthal equal-area (*right*) of the whole world centered on East Lansing, Michigan. The inner hemisphere surrounding the center point is well represented, but there is extreme distortion beyond the hemisphere

globe by positioning it with an arbitrary or preselected center, leaving a place of interest at the periphery of the visible hemisphere. Many globes are therefore available in cradles, rather than on fixed stands, because they are more useful when they can be turned to view any hemisphere.

The necessity of hemispheric viewing seems to be removed when we turn from the globe to projected maps. But one whole class of map projections is usually limited to showing a hemisphere in order to avoid excessive amounts of distortion. These are the planar or azimuthal and on which the directions (azimuths) from the center to all other points on the map are correct (Fig. 3.22). In addition, there are projections referred to as "globular" that project the hemisphere into a circle but are not azimuthal, i.e., central angles are not correct (Fig. 3.23). These projections were generally developed before 1700 and were commonly used in pairs to represent the whole Earth.

If we want to show more than a hemisphere, there are many non-azimuthal, non-globular projections from which to choose. Yet some of the azimuthal projections have properties that encourage us to put up with hemispheric representations, and, just as important, the representation of the hemisphere is often desirable in its own right. As to reasons for tolerating hemisphere representations, the azimuthal property (correct central angles) of all azimuthal projections is often desirable. The Azimuthal Equidistant projection also shows distances from the center point (or, with a modification, two points) correctly. In addition, most azimuthal projections are generally good for showing any compact area (hemisphere or smaller) with relatively small amounts of distortion.

Turning to reasons for wanting to show a hemisphere, rather than merely tolerating it as a limitation, we often want to simulate the experience of seeing a globe, the globe itself being an undistorted representation of the Earth. The Orthographic

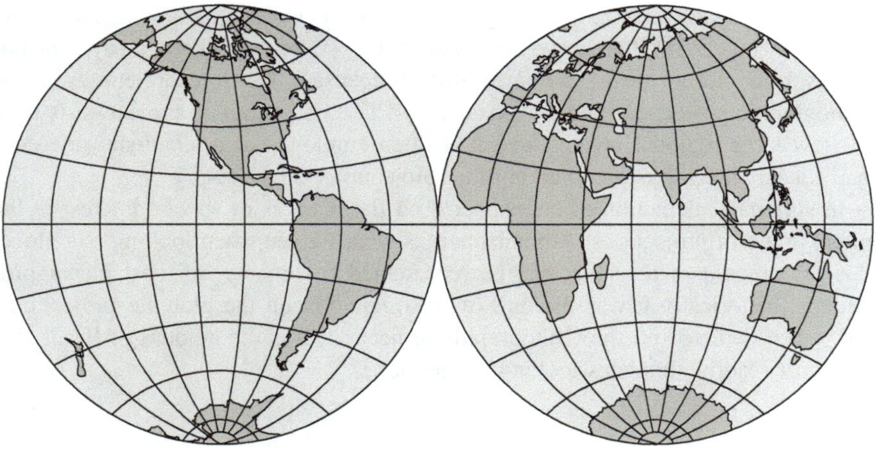

Fig. 3.23 The world on two Nicolosi globular hemisphere projections

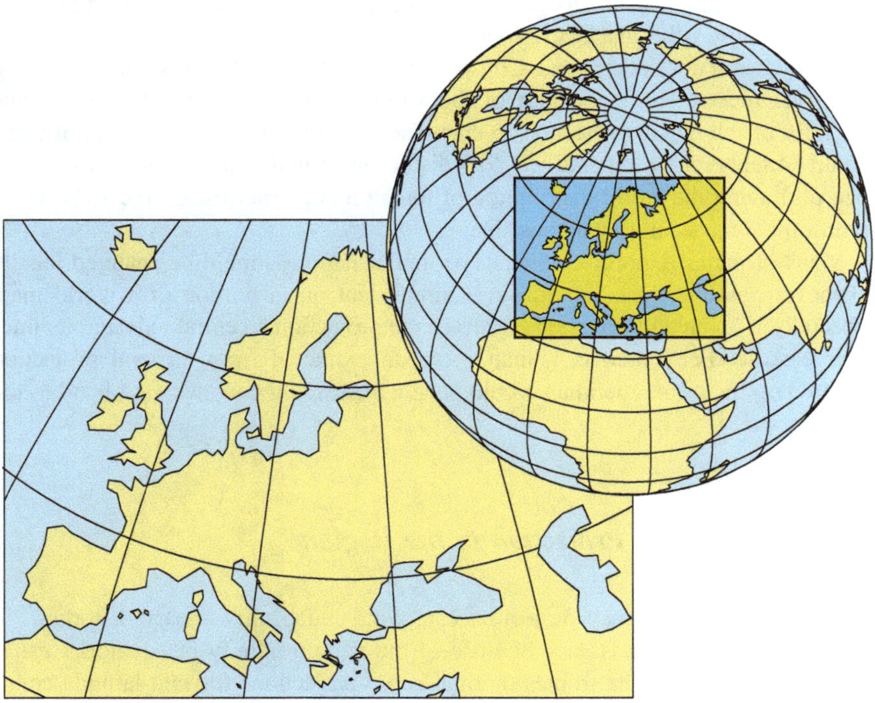

Fig. 3.24 The use of orthographic projection as a locator map. The area of the rectangular regional map is delineated on the orthographic *inset*. The shadowing helps to give us the impression that we are looking at a globe

projection is readily recognized as globe-like and is even perceived as having no distortion (Fig. 3.21). Just as we recognize that a person several feet away is just as tall as one closer to us, we mentally compensate for foreshortening in the Orthographic projection, and it "looks right." The Orthographic projection is used to show large regional layouts, and it is often employed as a locational reference map for larger-scale depictions on other projections (Fig. 3.24).

In summary, depiction of hemispheres on flat maps is of special interest in the employment of projections. A hemisphere is what we see when looking at a globe; it is the amount of area that can be represented reasonably on most planar projections; historically it was the unit of representation on the globular projections; and, when depicted on the Orthographic projection, it is the amount of Earth area that most readily simulates looking at the globe.

3.6 Centering a Map on the Point of Interest

Mark Monmonier

Because distortion is generally greater toward the margins of a map projection, each projection shows the distances, area, or angles of some parts of the world more correctly than it does other parts. Consequently, in choosing a projection for a world map, one must identify the part of the world to be shown or emphasized and carefully match a map projection to that area. A world map of agricultural regions would benefit from comparatively low distortion in mid-latitude zones, whereas a map portraying the world distribution of tundra and permafrost should show areas north of 70° as accurately as possible.

Maps of major regions, continents, or major nations should be centered locally whenever possible. One should never simply cut out a portion of a world map, especially if an area of interest is far from the projection's central axis and its lines of true scale. For example, a map of Japan extracted from an oval projection centered on the prime meridian would have a misleading and awkward bend to the left (Fig. 3.25).

3.6.1 Suiting the Projection to the Region

Careful choice of the map projection can avoid much unnecessary distortion. In some cases this choice is easy because of the natural affinity of equatorial *cylindrical* projections for the tropics, normal *conic* projections for mid-latitude continents and continental subregions such as Europe and Anglo-America, and polar *azimuthal* projections for Antarctica or the Arctic. Africa, which extends from roughly 37°N to 34°S, is well served by an equatorial cylindrical projection, with a line of true scale at the Equator or two lines of true scale, at 20°N and 20°S, say, for

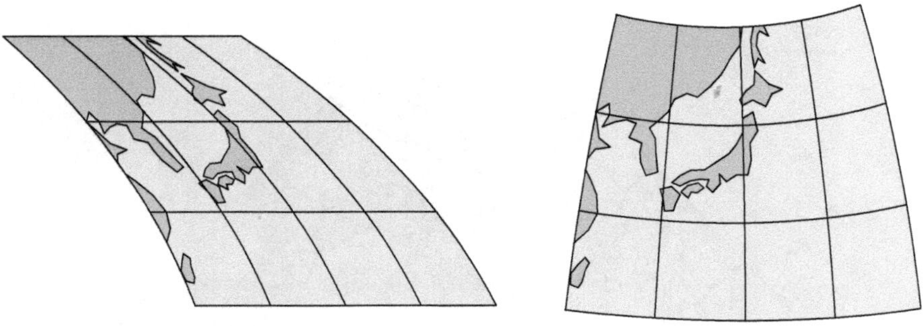

Fig. 3.25 Japan as a cut-out from a Mollweide equal-area world map centered on the Greenwich meridian (*left*) and on a locally centered transverse cylindrical equal-area projection (*right*)

Fig. 3.26 Africa on a cylindrical equal-area projection with true-scale standard parallels at 20°N and S

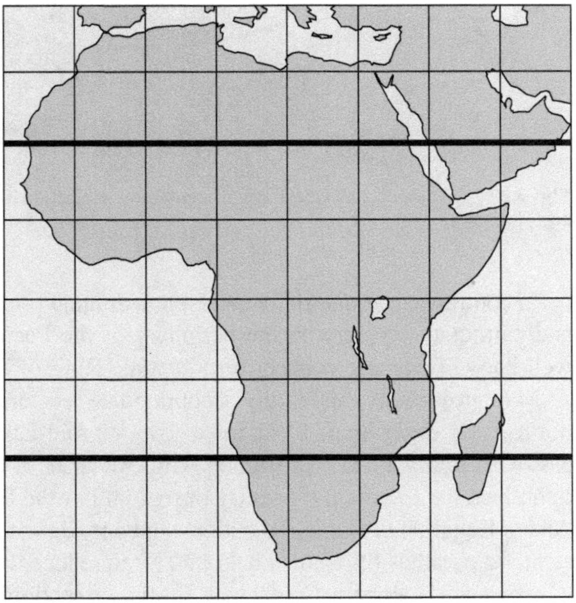

still greater fidelity (Fig. 3.26). The conterminous portion of the United States is similarly well represented on conic projections, which preserve the Canadian border from Minnesota to the state of Washington as a smooth, gentle arc. Other large, mid-latitude countries commonly portrayed on locally centered conic projections include Australia, Canada (Fig. 3.27), China, and the former Soviet Union.

Maps illustrating concepts in oceanography, long-distance navigation, and strategic defense often have special centering requirements. A map of shipping routes between Europe and North America would normally employ a conic projection, a map concerned with direct flights between North America and Asia would benefit from a polar azimuthal projection, and a map of circum-Pacific navigation could benefit from an equatorial cylindrical projection.

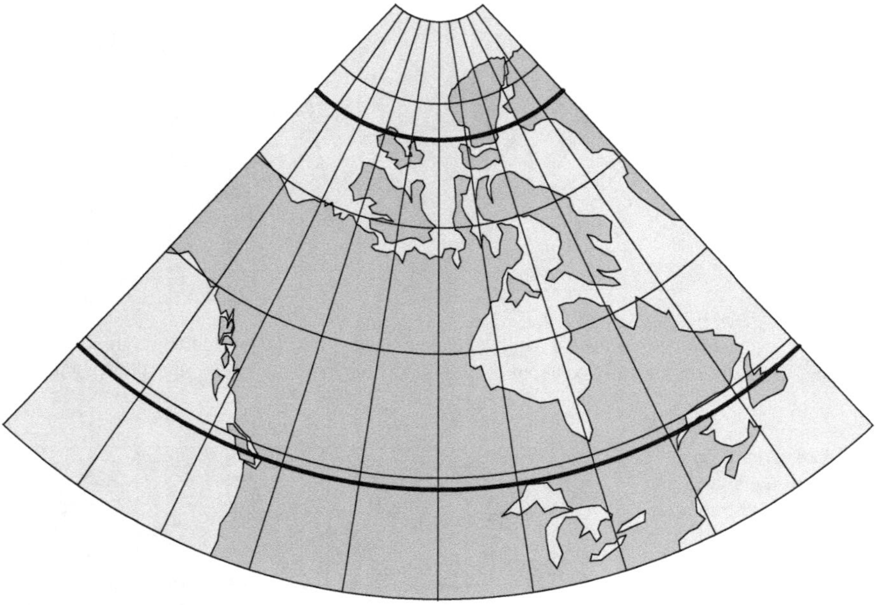

Fig. 3.27 Canada on an Albers equal-area conic projection with true-scale standard parallels of 49° and 77°N

In contrast, an equatorially centered azimuthal projection for a map of volcanoes and earthquake centers on the periphery of the Pacific Ocean would promote the well-known geologic concept of a "Pacific Ring of Fire" (see also Sect. 3.14).

Map projections can readily accommodate territories with a marked east-west or north-south elongation. Ideal for a low- or mid-latitude area with a pronounced east-west trend, a conic projection with two lines of true scale can be more or less optimized for a particular country or region. For the United States, as examples, the Albers Equal-Area Conic projection can be made with two lines of true scale called standard parallels (29° 30′ and 45° 30′N) to reduce the effects of angular distortion, whereas the Lambert Conformal Conic projection (Fig. 3.28) employs similar *standard parallels* (33° and 45°N) to control the effects of area distortion.

In contrast, transverse cylindrical projections provide low-distortion maps of areas with a wide range of latitude and a narrow range of longitude, such as Argentina (Fig. 3.29), Chile, and New Zealand. These projections are often made with two "lines of strength" (lines of true scale) parallel to a vertical central meridian chosen to minimize distortion across the territory of interest.

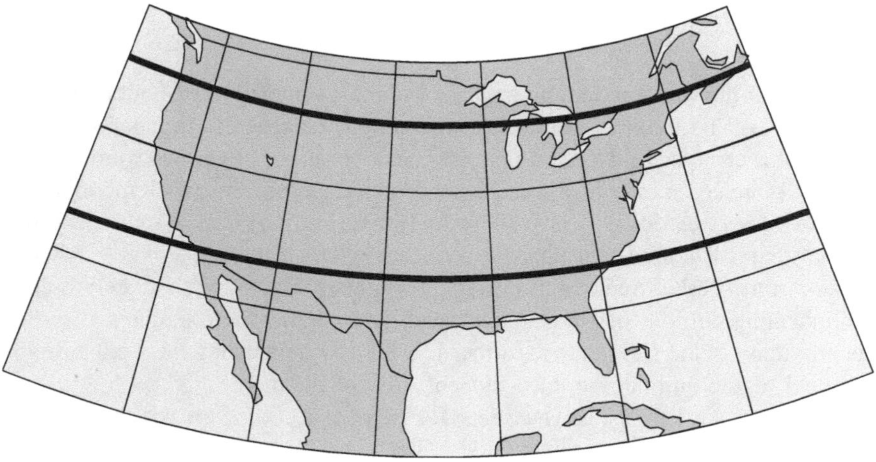

Fig. 3.28 A Lambert conformal conic projection of the U.S. based on true-scale standard parallels at 33° and 45°N

Fig. 3.29 Argentina portrayed on a transverse Mercator projection with true-scale lines parallel to the central meridian (not shown)

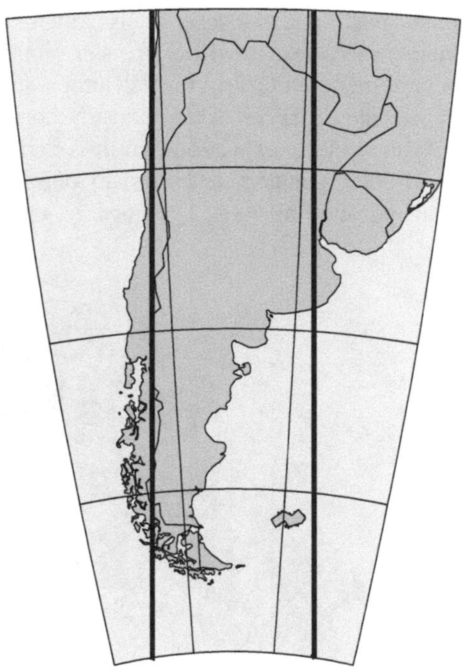

3.6.2 Centering a World Map

On a map of the whole world, how the projection is centered affects not only where distortion will be minimal but which areas might be severed. Equatorially based world maps centered on North America are rare because of their awkward partition of Asia. More commonly, world maps are centered on the Greenwich meridian (0°) and extend symmetrically from 180°W to 180°E. Yet if strictly symmetrical, they not only cut off the eastern tip of Siberia but separate the westernmost Aleutian Islands from Alaska. Some map makers solve this problem by alternately bulging and indenting portions of the map's left and right edges, whereas others also show the proximity of the U.S. and the former U.S.S.R. by duplicating the Bering Strait's marginal region on both the left and right sides of the map.

Centering can be especially critical for oval projections, on which maximum distortion occurs in the outer extremities. Greenwich-centered oval projections thus provide particularly distorted portrayals of Alaska, northeastern Siberia, Australia, and New Zealand. In 1953, Briesemeister, of the American Geographical Society, partially overcame this difficulty with an oblique equal-area projection that groups Europe, Asia, Africa, and the Americas toward the center, assigns the oceans to the outer quadrants, and severs only Antarctica (Fig. 3.30). Yet, like other full-world map projections, the Briesemeister cannot treat all regions equally. Its vertical central meridian clearly favors Europe and Africa with only minimal distortion of angles and distances, whereas nearly horizontal meridians yield less familiar, more deformed views of western North America and eastern Asia.

Military planners and tourism officials might favor an azimuthal projection centered at an air base, a major city, or a resort. Such projections are commonly

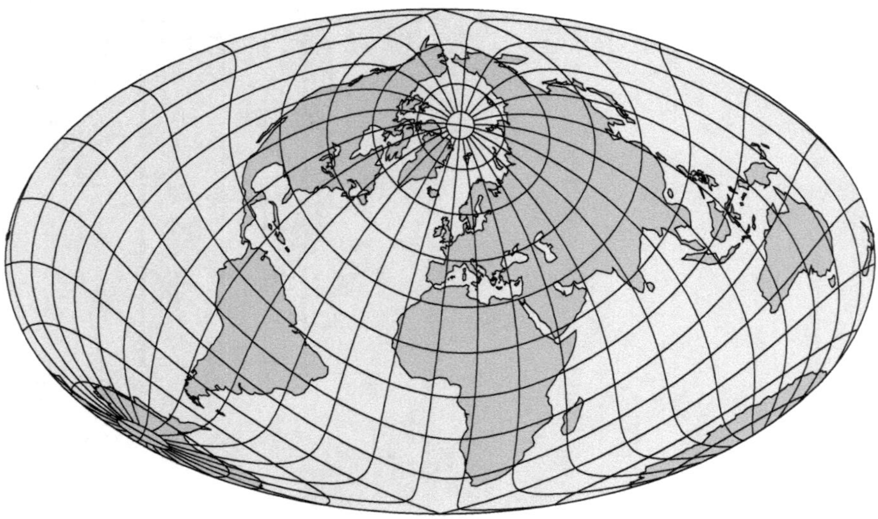

Fig. 3.30 A Briesemeister equal-area projection of the world

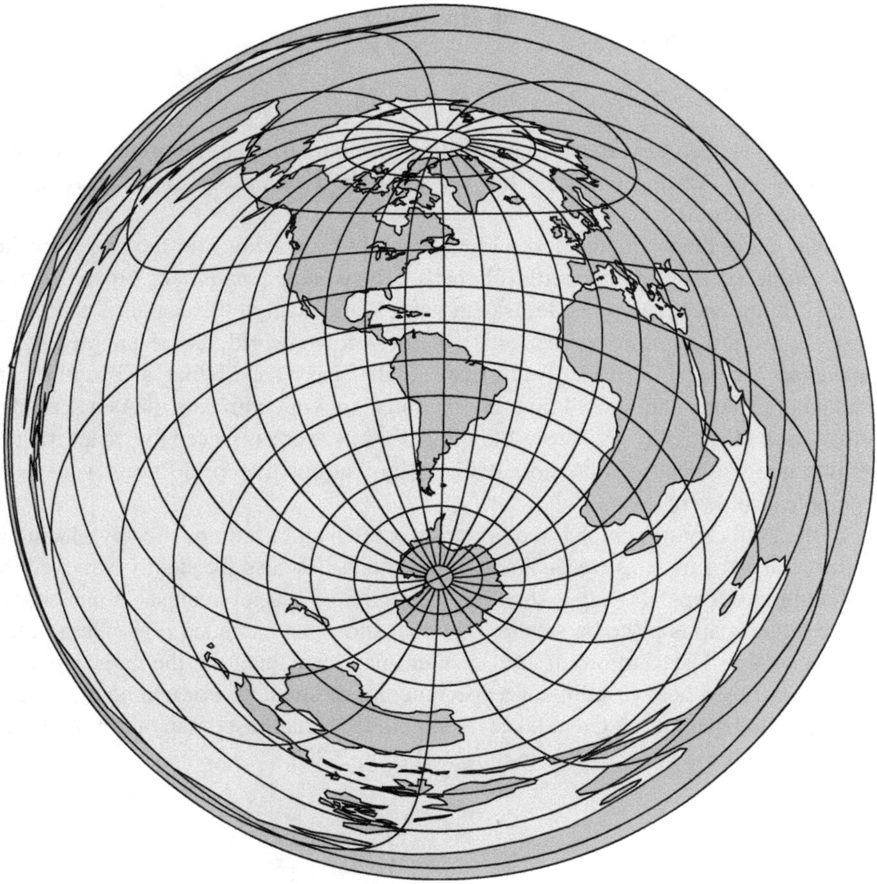

Fig. 3.31 An azimuthal equidistant projection of the world centered on Buenos Aires, Argentina

equidistant, to allow viewers to compare relative distances to other places (Fig. 3.31). (An interesting and often useful variant is the *linear cartogram*, which omits coastlines and political boundaries and physical features but shows cities with distances to the center adjusted to portray travel time, transport cost, or some other measure of accessibility or relative distance.) If relative area is important, a locally centered equivalent (or equal-area) azimuthal projection is appropriate. In either case, distortion will be minimal near the center and extreme at the margins.

3.7 Enlarging the Heart of the Map

John P. Snyder

Frequently, instead of wanting to give all parts of a map equal importance, we wish to emphasize a certain part of it but yet show the relationship of that part to the surrounding regions.

Many early maps emphasized a region merely by placing it in the center of the map. Some of these were medieval maps called *mappaemundi* or "maps of the world." In the western world, Jerusalem was often placed in the center of a circular map, signifying a philosophical unity of the known world. Some simpler *mappaemundi* are often called "T–O" maps, with seas shaped like a T within an enclosing O, surrounding and dividing the three known continents of Asia, Europe, and Africa (Fig. 3.32). As these allegorical maps were replaced by maps representing geographical regions more realistically, various map projections were used to emphasize the regions.

In the 20th century, emphasis of a region is often still accomplished by placing it in the center of a map, as we shall see later. Emphasis is also frequently achieved by including two maps, the larger showing the region of special interest. In one corner of the larger map is placed a smaller map that shows an extended region at a much smaller scale. The portion of the smaller map contained in the larger one is emphasized by a heavy outline or a special color or shading. For example, a larger more detailed map of the Great lakes may include a small inset map outlining North America, with the Great Lakes highlighted (Fig. 3.33).

Fig. 3.32 The Isidore T–O map of 1472

Fig. 3.33 The Great Lakes with a North American *inset*. Lambert azimuthal equal-area projections

3.7.1 Emphasis Using the Orthographic Projection

Emphasizing an area of special interest in the midst of its surroundings can also be accomplished by using a single projection instead of two separate maps. The Orthographic projection is sometimes appropriate for this purpose. This very old projection shows a hemisphere in much the way that a globe appears to us. It is more suitable for emphasizing regions that make up a substantial part of the hemisphere, such as North America, because the surrounding regions that are to be shown should lie near the edges of the hemisphere, in this case one centered on the Great Lakes (Fig. 3.34). It is not very satisfactory for emphasizing, say, the relatively small Great Lakes themselves.

The Orthographic is one of the group of azimuthal projections. When displaying regions that are nearly circular, members of this group generally have a more suitable arrangement of distortion than corresponding conic and cylindrical projections. Since highlighted regions often tend to be more circular than lengthwise, it is appropriate to consider other azimuthal projections for enlarging the central region of a map.

Fig. 3.34 The orthographic projection centered on the Great Lakes, 85°W, 45°N, as are Figs. 3.35, 3.36, 3.37 and 3.38

3.7.2 Zooming in with Perspective Views

One choice to consider is the general perspective view of the globe. The Orthographic is a view from an infinite distance, but by bringing in the viewpoint to a few thousand or a few hundred kilometers from the surface of the Earth, we can get other perspectives, like that of a camera viewing Earth from space (Fig. 3.35). When astronauts photographed portions of the Earth during the Apollo and Gemini missions, they were in effect using a tilted form of a perspective projection, with the regions that surround the points of interest fading into the horizon.

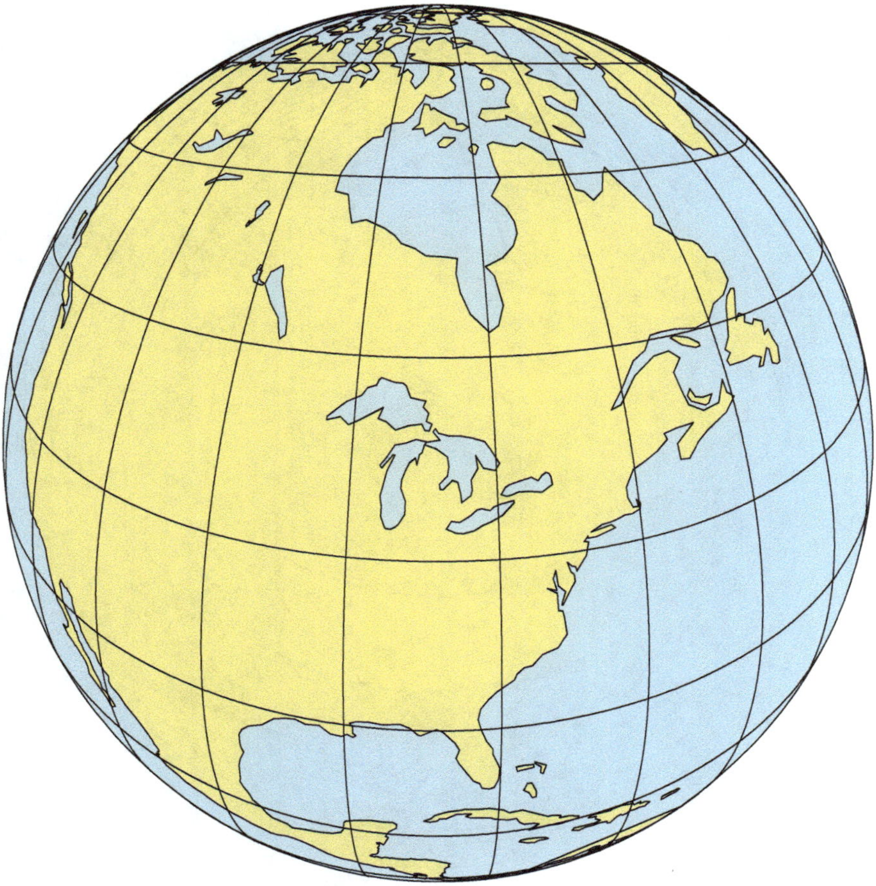

Fig. 3.35 A vertical perspective projection from a height of 2000 km

3.7.3 Newer Projections that Enlarge the Center

About thirty years ago, Swedish geographer Torsten Hägerstrand used a specially tapered azimuthal projection to show migration patterns near a Swedish town. This type of projection, called a Logarithmic Azimuthal, can be used to enlarge a circular region of almost any size, with the surrounding region subordinated to an adjustable extent. The effect can resemble a "fish-eye" photograph (Fig. 3.36).

Recently, another type of projection has been developed. The series, called "Magnifying-Glass" Azimuthal projections, includes several different forms. One type consists of a circular region of any chosen extent and at true area scale, surrounded by a region also at a constant area scale that is smaller by a predetermined amount than the scale in the central portion. Thus, it is as if a round magnifying glass were held over one part of an equal-area map so that the user can also

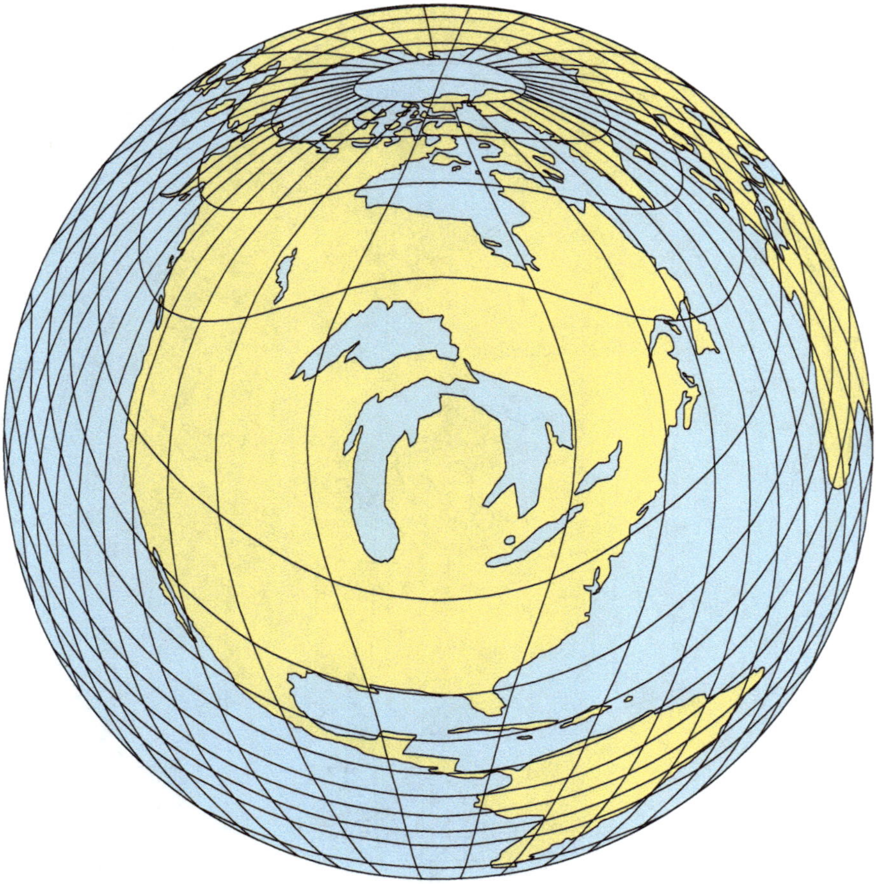

Fig. 3.36 A logarithmic azimuthal projection showing one hemisphere

see the surrounding regions outside the magnifying glass, but at a smaller scale. With an actual glass, some of the map would be hidden under the edge of the glass, but with this projection the entire surrounding region can be seen (Fig. 3.37). Other possibilities include the use of the Azimuthal Equidistant projection instead of the Azimuthal Equal-Area, or tapering the surrounding regions instead of maintaining a constant scale (Fig. 3.38).

Fig. 3.37 A "magnifying-glass" azimuthal equal-area projection, with an inner circle 7° in radius, and outer circle 20° in radius, and an outer area scale 1/6 the area scale of the central portion

3.8 Viewing the Earth from Space

Henry W. Castner

The selection of an appropriate map projection includes arranging the projection's pattern of deformation, i.e., its pattern of scale changes, in some optimum way. When we envision the Earth from space, we are above only one point on its surface, which we see essentially in plan and at a maximum scale, while all other points are seen at decreasing scales and in some increasing degree of obliquity out to the surrounding horizon line. Features there, if greatly exaggerated, could be seen only in profile. To represent this view of the Earth, cartographers may utilize an oblique aspect of the Orthographic projection, a perspective projection derived from some

Fig. 3.38 A tapered "magnifying-glass" equal-area projection, with an inner circle 30° in radius, and outer circle 120°; the radius of the inner circle is 0.65 that of the outer circle

specific altitude above the Earth, or any oval or azimuthal projection that preserves some desired quality of the Earth's surface.

Normally the chosen projection would be centered on some specific area of interest as described in the preceding two chapters. Scale decreases progressively outward and deformation increases from the centers to the peripheries of Orthographic and general perspective projections (Fig. 3.39).

The Orthographic projection views the Earth from infinity, hence half the Earth can be displayed on any one map, but perspective projections can be constructed from any elevation so that their limits extend from a great circle, as with the Orthographic, down to a very small circle centered on the point of interest. Other projection choices may allow viewing more than half of the Earth. Consequently, the map designer can control two important aspects of the finished map: its direction of viewing and extent of coverage.

Fig. 3.39 The pattern of deformation of the orthographic and all perspective projections. The darker the shading, the greater the deformation

3.8.1 The Direction of Viewing

We often insist, unnecessarily, that maps be oriented with north at the top. There are good reasons for this, particularly when the area being presented is not a familiar one. But there are often compelling reasons to use other orientations. These may relate to the desire to match a known viewpoint, to force the viewer to consider a geographic relationship from another perspective, or to suggest a change of importance or value from one area to another.

To present a specific direction of viewing, the projection is rotated so that the direction of interest is at the top. The projection can also be cropped below the center point so as to place the horizon in its natural position at the top or far side of the "scene" (Fig. 3.40).

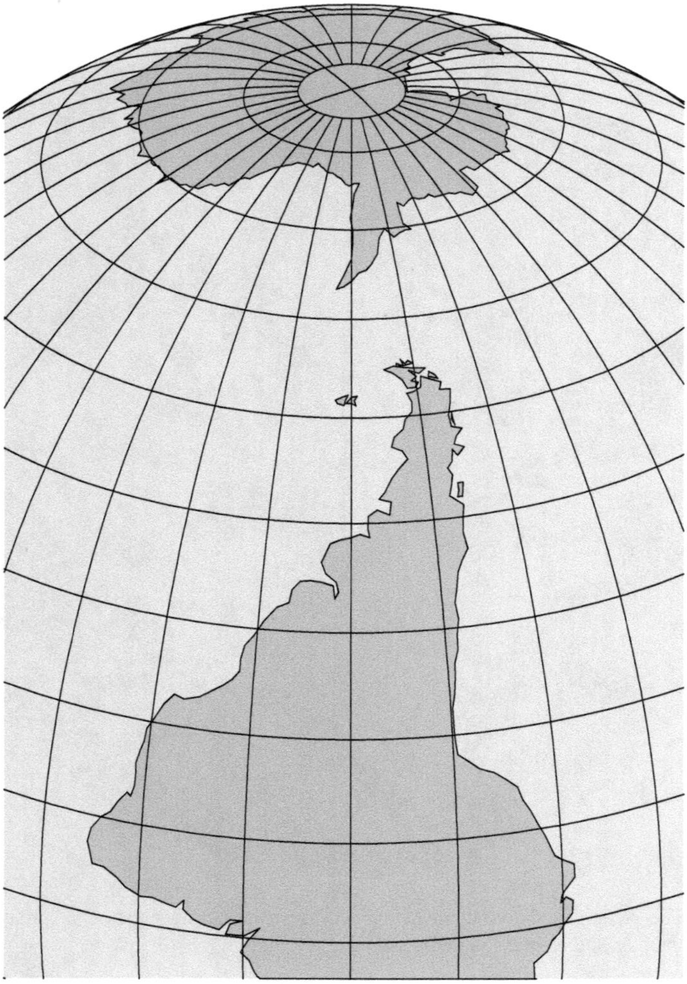

Fig. 3.40 A framed, oblique view of a portion of the globe looking southward across South America toward Antarctica. The remainder of this orthographic projection need not be used

If the projection is not cropped, the cartographer can place any lettering so that it is easily read in the preferred orientation. While this signals the intended orientation for viewing, it does not prevent a map user from rotating the map to a more familiar orientation so as to better appreciate the geographic area being viewed.

Because of the decrease of area scale away from the projection center, areas near the center can carry greater detail than those toward the horizon line. This, together with the change in perspective from orthogonal at the center to oblique at the periphery, suggests a parallel change in importance or significance for those areas away from the center. This effect can be used to imply a reduction in some phenomenon other than geographic area.

Fig. 3.41 A perspective view of eastern North America and northeast South America that reflects the limited understanding of the New World by southwest Europeans during the early years of the 16th century

For example, a perspective view of North America from over a point in southwestern Europe could be used to suggest the difference in knowledge about the New World held by Europeans in the middle of the 16th century (Fig. 3.41). The coastal areas were better known than the continental interior; but beyond some horizon, nothing was known—the known world essentially disappeared from view. From such an image it is clear why early exploration focused on the various inlets and breaks in that coastline in search of waterways to the Orient.

3.8.2 The Extent of Coverage

The area covered in a view from space can be varied by changing the altitude above the Earth's surface at which the Earth is being viewed. The lower the vantage point, the smaller the area of coverage (Fig. 3.42); the higher the viewer's altitude, the greater the coverage (Fig. 3.43). Being able to change the altitude of the vantage point allows one to illustrate such things as the increase in coverage or range of different communication or transportation technologies, the hinterlands around various-sized urban areas, or the surrounding regions of political or military hegemony of particular countries.

It is also possible to simulate a global view with an oblique aspect of any world projection that has a circular or oval perimeter. This might be done to preserve, for

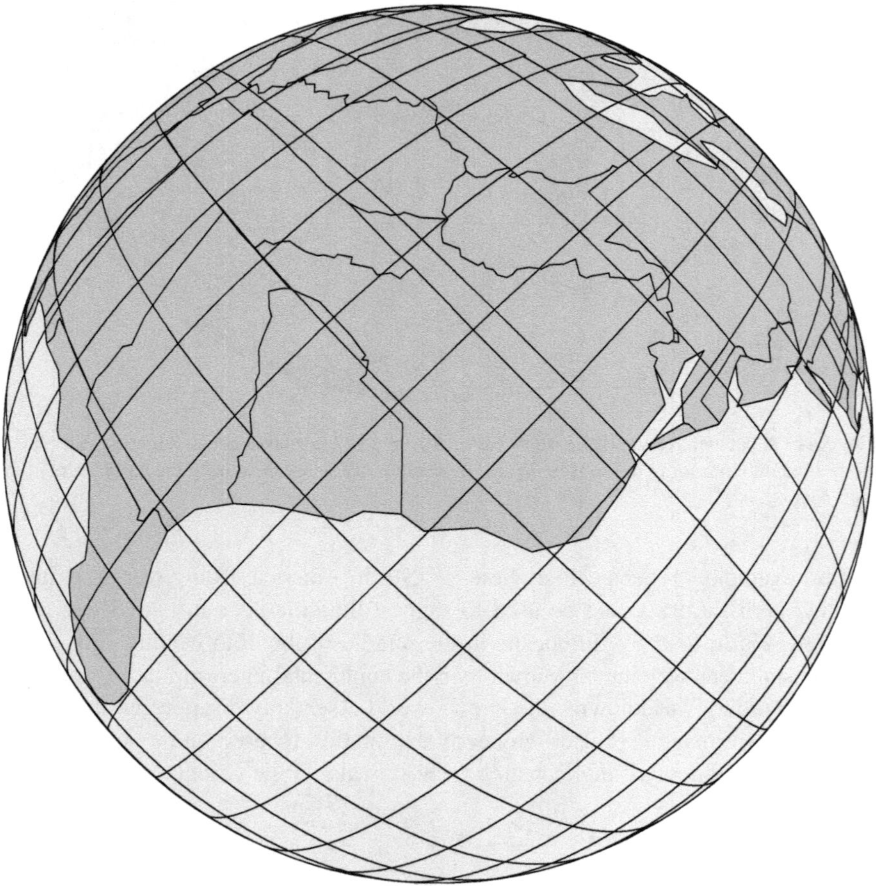

Fig. 3.42 A perspective view of the U.S. looking northwestward from a point 200 miles above Raleigh, N.C

Fig. 3.43 A perspective view of the U.S. looking northwestward from a point 800 miles above Raleigh, N.C

example, equivalence among areas in the projection (Fig. 3.44), something that could only be done with very small areas in a perspective projection from a very low altitude.

3.9 Showing Routes for Globe Circlers

Hsu Mei-Ling and Philip M. Voxland

Throughout the centuries, curiosity about the Earth has led explorers to embark on journeys around the world. Adventures that were undertaken by celebrated global circlers like Ferdinand Magellan and Sir Francis Drake are well known. More

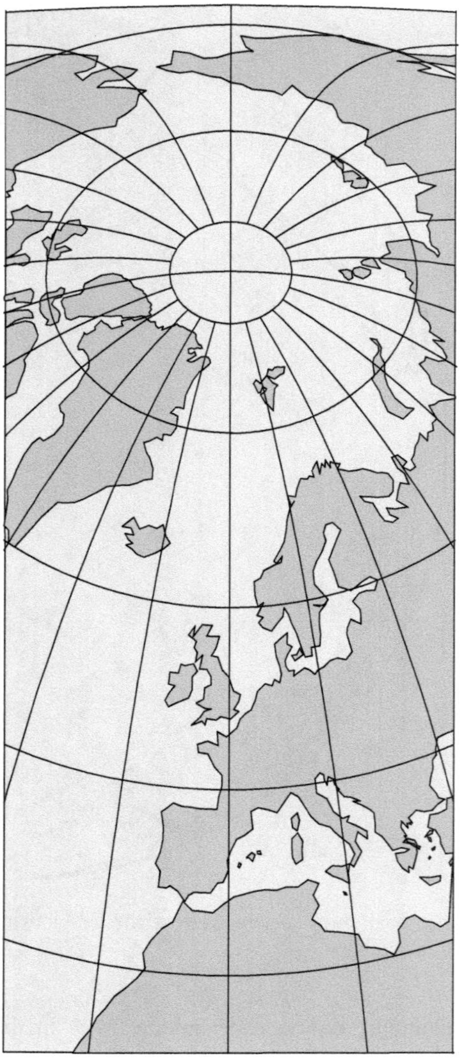

Fig. 3.44 A portion of an oblique Mollweide projection that preserves area relationships while giving the appearance of a perspective view of the earth

recently, global and space flights such as the Voyager airplane and the first artificial satellite, Sputnik I, also roused much public interest.

In this chapter we portray the routes of three of the above circumnavigations. It is a challenge to show routes on a small-scale world map. We begin by asking the following questions:

What areas on the Earth do the route predominantly transverse (e.g. equatorial, northern or southern hemisphere)? The distance between and areas along the route should be well preserved on the map.

What information should be included on the map so that the route can be interpreted correctly?

What projection(s) should be selected and specially designed to help the map reader visualize the spherical character of the global circler's journey? Because the route is continuous and global, it imposes limitations on projection design. For example, the trip should be shown on a world map without interruption; also, the entire route is important and no one segment can be slighted.

3.9.1 Magellan's Expedition—The First Circumnavigation

Ferdinand Magellan sailed with five ships and some 250 men in September 1519 from the port of Seville, Spain. He continued to the Cape Verde Islands, crossed the Atlantic Ocean, discovered the turbulent strait that now bears his name in October 1520, entered the calm ocean that he named Pacific in November 1520, passed Guam to reach the Philippines in the spring of 1521, and was killed in a local conflict shortly afterward. Elcano led the only surviving ship and 17 men returning to Seville in September 1522 via the Indian Ocean and around the Cape of Good Hope. This was the first circumnavigation.

Magellan's trip began in the northern hemisphere. While the crew sailed primarily along middle and lower latitudes, they also reached the southern extremities of Africa and South America. A projection that emphasizes the spherical nature of the trip in these regions is desirable.

The Raisz Armadillo projection has been chosen to depict the Magellan-Elcano route because it looks three-dimensional. It compresses areas at the periphery of the map and hides some areas altogether (Fig. 3.45). Raisz tilted the equator down (see Fig. 3.12) to emphasize land, but the tilting can be upward, as is done here.

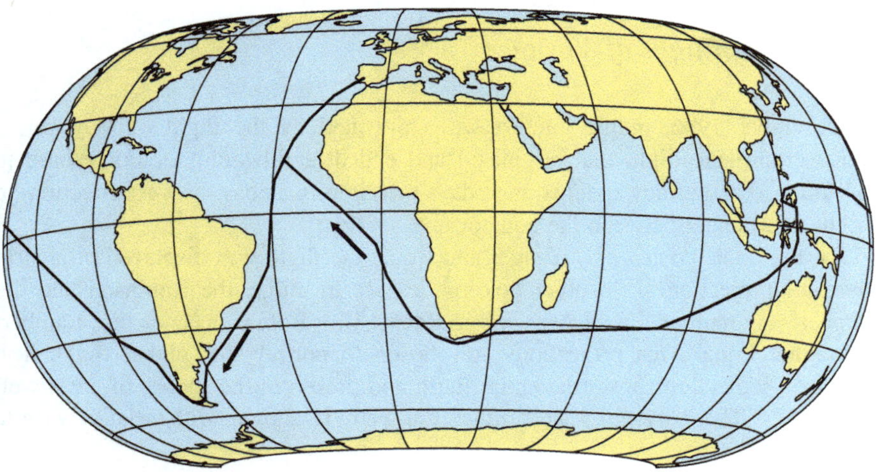

Fig. 3.45 The Magellan-Elcano route, Raisz Armadillo projection

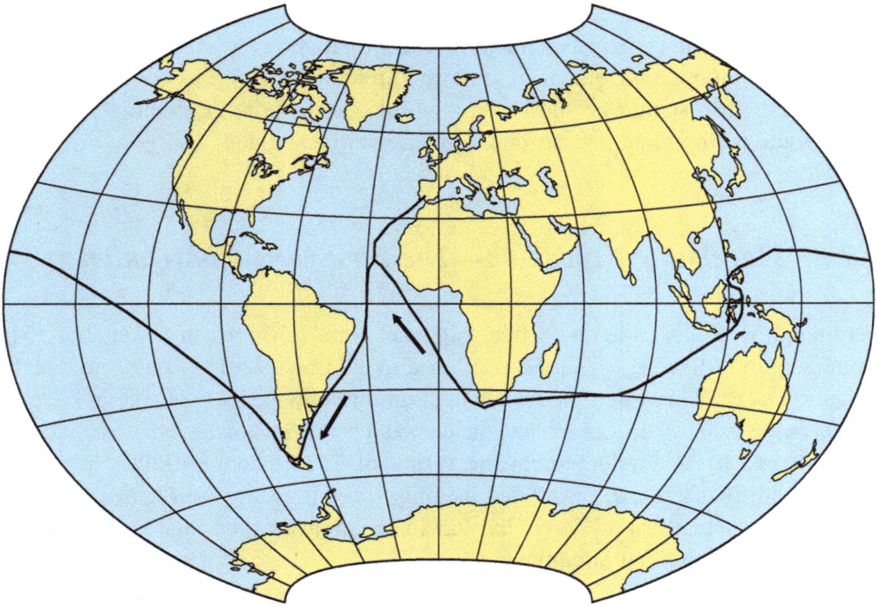

Fig. 3.46 The Magellan-Elcano route, Canters minimum-error projection

In some cases, it is important to show the entire sphere; one of Canter's projections can be used to reduce shape distortion of continents. Figure 3.46 shows his 1989 projection termed the "minimum-error polyconic projection with equally spaced parallels and pole line." Like the Raisz projection, it gives the impression of a sphere spread out onto a flat surface.

3.9.2 The Flight of Voyager

In December 1986, public interest was captivated by the flight of Voyager, a non-stop airplane flight that originated and ended at Edwards Air Force base in California. The journey spanned nine days traveling in an east-to-west direction on a path that ranged between the equator and 35°N.

For the book *Voyager* by Yeager and Rutan the flight was displayed on a strip Mercator projection. Two other possibilities are to utilize the Robinson and the Lambert Azimuthal Equal-Area projections. The Robinson and the Lambert Azimuthal Equal-Area projections are chosen to portray this global flight. The Robinson projection shows the entire Earth and preserves the shapes of areas well (Fig. 3.47). The Voyager's near-equatorial path is shown with relatively little distortion.

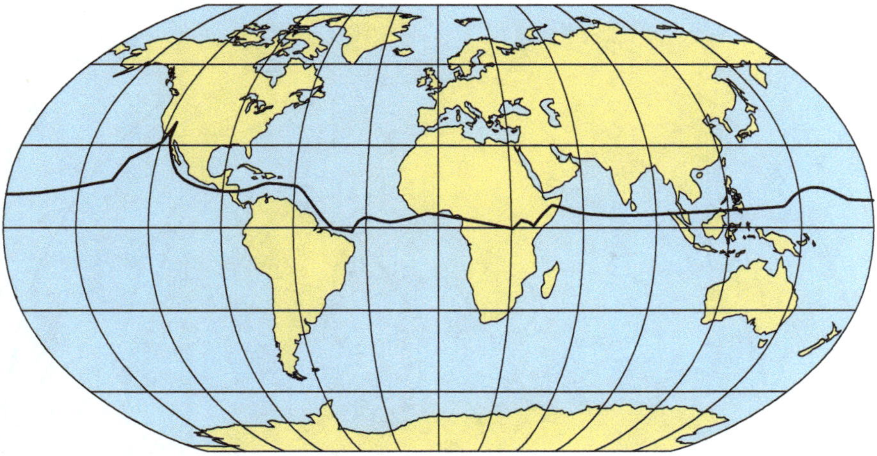

Fig. 3.47 The Voyager flight, Robinson projection

If we choose to emphasize the circular nature of the flight of the Voyager, a good choice is a Lambert Azimuthal Equal-Area map centered on a point near the pole to depict the uninterrupted flight path realistically (Fig. 3.48). This design shows the entire Earth and also gives an impression of three dimensions.

3.9.3 Sputnik I—The First Artificial Satellite

While east-west circumnavigation captured the human imagination in the past, north-south circling is of great commercial and strategic importance in modern times.

The launching of the Soviet Sputnik I in October 1957 began a new era of high technology. Sputnik [meaning fellow traveler (of the Earth)] was the first artificial satellite. It traversed a north-south path ranging between parallels 65°N and 65°S; each orbit of the Earth lasted approximately 92 min (Fig. 3.49).

Orbiting satellites travel steadily along what would be great-circle routes except that the Earth moves along its orbit and also spins beneath them, so the satellite *groundtracks* spiral around the Earth. Their circumnavigation may journey from almost pole to pole, incline relative to the equator, or align with the equator.

Generally, cylindrical or rectangular projections are not ideal choices for showing the entire Earth because they present significant distortion in the polar areas. However, they can show effectively the north-south traversals of satellites, which follow consistent cyclical patterns. The projected Sputnik path on the map should emphasize this regularity visually.

The Miller Cylindrical projection shows the entire Earth in a rectangle (Fig. 3.49). The successive orbits of Sputnik I are identical and repeating sinuous

Fig. 3.48 The Voyager flight, oblique Lambert azimuthal equal-area projection

curves. The path of one orbit is highlighted. On this map, the highlighted orbit begins at the equator near Borneo, moves toward the north, and then moves southward. When it returns from the south to cross the equator, the satellite is about 24° west of its previous crossing, since the Earth has rotated relative to the satellite. The successive northbound paths are omitted to simplify the map.

Lastly, the Snyder Cylindrical Satellite-Tracking projection is a good choice to depict the satellite path as a set of straight lines (Fig. 3.50). Areas beyond the northern- and southernmost points of the satellite track cannot be shown by this projection. The same orbit highlighted in Fig. 3.49 is emphasized. As in Fig. 3.49 the northbound paths are omitted.

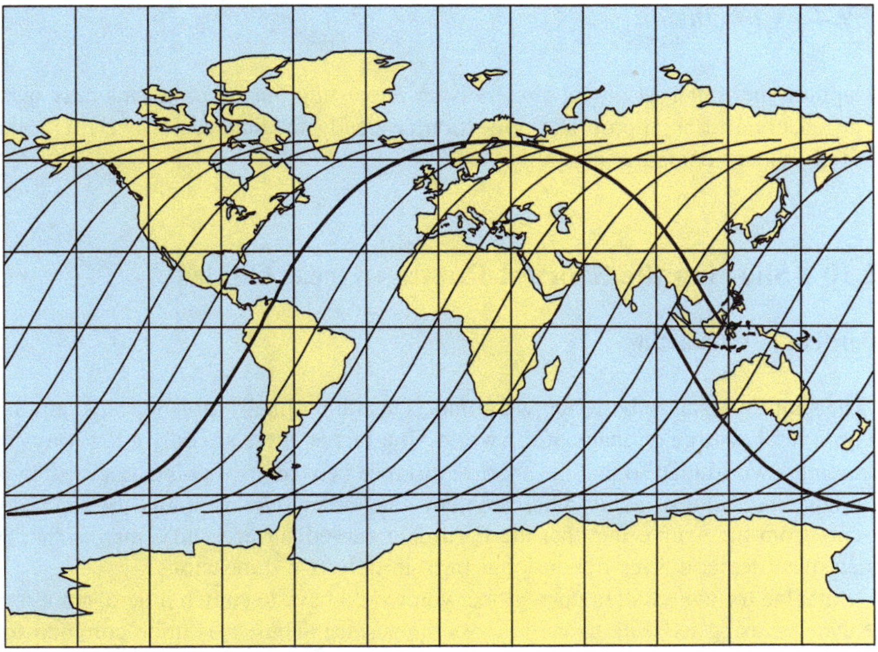

Fig. 3.49 Sputnik I, Miller cylindrical projection

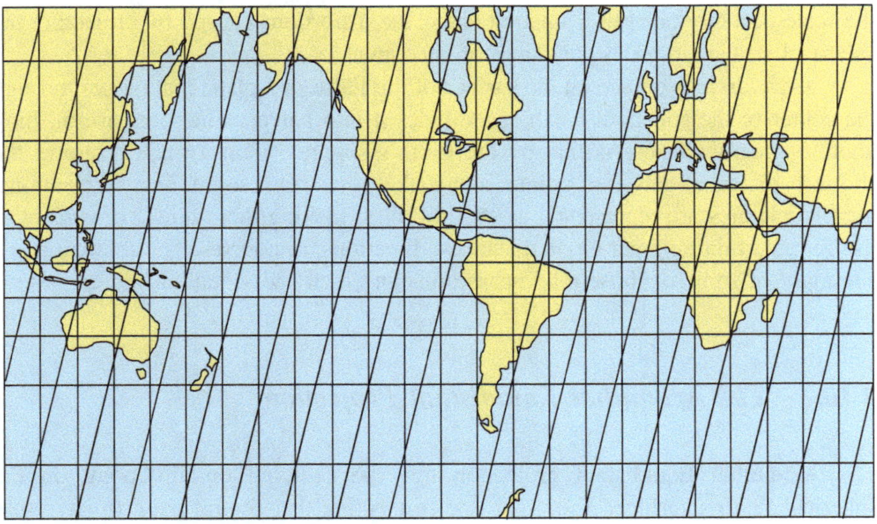

Fig. 3.50 Snyder cylindrical satellite-tracking projection

3.9.4 Conclusion

Mapping the routes of global circlers is an interesting but by no means easy task. The selection of an appropriate projection is greatly facilitated, however, with the use of modern computer programs.

3.10 Showing the Shortest Routes—Great Circles

Patricia P. Gilmartin

"The shortest distance between two points is a straight line"—or is it? This familiar axiom rarely is true on maps of the world. The fact is, there are only a few ways of creating a world map so that the shortest distance between two points is shown by a straight line. Like so many other problems encountered in mapping, the difficulty stems from the simple fact that the Earth is a curved surface and a map is flat: A map must depict a three-dimensional form in only two dimensions.

Imagine trying to flatten out a globe; you would have to stretch it here, compress it there, causing its scale to vary across the surface. Thus, it is quite common for directions and distances between pairs of locations to be represented unrealistically on maps. The amount of such distortion is significant only if the map shows a large portion of the spherical Earth's surface, such as a hemisphere or the entire world. At the scale of a city or even a small country, the distortions caused by projecting the spherical surface to a flat page are not very great.

A *great circle* is a trace on the surface of the Earth of a plane that passes through the center of the spherical Earth and divides it into halves. The most useful thing about a great-circle arc is that on the Earth's sphere, or on a map, it shows the shortest distance between points along that line. On some map projections, great-circle arcs are represented as straight lines, making them quite convenient to use for determining great-circle distances, directions, or courses. The following is a summary of some of those map projections and their individual characteristics.

3.10.1 The Azimuthal Equidistant Projection

The Azimuthal Equidistant projection may be centered on any point on the spherical Earth's surface. A straight line connecting that central point to any other point on the map will represent the shortest route between the two locations. Distances and directions along the line also will be correct; hence the term "equidistant" in the name of the projection.

If the central point is made the North or South Pole, then the parallels would be equally spaced concentric circles. It is usually more useful, however, to center the

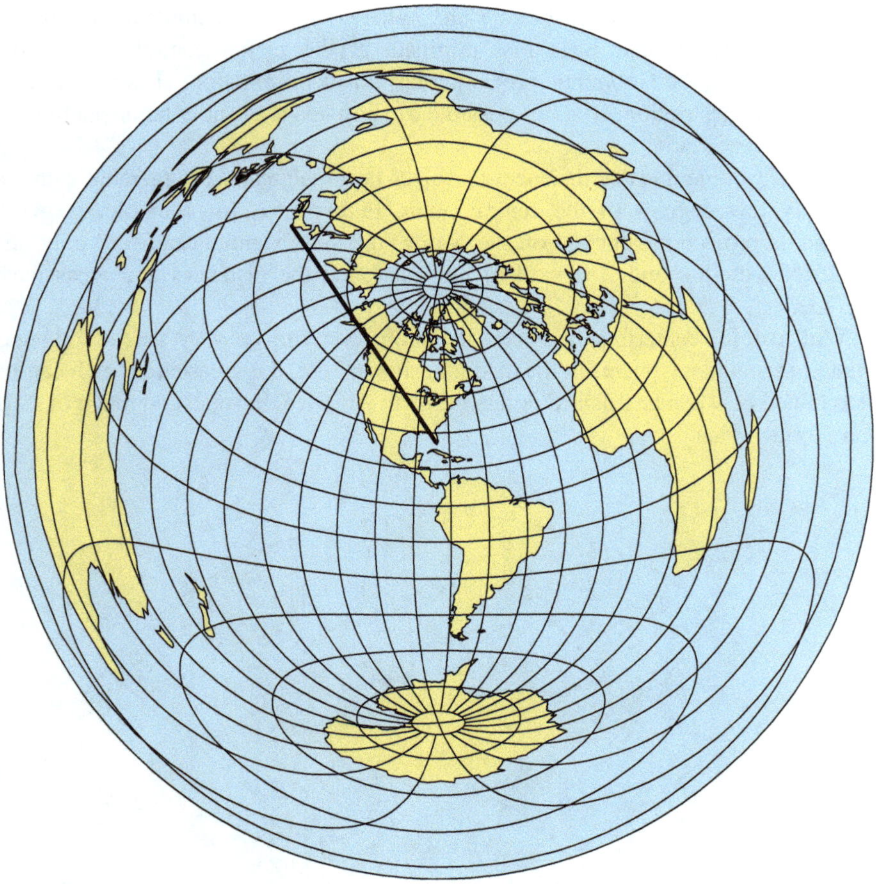

Fig. 3.51 The azimuthal equidistant projection centered on Miami, Florida

projection on a particular city or other point of interest, as has been done in Fig. 3.51. A straight line from the map's central point (Miami, in this case) to Tokyo (or any other place) shows the shortest route and the correct distance and direction between the two. The central point could be any other significant place such as an airport, radio transmission facility, missile launch site, or seismographic station. The entire world can be shown on this projection, although areas and sizes become very distorted near the outer edges.

3.10.2 The Gnomonic Projection

The Gnomonic projection (Fig. 3.52) is another member of the azimuthal projection family, and it has the distinction of being the only map projection on which *any* straight line represents a great-circle arc. Whereas on the Azimuthal Equidistant projection only straight lines that originate at the map's central point show great-circle arcs, the Gnomonic projection has no such limitation. However, it does have a major restriction; it is not possible to show even a whole hemisphere on a Gnomonic projection.

Another problem is that distances and sizes become greatly exaggerated within a relatively few degrees of the central point. Figure 3.52, for example, shows a Gnomonic projection centered on the North Pole and extending outward only 65° (to 25°N latitude), and it is evident how distorted the distances and areas have become.

With careful centering, however, the projection can be very useful for areas smaller than a hemisphere. In the example shown here, the great-circle path can be determined between any pair of points in North America, Europe, and much of Asia and North Africa.

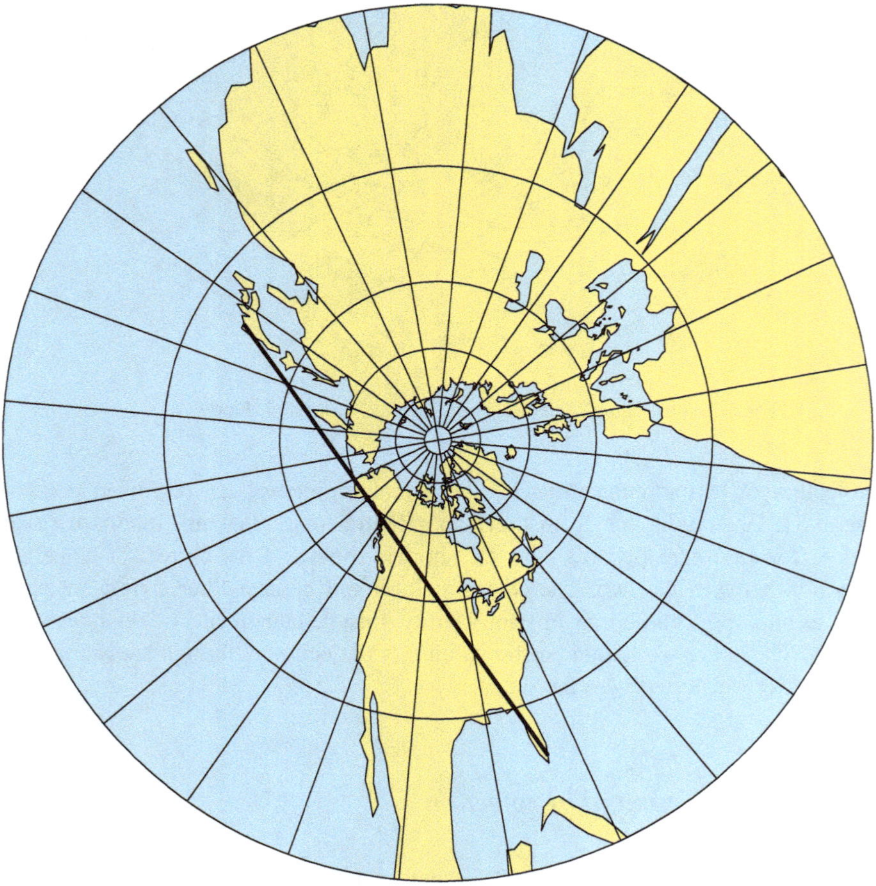

Fig. 3.52 The gnomomic projection centered on the north pole

These two projections, the Azimuthal Equidistant and the Gnomonic, probably are the most versatile and commonly used projections for representing great-circle routes. They are not the only alternatives, however.

3.10.3 Great Circles on Cylindrical Projections

Some projections show great-circle arcs as straight lines in more limited ways, such as in a single direction or along just one or two lines on the map. For example, the basic conic and cylindrical projections show north-south, great-circle courses (meridians) as straight lines, and on cylindrical projections the great circle formed by the equator also appears as a straight line.

It is possible to "force" two points to lie along one of the straight lines that is a great-circle arc on the projection. This necessitates "tilting the Earth" so that the map is transverse (90°) or oblique (less than 90°) to the normal equatorial orientation of the projection. Figure 3.53, for example, shows a form of a cylindrical projection, the Plate Carrée. In this case the globe has been rotated at an oblique angle until Miami and Tokyo were aligned along the projection's "central meridian." On the normal aspect of the Plate Carrée, the great-circle arcs of the equator and all meridians are represented as straight lines that are true to scale.

As a further illustration of how different projection systems can affect the representation of shortest routes, look at Fig. 3.54. This shows the normal aspect of another cylindrical projection, the Mercator, with the great-circle route between Miami and Tokyo plotted on it. The equator and the meridians are the only great-circle arcs this projection represents as straight lines.

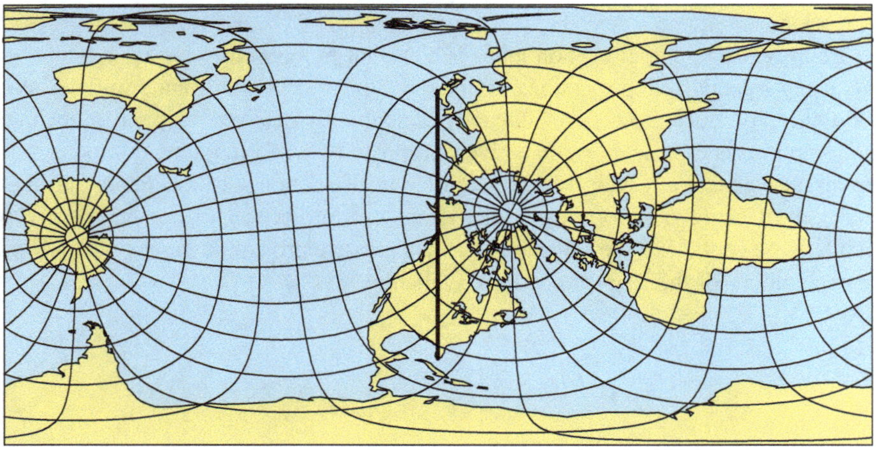

Fig. 3.53 The oblique Plate Carrée projection with Miami and Tokyo aligned along the projection's "central meridian"

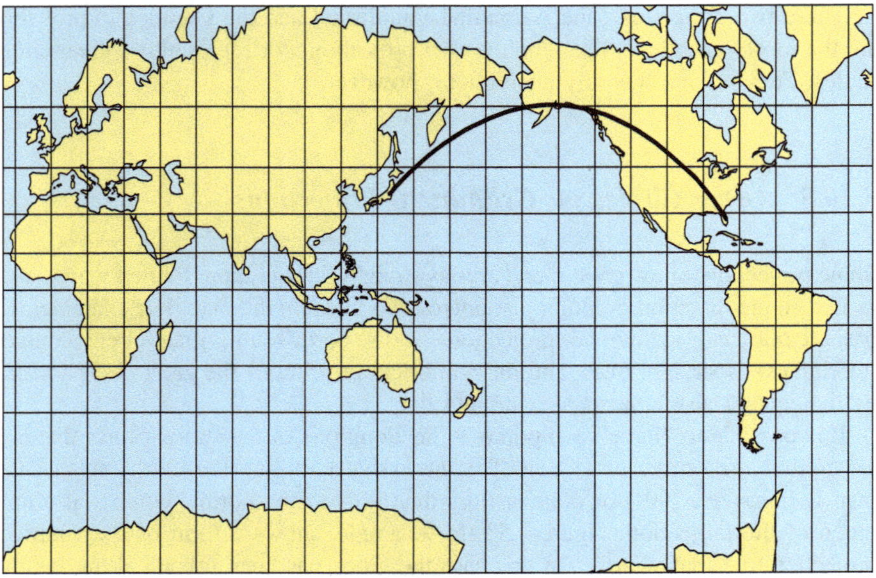

Fig. 3.54 The normal aspect of the Mercator projection showing the great circle between Miami and Tokyo

Obviously, in its normal aspect, this is not a good projection for showing the shortest route between points on the Earth's surface unless they happen to lie on the equator or a meridian. Otherwise, great-circle routes do not look like the most direct ones at all. To achieve a straight line for the great-circle arc between Miami and Tokyo with this projection, one could rotate the projection to a position where Miami and Tokyo lie on the "equator" of the projection, much as was done along the central meridian in Fig. 3.53. Figure 3.55 shows the results of such an operation.

Representing great circles on maps is a special problem that can be solved only by using specialized projections. Formerly, the calculations and drafting skill required to produce projections manually undoubtedly deterred people from using a more appropriate projection for a given purpose. Now, fortunately, computer programs that generate a wide variety of different projections are available, and they make the task of transforming a map from one projection to another relatively painless. Nevertheless, the selection of an appropriate projection begins with a basic understanding of the map projections available!

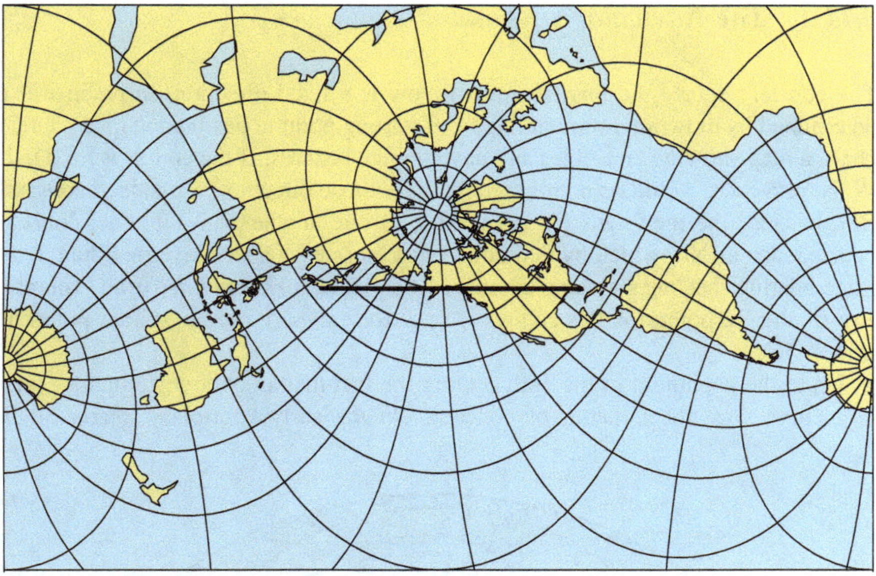

Fig. 3.55 The oblique aspect of the Mercator projection with Miami and Tokyo aligned the projection's "equator"

3.11 Straightening a Rhumb

Arthur H. Robinson

On a clear day on a small body of water a sailor can aim for a point visible on an opposite shore. But when a destination is not visible, as on the open sea, the mariner must try to proceed along a constant, predetermined, directional course. A constant course is a practical necessity since it is much more difficult to navigate a curved course.

Determining a constant course is more complicated than it might seem at first because our system of "compass" directions, N, S, E, W, etc., is defined by the coordinate system of meridians and parallels fitted to a sphere. (The needle of a magnetic compass usually varies somewhat from the true direction.) As pointed out in Sect. 3.3, meridians are lines that converge to each pole and define the directions N or S. A second set of lines, called parallels, is perpendicular to the meridians at every intersection. They form a set of circles that define the directions E or W. This spherical coordinate system is not spaced at equal distances as is a rectangular system on a flat surface, and therein lies the problem.

3.11.1 The Navigator's Rhumb

The constant course the mariner must follow is a fixed direction. In practice it is determined by drawing a straight line from starting point to destination on a nautical chart, a map showing coastlines and navigational hazards. If the course were due E, W, N. or S there would be no problem because those directions are indeed constant. But, because the meridians lie at an angle to one another, any oblique, constant course, such as NE, which by definition must cross each successive meridian at the same bearing, becomes a complex curve. It is called a *rhumb line* (from "rhumb": any of the 32 points of the compass) or *loxodrome*. A rhumb spirals poleward (Fig. 3.56).

Up to the beginning of the 16th century the fact that an oblique compass course is a curve was not a matter of concern. On the early "portolan" charts of the

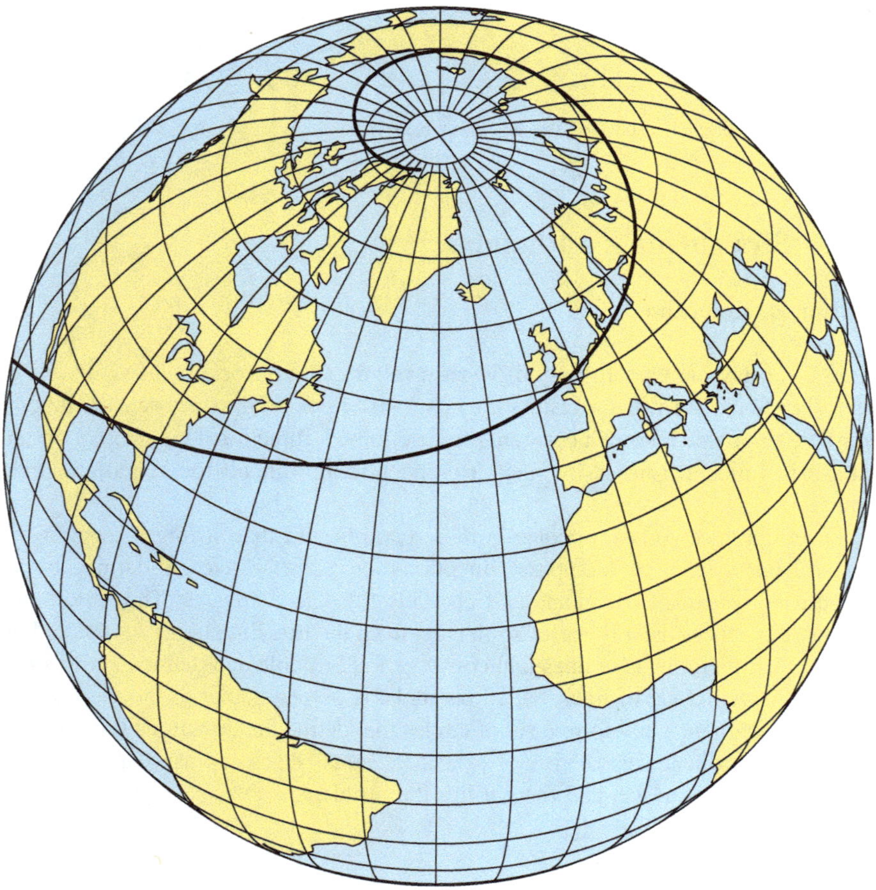

Fig. 3.56 A rhumb line on the spherical earth

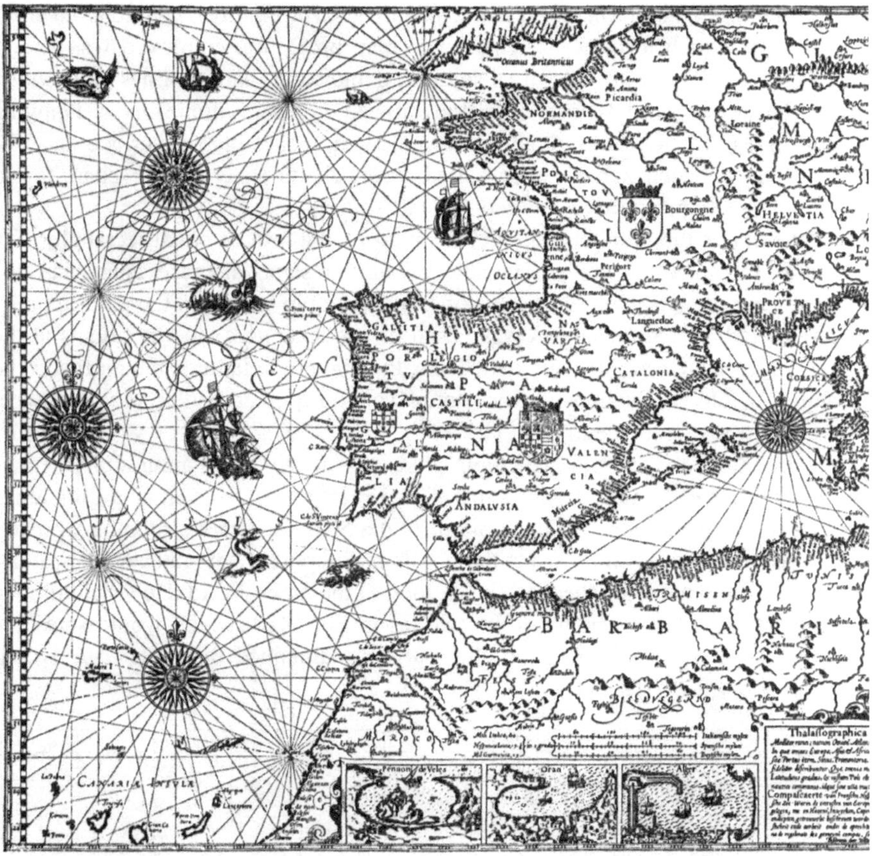

Fig. 3.57 A portion of W. Barent's chart of the Mediterranean showing radiating rhumbs. From Nordenskiöld, A.E. 1889. Facsimile-Atlas... New York: reprinted Dover Publications, 1973, p. 39

Mediterranean a base of parallels and meridians was not needed since frequent passages and relatively short distances within the enclosed area led to accurate relative locations. Courses were not long enough to cause problems. As an aid to navigation these early charts were crisscrossed by numerous sets of straight lines radiating from compass roses (Fig. 3.57). These "rhumb lines enabled the navigator to determine the correct compass course easily. Apparently the system worked well in the confined Mediterranean area.

3.11.2 Plane Charts

Early in the 16th century mariners began regularly plying the much more extensive oceans. A chart of these larger regions had to be based on some systematic transformation (a map projection) of the spherical surface of the Earth to a flat sheet as a base on which to draw the chart. The map projection employed was one in which the meridians were drawn as straight, equidistant, parallel lines; perpendicular to these, the parallels were also drawn as equidistant, parallel lines. Drawing the converging meridians as parallel lines, in effect, makes the sea surface flat. The charts made on such a base were called *Plain* charts (later, *Plane* charts), and navigating by them was called *Plain* (*Plane*) *sailing*.

On the charts the parallels and meridians formed a grid of equal rectangles. On the Earth the parallels are everywhere about the same distance apart, but the meridians are progressively closer together as they draw closer to the pole. Ideally, the E-W width of the rectangles was based on the spacing of the meridians at the latitude of the center of the chart area. Thus, the higher the latitude the narrower the rectangles would be. Such projections are all called Equirectangular no matter what parallel is chosen to be central.

The tradition of covering the charts with crisscrossing, radiating straight lines and the practice of "sailing by the rhumb" were continued. This created difficulties since on these charts a straight line oblique to the meridian is not a rhumb. A greatly exaggerated example will illustrate the problem. Suppose a navigator were heading for Bristol, England, from the Carolina colony in the New World. He would draw a straight line on his chart between the two, as in Fig. 3.58, which would show his course to be about N 62° 30°E. If the mariner maintained that course he would

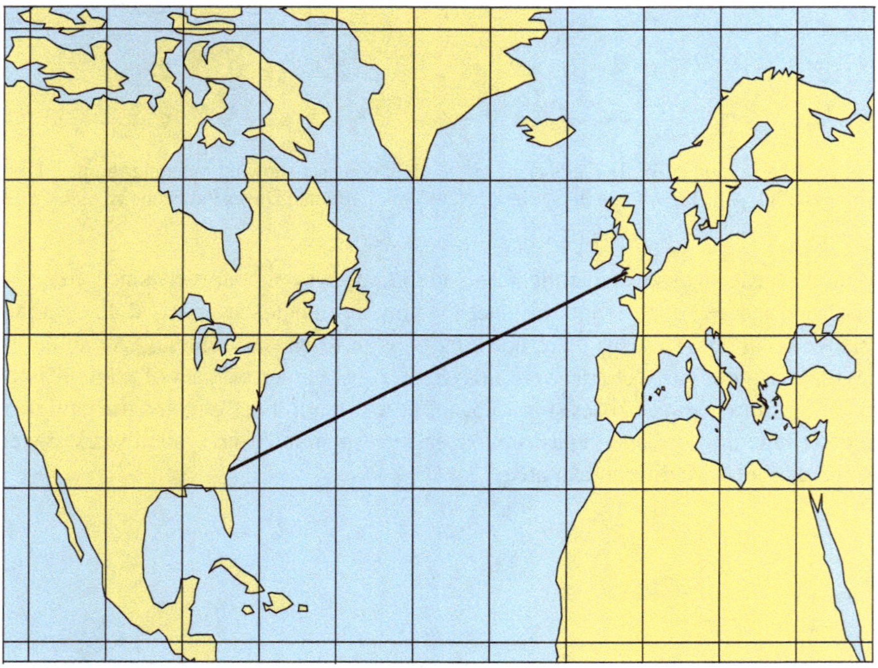

Fig. 3.58 An Equirectangular projection of the North Atlantic with an incorrect "rhumb"

actually miss England entirely, going far to the north. If the navigator's chart were based on another Equirectangular projection, the indicated course would be different, and wrong. In the 16th century sailing on the open ocean was adventurous, to say the least. The couplet of the 19th-century poet Arthur Hugh Clough captures its essence:

Where lies the land to which the ship would go?
Far, far away is all her seamen know.

3.11.3 Mercator's Solution

The fact that a rhumb is a complex curve on the Earth globe was recognized by cosmographers early in the 16th century, but the problem of how to construct a chart that would show rhumbs as straight lines was not solved until 1569. In that year the Flemish cartographer, Gerardus Mercator, published a large world map on a projection he had devised for the purpose of navigation. Instead of spacing the parallels equally, Mercator expanded the distances between them progressively

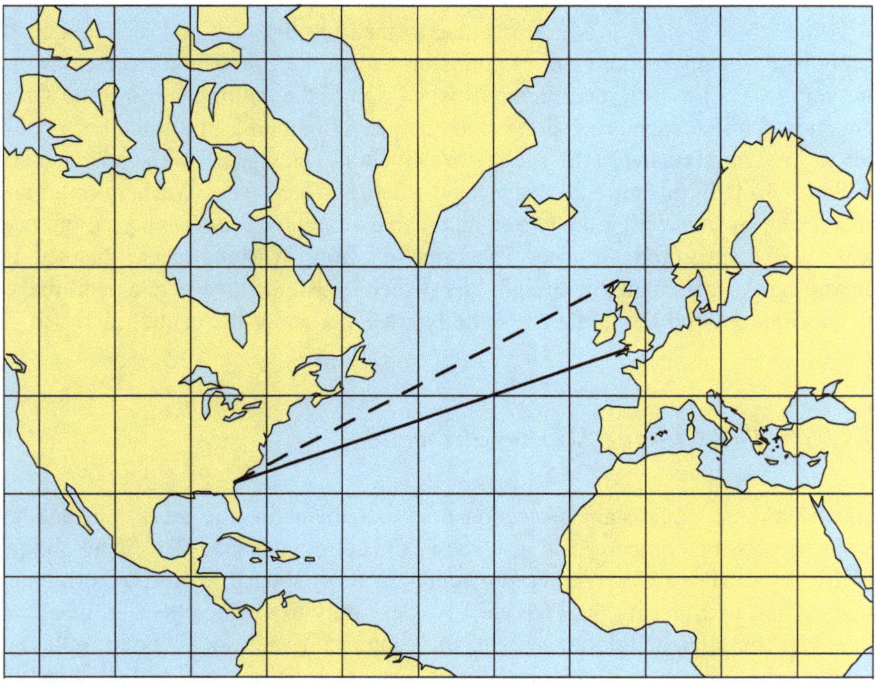

Fig. 3.59 A Mercator projection of the region in Fig. 3.58 with the correct rhumb. The *dashed line* shows where the "rhumb" on Fig. 3.58 would have led

poleward to match the rate at which the spacing of the meridians had been expanded by making them parallel. On a chart made on a Mercator projection any rhumb plots as a straight line with the correct compass angle. Figure 3.59 shows a Mercator map of the same area as Fig. 3.58 and the correct compass course, about N 72°E.

Although the projection solved the problem of sailing by the rhumb, it was not adopted immediately. Determining distance on a Mercator chart was more cumbersome than on a Plane chart, and the latter remained in use well into the 17th century. Since then, however, nearly all navigational charts have been made on the Mercator projection.

3.12 Measuring Directions and Distances for the Shortest Routes

John P. Snyder

For some maps, we may be less concerned about the great-circle path between two points than we are about the direction and distance from one point to another. For radio signals, for example, these values affect the pointing of a directional transmitter or receiver as well as the strength of the signal.

Although in general it is not practical to measure distance and direction accurately between any random pair of points on a map, there are some projections that are very useful for such measurements when one of the points is always the same. The most familiar such projection is the oblique Azimuthal Equidistant, centered at any desired point on the Earth, such as Washington, D.C. (Fig. 3.60), and discussed in Sect. 3.11. In this case all distances can be measured from Washington to any other point by connecting the two points with a straightedge and applying the map scale to the measured distance. The direction from Washington is obtained by measuring the angle that this straight line (which is also the great-circle path) makes at Washington with the direction to the North Pole, along the central meridian.

3.12.1 The Oblique Werner Projection

The Azimuthal Equidistant projection does not maintain true areas, but another projection has this property and also shows the correct distance. This is an oblique form (Fig. 3.61) of the Werner projection, a heart-shaped projection originally devised and used during the 16th century, but rarely used since then. In this case, however, the straight line connecting the center and some other point, although having the correct length, does not follow the great-circle route, and its direction cannot be determined readily.

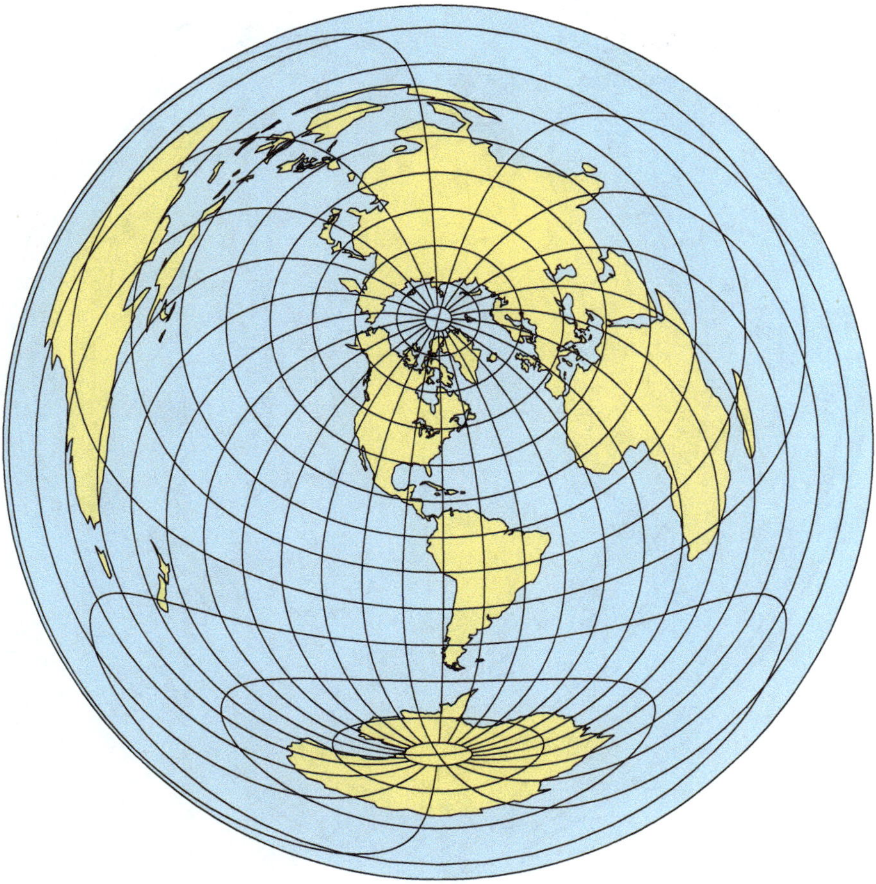

Fig. 3.60 The oblique azimuthal equidistant projection of the world centered on Washington, D.C

Another way to measure distance and direction is to treat them both as rectangular coordinates by plotting an oblique aspect of another very old projection called the Plate Carrée or Plane Chart (Fig. 3.62), discussed before (see Sects. 3.11 and 3.12). In the normal case all meridians and parallels are equally spaced straight lines, and the scale is correct along all meridians and the equator. If the point of interest, such as Washington, is placed at the "north pole" of this projection (Fig. 3.63), that point becomes the entire top edge of the map, and its antipode, the bottom edge. Then the great-circle route from Washington to any second point is the straight vertical line intersecting the second point, while the distance is that directly measured from the top edge.

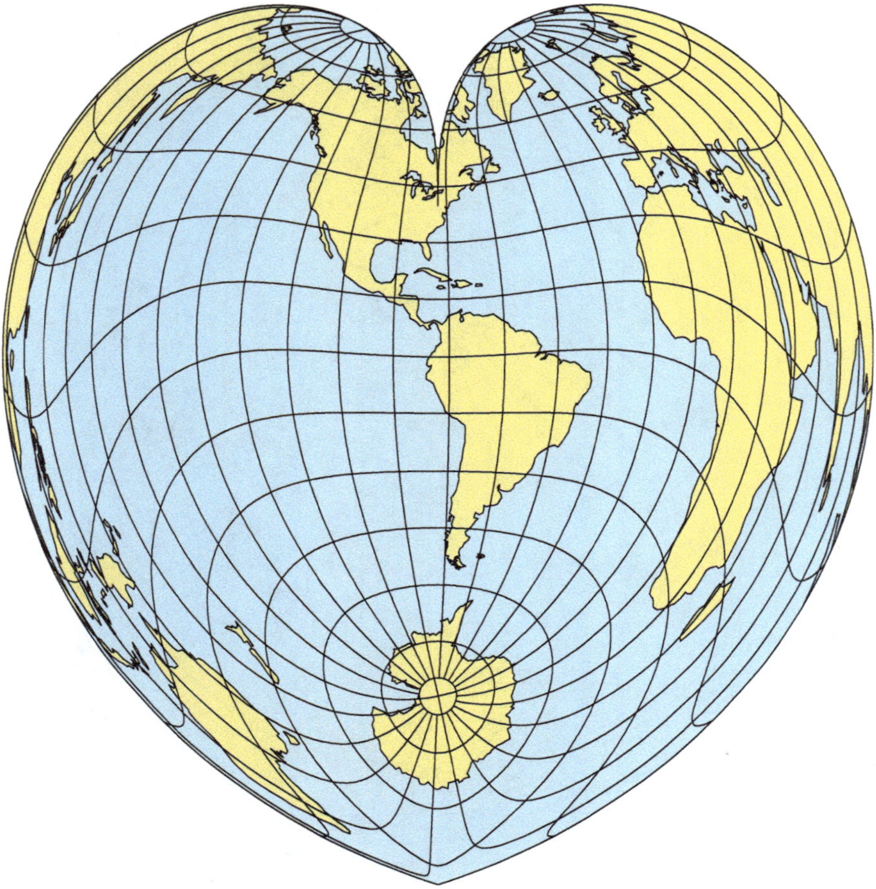

Fig. 3.61 The oblique Werner projection of the world with its pole at Washington, D.C

3.12.2 Retroazimuthal Projections

It is also possible to have a projection on which a straight line connecting the center and any other point shows the correct distance, and on which the direction from the second point to the center relative to north is the same as the angle between the connecting line and a line extending straight up from the second point. The direction from the center to the second point, however, will generally not be correct. This is called a *retroazimuthal* projection, and this form was devised by Hammer in Germany in 1910. In 1929, it was extended in England to a world map, inspired by the desire for a map to allow persons throughout the world to determine the directions to point antennas on radio receivers in order to pick up transmission from a powerful station in Rugby, England. The far side of the Earth must be plotted

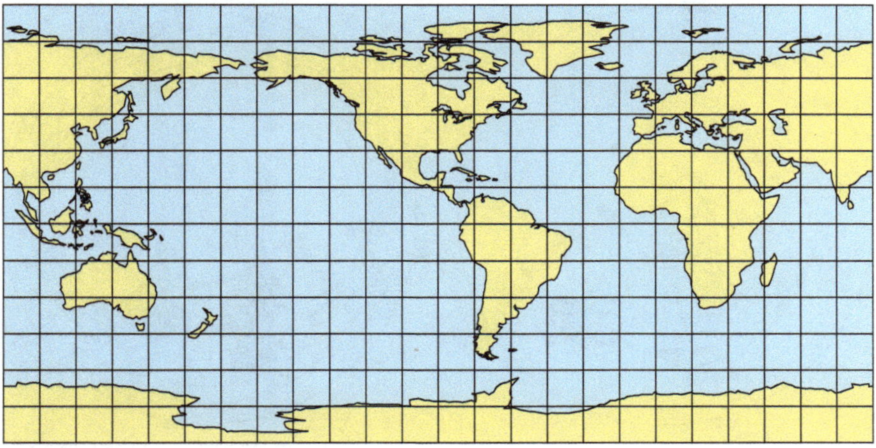

Fig. 3.62 The direct Plate Carrée projection of the world

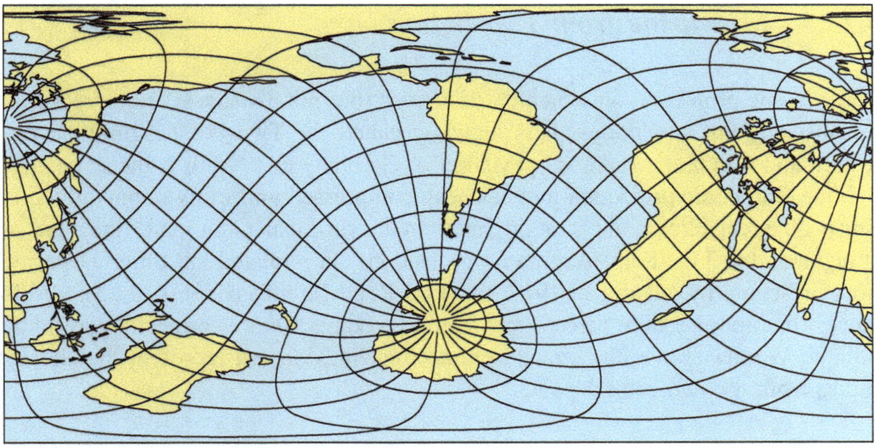

Fig. 3.63 The oblique Plate Carrée projection of the world with its upper pole at Washington, D.C

backwards, and the entire projection has a strange shape. Figure 3.64 shows the two parts of a world map centered near St. Louis, Mo. They should be superimposed.

The first retroazimuthal projection was devised by Craig in Egypt in 1909 as the Mecca projection, so-called because with it Moslems anywhere could determine which way to face toward Mecca (the center) for prayers. Craig's retroazimuthal projection is different from Hammer's in that meridians are straight and vertical instead of curved, and distance cannot be measured readily, but direction is measured as on Hammer's.

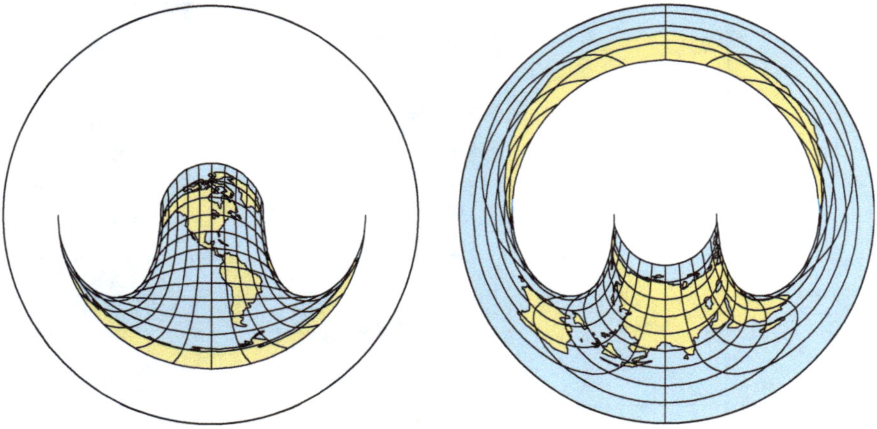

Fig. 3.64 The oblique Hammer retroazimuthal projection of the world, centered near St. Louis, Mo., at longitude 90°W, latitude 40°N. Two hemisphere bounded by meridians at 0° and 180°

3.12.3 Measuring from Two Points

There is one projection on which all directions (but not distances) are correct from not just one but two points. It is logically called the Two-Point Azimuthal projection and was first presented by Maurer of Germany in 1914. It is the same as an oblique Gnomonic projection that has been compressed uniformly in one direction. Since all great circles are straight lines on the original Gnomonic, they are all straight on the Two-Point Azimuthal projection. On a related but different projection, presented by Maurer in 1919, the Two-Point Equidistant (Fig. 3.65), all distances (but not directions) are correct from two points. On this projection, as on the oblique Werner above, the straight connecting lines do not represent the shortest routes from the two central points.

3.13 Showing Ranges and Rings of Activity

Phillip C. Muehrcke

Sometimes we wish to preserve the shape of environmental features on a map. In special cases this shape is a circle on the ground, or a phenomenon that spreads outward in circular ring-like fashion from a source area. A special class of projections makes it possible to capture the circular character of these features.

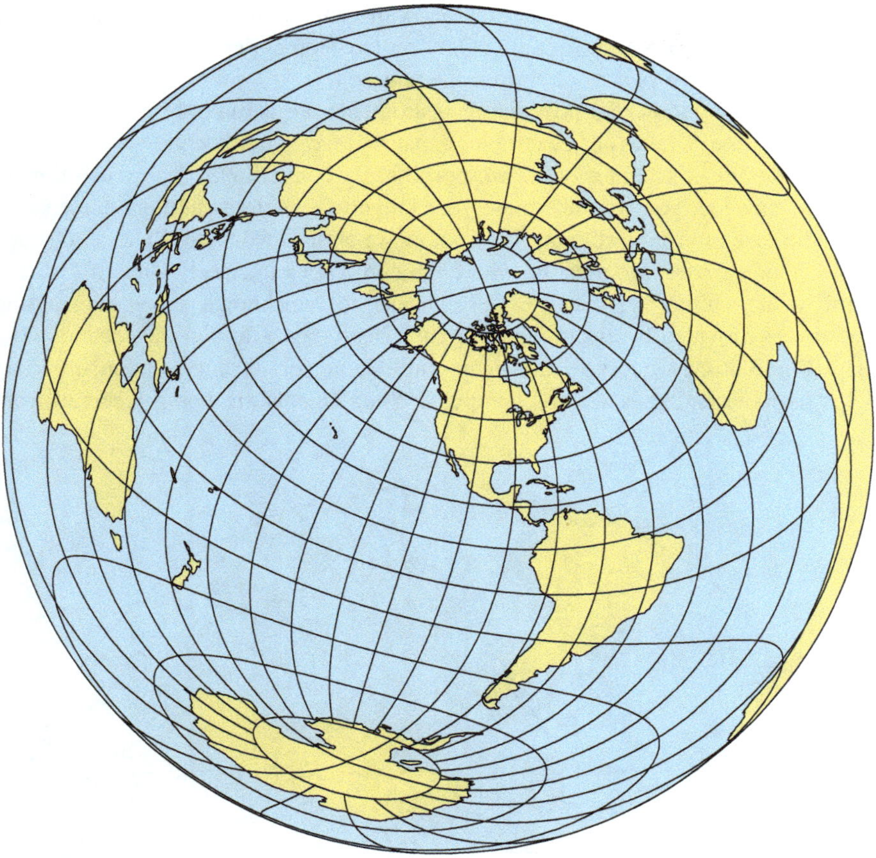

Fig. 3.65 The two-point equidistant projection of the world with centers at Washington, D.C., and Honolulu

3.13.1 Ranges and Positions

Most circles on the Earth's sphere are created when something extends outward a specified distance from a central point. Radio, television, and cellular telephone signals all exhibit this characteristic. Similarly, an airplane can only fly a limited number of miles or hours from an airport before it runs out of fuel. In each case the effective range is a circle on the ground.

To illustrate the range of something, we need to choose a projection on which ground circles are preserved as circles on the map. *Conformal* projections have this special property if the circles are relatively small. But since we cannot preserve both distances and area on the same map projection, circles of constant size will plot in different sizes from one part of a conformal map to another. The azimuthal family of projections will preserve a circle of any size as long as the projection is centered on that circle. But this means that circles with centers elsewhere will not plot as circles on most of these projections. Only on the Stereographic projection will all circles on the globe plot as circles on the map.

Modern navigation and positioning instruments are commonly based on measuring the time it takes an electronic signal to reach your location from several ground stations or satellites of known position. By converting signal transit time to distance traveled, you can determine your range to each transmitting source.

In the case of ground stations, if we know we are located somewhere along each of two range circles, we must be where the two circles intersect. Thus, if lost in a storm while sailing, we can use the distances to two known shore transmitting stations, as provided by electronic navigation instruments, to plot our location on a map (Fig. 3.66). Notice how the range circles intersect at our location; the map helps us to visualize as well as compute. When signals are being received from

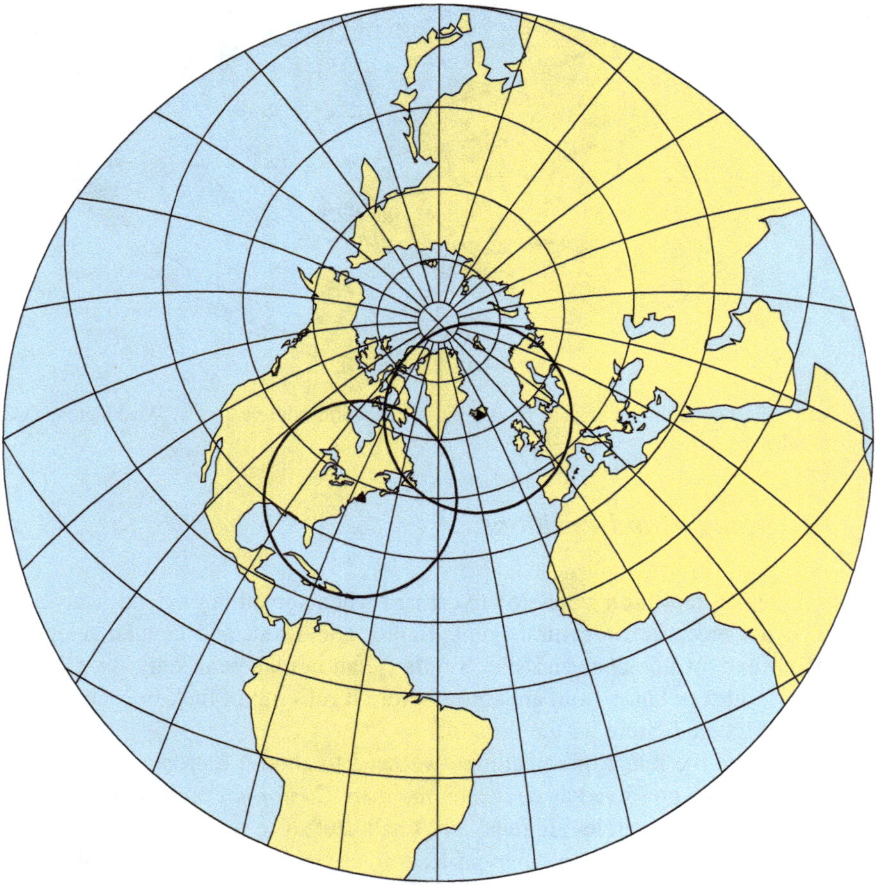

Fig. 3.66 The stereographic projection of a hemisphere centered at latitude 60°N, longitude 45°W, with range circles centered on Boston and Reykjavik

satellites, we now have a third dimension and must deal with intersecting spheres. The same principles apply in both cases, however, since the intersection of a satellite range sphere with the Earth sphere is a circle.

3.13.2 Rings of Activity

The distribution of landscape forms created by environmental processes on spherical Earth can be difficult to visualize. The pattern of earthquake and volcanic activity around the Earth illustrates this point. A thoughtless choice of map projection that pays no attention to the geographical character of the phenomenon being mapped might look like Fig. 3.67. The geometry of this map provides little insight into process and structural relations and, as a result, the pattern shown is complex and confusing. The band of intense tectonic activity known as the "Ring of Fire" surrounding the Pacific Ocean is apparent but its understanding is obscured more than enhanced by this map.

To choose a projection well, we must take advantage of what is known about a phenomenon. In this case, tectonic activity is said to be associated with the slow breakup of a single ancient land mass centered approximately at the present location of the southern tip of Africa, and the subsequent drift of the pieces (tectonic plates) holding the continental "islands" that make up the surface layer of the Earth. It is believed that tectonic activity is greatest along the leading edge of the continental plates as they move outward from the position of the original landmass. Thus, if we choose an Azimuthal Equidistant projection and center it on the tip of South Africa, we obtain a most revealing pattern of tectonic activity (Fig. 3.68). The intensely active "Ring of Fire" becomes an encompassing circle that coincides with the

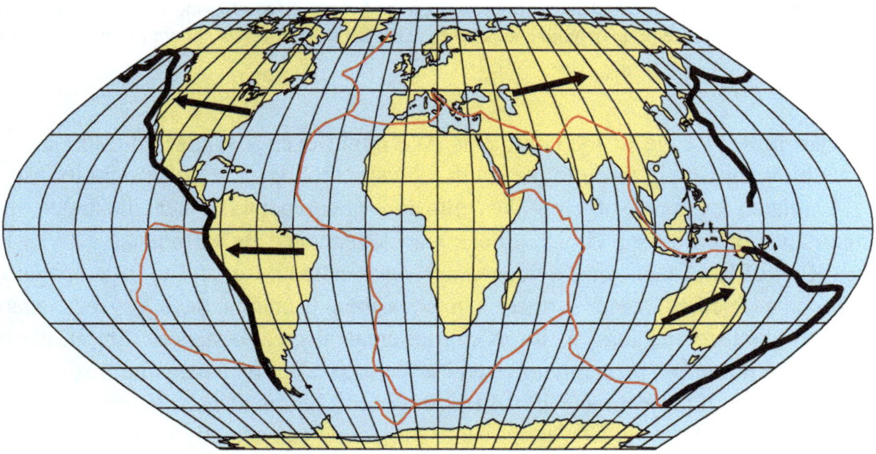

Fig. 3.67 The Eckert VI equal-area projection of the world with tectonic plate boundaries and regions of volcanic activity. The "Ring of Fire" is shown by the very *thick line*

Fig. 3.68 The azimuthal equidistant projection centered at the Cape of Good Hope, with tectonic plate boundaries and regions of volcanic activity. The "Ring of Fire" is shown by the very *thick line*

forward-moving edge of the continents. We might even visualize ship-like continents plowing slowly through the Earth's crust, creating zones of instability.

A related example of ring-like diffusion phenomena further illustrates the importance of projection choice. Shock waves generated by earthquake activity in the "Ring of Fire" can cause a huge water wave called a tsunami to be propagated across the Pacific Ocean surface. On occasions tsunamis have caused severe property damage and loss of life in coastal areas at great distance from an earthquake site. If our aim is merely to visualize the path taken by a tsunami, maps based a wide variety of projections will suffice. But in this circumstance we might want to go beyond visualization. Clever choice of projection can actually turn the map into a powerful nomographic tool for geographical analysis.

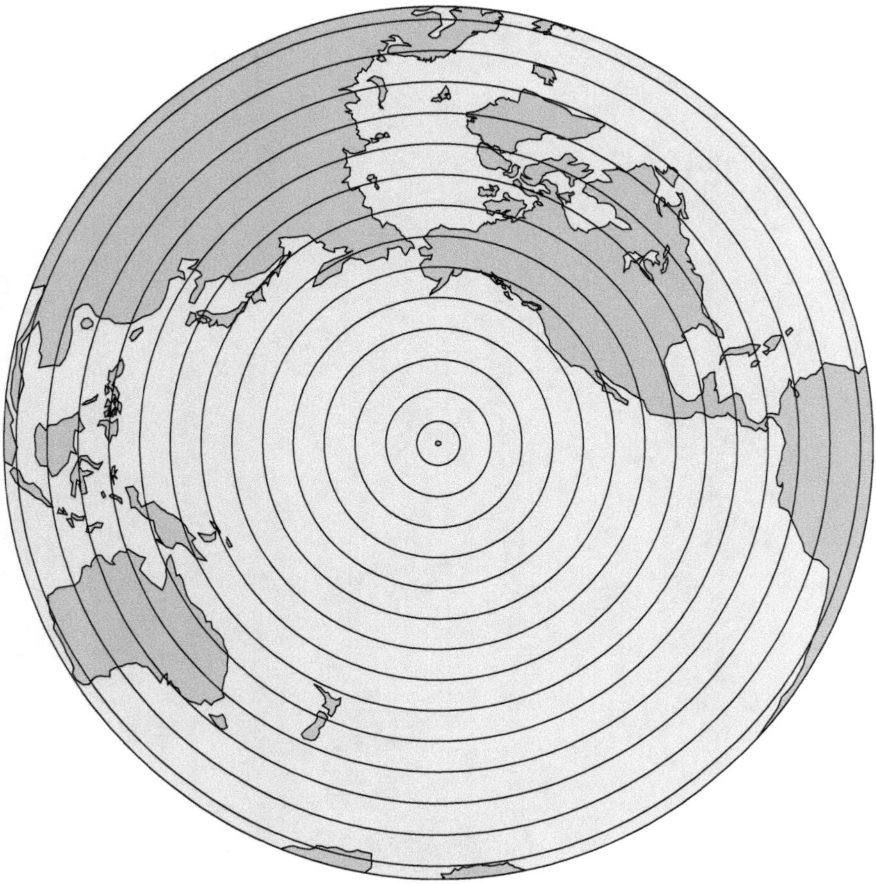

Fig. 3.69 The azimuthal equidistant projection centered on the island of Hawaii and extending 100° with equally spaced circles

Consider, for example, the fate of residents living in coastal lowlands on the island of Hawaii. When a tsunami is caused by an earthquake occurring somewhere in the Pacific basin, they would like to know how big a water wave has been generated and how long it will be before this wave hits their shores. Since the wave will travel across the ocean surface at a nearly constant rate of speed, the important thing to know is distance to the earthquake site. An Azimuthal Equidistant projection centered on Hawaii and marked off to the Pacific Ocean rim with equally spaced tsunami travel time lines admirably serves this purpose (Fig. 3.69). A glance at the map would suffice to determine the estimated time of arrival of the potentially dangerous water wave, regardless of its origin in the basin. Preparations can be made accordingly.

Chapter 4
Map Projections and the Internet

Fritz C. Kessler, Sarah E. Battersby, Michael P. Finn and Keith C. Clarke

Abstract The field of map projections can be described as mathematical, static, and challenging. However, this description is evolving in concert with the development of the Internet. The Internet has enabled new outlets for software applications, learning, and interaction with and about map projections. This chapter examines specific ways in which the Internet has moved map projections from a relatively obscure paper-based setting to a more engaging and accessible online environment. After a brief overview of map projections, this chapter discusses four perspectives on how map projections have been integrated into the Internet. First, map projections and their role in web maps and mapping services is examined. Second, an overview of online atlases and the map projections chosen for their maps is presented. Third, new programming languages and code libraries that enable map projections to be included in mapping applications are reviewed. Fourth, the Internet has facilitated map projection education and research especially with the map reader's comprehension and understanding of complex topics like map projection distortion is discussed.

F.C. Kessler (✉)
John A. Dutton e-Education Institute, Department of Geography,
Pennsylvania State University, State College, USA
e-mail: fck2@psu.edu

S.E. Battersby
Tableau Software, Seattle, WA, USA

M.P. Finn
Center of Excellence for Geospatial Information Science,
U.S. Geological Survey, Denver, CO, USA

K.C. Clarke
Department of Geography, University of California, Santa Barbara, USA

M. Lapaine and E.L. Usery (eds.), *Choosing a Map Projection*,
Lecture Notes in Geoinformation and Cartography,
DOI 10.1007/978-3-319-51835-0_4

4.1 Introduction

This chapter presents an overview of how map projections have been used for Internet cartography. The first section provides a general overview of map projections. The second section covers the map projections that are used by Web maps and map services. This section is divided into two parts and covers map projections adopted for both global and local-global mapping scales. The third section presents a review of online atlases. Specifically, this section examines the common map projections that are selected for thematic and general purpose maps appearing in online world atlases. The fourth section considers online map projection software. For instance, there are several useful code libraries that can be incorporated into a Website that facilitate map creation and projection. The fifth section discusses topics related to map projection research and education. Of particular importance in this section is the level to which map readers understand map projection distortion and how this understanding enhances or detracts from map learning.

4.1.1 Map Projection

Map projection is the mathematical way the Earth's two-dimensional curved surface is transformed and mapped onto a planimetric surface. Hundreds of map projections have been developed. Many were created from mathematical curiosity or as a means to reduce the inevitable distortion that results from the projection process (Snyder 1993). For instance, the Eisenlohr projection is conformal (preserves local angles) and rarely used. The Eisenlohr was mathematically contrived so that conformality does not fail at any point (Eisenlohr 1870). Despite the many projections that have been developed, only a handful have seen wide-spread use. The Gall stereographic is one projection that has appeared in many atlases as the basis for world maps of thematic data. In some instances, a few map projections were developed to meet the specific needs of a map purpose. For example, the American (or ordinary) polyconic projection was developed by Hassler (1825) for large-scale topographic mapping of the United States. On the other hand, the Robinson projection, developed by Robinson (1974), was designed specifically to show the appearance of the world's landmasses as if looking on a globe.

Map projections can be generally categorized based on the properties that are preserved in the transformation process. The properties that are preserved make particular types of projections useful for specific purposes. For instance, equal area projections preserve areas measured on the Earth's surface. When using symbolization methods, such as dots, to show the distributions of a phenomenon across space, an equal area projection is preferred for thematic mapping. Using an equal area projection in global-scale mapping ensures that each country is shown with the same area as on a globe. Preserving area allows for visual comparisons of a distribution of a data set across the geographic area of interest.

If the map purpose calls for preserving distances, then an equidistant projection should be considered. This property allows distances to be preserved along all lines of longitude or from a single point to all other points following great circles. Like azimuthal projections, equidistant projections are beneficial for navigational purposes. If an angle is measured on the Earth's surface and that angle needs to be preserved on a map (e.g., for navigational purposes), then a conformal projection is appropriate.

Some projections do not preserve any specific property. These projections are referred to as "compromise" and balance the distortion levels across the projected area. The result is a projection that attempts to preserve the visual appearance of Earth's landmasses as much as possible. Due to this characteristic, compromise projections are suitable for general reference or political maps. As will be described through this chapter, not all projections and their properties are correctly matched to the map purpose.

4.2 Map Projection Use in Web Map Services

In this section we will discuss common map projections as used in Web maps and map services. We will start with a discussion of global-scale mapping options (largely focusing on those tailored for tiled map services to be used across scales), and then present general discussion of projections for online local-scale mapping.

4.2.1 Global-Scale Maps

For general purpose online maps designed to be panned and zoomed across scales, from global to local, we generally see use of cylindric map projections. Cylindric projections allow for easy tile creation and the rectangular shape makes it easier to facilitate continuous panning without discontinuities on the map periphery. Though projection choice for online maps has varied over the years (Battersby et al. 2014), most online maps now utilize Web Mercator. Though we do not know the exact cartographic and/or computational reasons behind the choice of Web Mercator for online tiled map systems, it has become the de facto standard for global-scale tiled maps since Google Maps was introduced in 2005 using this system. For reference, the Web Mercator projection may also be referred to as Spherical Mercator, Google Mercator, or WGS84 Web Mercator, and is formally referenced by the European Petroleum Survey Group (EPSG), now the International Producers of Oil and Gas) definitions as EPSG:3857 (Battersby et al. 2014).

Web Mercator is a projection related to the cylindric projection introduced by Gerardus Mercator in 1569. The significant difference with the Web variant of the projection is that it uses the spherical form of forward equations with an Earth model radius of 6,378,137.0 m. Ellipsoidal latitude and longitude values from the

World Geodetic Survey of 1984 (WGS84) ellipsoid are then projected using the spherical form of the equations. Using the spherical equations with the ellipsoidal values (namely, the semi-major axis) makes the Web Mercator non-conformal (it does not preserve local angles), unlike the conformal Mercator projection), though visually this difference at the global scale is not detectable.

4.2.2 Web Mercator Issues and Controversies

The adoption of Web Mercator has been fairly controversial for Web mapping. Much of this controversy is related to the general need to use a map projection that is appropriate for the scale, data, and application. Unfortunately, for general purpose mapping (thematic and reference) across scales that we see in Web map services there is *no* single best projection that will be appropriate for all uses. Though we discuss several issues and controversies surrounding the adoption of Web Mercator, we note that these issues would not be eliminated with the selection of any other single projection. We also note that we are only presenting a few notable issues with the Web Mercator projection; for reference, Battersby et al. (2014) provide a more detailed discussion of the challenges of the Web Mercator projection for online mapping.

4.2.3 Mapping Accuracy

While there is no visual difference between Mercator and Web Mercator in global-scale mapping, from a geodetic standpoint the difference in calculated measurements can be substantial—up to 50 km at the extreme north and south latitudes (Battersby et al. 2014). Because of these differences, agencies such as the US National Geospatial-Intelligence Agency (NGA 2014) have indicated that the Web Mercator is not acceptable for global-scale intelligence mapping. Related to this, NGA (2014) has issued an advisory notice indicating that "this erroneous geospatial positioning information poses an unacceptable risk to global safety of navigational activities… that require accurate and precise positioning and navigation information" and has explicitly stated that it does not endorse or support the Web Mercator projection for any geospatial intelligence activities (NGA Office of Geomatics 2014). The NGAs concern about the distortion of the Mercator projection is not a new development in discussion of appropriate global-scale projections, as can be seen in the 1989 resolution (American Cartographic Association 1989) rejecting all rectangular projections (such as the Mercator projection) for global-scale mapping.

Another map distortion consideration is how scale distortion exists across the Web Mercator projection and how Web mapping services account for and communicate this distortion. As discussed earlier, the Web Mercator is approximately conformal; no scale distortion is present along the equator. Thus, distances measured between two points that fall exactly on the equator are identical to the true Earth distance. As one moves away from the equator, scale distortion begins to increase gradually. As one nears the poles (around 60°N/S) scale distortion increases rapidly. Although scale distortion increases with higher latitudes, distortion is constant along all lines of latitude. The way in which scale changes across the Web Mercator projection is an important consideration when using Web mapping services to measure, for example, distances.

Web mapping services such as Google Maps and OpenStreetMap use zoom levels (0 to 20) to allow users control of the geographic extent shown on screen. Zoom levels range from very large map scales showing street-level views (approximately 1:1000) to very small map scales showing the entire world (approximately 1:600,000,000). Associated with each zoom level are graphic scale bars (e.g., showing miles and kilometers) that appear inside the mapping service window and change in size according to the zoom level. Using the scale bars as reference, users can take approximate measures of distances or areas on the map. As users zoom in on and out from a single point the scale bars automatically adjust in size to account for the changing zoom level and thus map scale. Additionally, as one pans north or south across the map (not changing zoom levels) the scale bars also adjust to accommodate the changing scale distortion across the Web Mercator's surface with different latitude. However, as one pans east or west along a line of latitude (not changing zoom levels), the scale bars remain fixed in size accounting for the fact that scale distortion is constant along all parallels. By incorporating the Web Mercator and changing scale bars Web Mapping services have minimized the problems of accounting for scale distortion that was problematic by the presence of a single scale bar commonly found on printed map. Note, however, that the accuracy of the scale bar across the entire viewing window and thus the geographic extent shown will vary depending on the map's zoom level. Since scale distortion varies with changing latitude the scale bar will be correct only along a single parallel in the view. For large scale mapping changes in scale across a small geographic extent are generally inconsequential, however for small scale mapping, the changes in scale will be substantial.

Additionally, it is important to note the relative importance of the map projection and scale as an influence on visual analysis or when computing a spatial metric. For large scale mapping, the impact is likely to be minimal regardless of projection as the relative change in distortion across the area will be small; for small scale mapping the distortions present significant issues (e.g., a tightly clustered set of points at high latitude on the Earth may appear as a widely distributed set of points in Web Mercator; distances between these points as measured in projected space will be exaggerated).

4.2.4 Map Distortion

While projection distortion in general is an accuracy issue, the controversy surrounding distortion in the Mercator and Web Mercator projections is a cognitive issue as well. Numerous scholars have suggested that frequently seeing distorted maps will warp our global-scale cognitive maps (e.g., Robinson 1990; Monmonier 1995) providing worldviews in which regions at higher latitudes are given a "false connotation of importance" (Vujakovic 2002).

In addition to the general issue of the distortion patterns potentially influencing our cognitive understanding of global-scale geography, we must also consider that most map users do not recognize the influence of distortion on areas or angles across a map. This presents an issue for both designers and map readers. Designing online maps using the application programming interfaces (APIs) from Web map providers is often seen as more an issue of understanding the coding to control map symbology than understanding the nature of map distortion. This means that unknowingly a map designer may introduce significant geographic errors. For instance, we may see "shortest" routes that are straight lines connecting locations (instead of a great circle route) or buffers around points that are a consistent circular shape and size (instead of the varying size and shape expected due to distortion in Web Mercator). The map reader may make the same mistake—interpreting these symbols as truth in their representation of spatial relationships. On the other hand, readers may interpret *accurate* representations on the map as incorrect because they do not follow their intuitive, but incorrect, understanding of spatial relationships because the spatial relationships that are expected to be true in a Cartesian coordinate system are not always true in a map projection because the projection transforms the Earth's curved surface to the 2-dimensional plane. This educational issue is a significant problem that must be overcome as online maps with a Web Mercator projection become our most familiar global-scale map.

4.2.5 Non-web Mercator Options

Web Mercator is not the only option for zoom and pannable multi-scale maps. As we stated earlier, no single projection can meet all mapping needs at all scales and for all purposes. To address this problem, Jenny (2012) has suggested that we need adaptive composite map projections that can seamlessly morph between projections as a user pans and zooms around a map. Figure 4.1 depicts the variety of projections used based on zoom level and central latitude of a map. Though this system for Web maps to adapt on-the-fly based on user location and zoom level (and, one would assume, map purpose) presents benefits, it has not yet been adopted for widespread use.

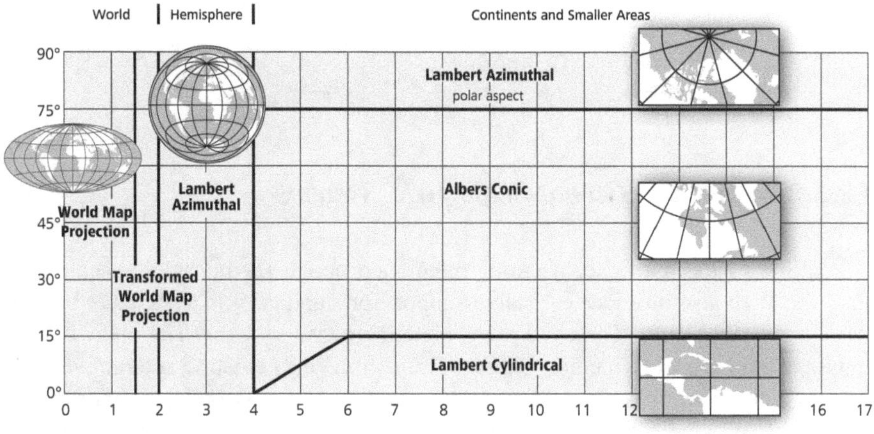

Fig. 4.1 Bernhard Jenny's adaptive map projection matrix. The horizontal axis shows zoom factor and the vertical axis shows central latitude of map. (Figure courtesy of Bernhard Jenny)

4.2.6 Non-global Scale Map Projections

At the global scale map projections, such as Web Mercator, are usually based on treating the Earth as a sphere and simply converting latitude and longitude pairs into the projected coordinate system of the map display window. An alternative to this is use of three axis (x, y, z) coordinates or Earth Centered Earth Fixed, as is common in geodesy, space geometry and computer graphics. This type of transformation would provide a more accurate view of local-scale geographic data.

By far the majority of Web maps have a regional or continental focus, especially in thematic mapping. This is true of Web mapping applications with a national focus such as Baidu Maps for China (http://map.baidu.com/) and Bhuvan for India (http://bhuvan.nrsc.gov.in/bhuvan_links.php), and for more local applications such as state geographic information systems (GISs) (http://www.kansasgis.org/). In most cases, projections are chosen that suit a local map extent, meet various distortion minimization criteria for the region, or fit a local grid. General guidelines for scales larger than global-scale map projection exist, for example Snyder (1987, Appendix B) provided a list of projections used in National mapping for the United States and its territories, and Bugayevskiy and Snyder (1995, pp. 238–9) provided such a list for Russia and the former Soviet Union. Practically, these guidelines seem little used in Web mapping with the exception of national reference maps and some demographic mapping. For example, the Lambert conformal conic has long been standard for mapping demographic data from the United States Bureau of the Census, for example on the Census Data Mapper (http://tigerWeb.geo.census.gov/datamapper/map.html). It also does not seem that much attention in Web mapping has been paid to the research on projection optimization based on geographic extent. Hopefully, the use of better matched projections will expand as Web-based map dissemination becomes more widespread, both geographically and intellectually.

4.2.7 Local- and Global-Scale Grid Systems

In addition to the named projections mentioned above for non-global-scale mapping, the Web also provides extensive support for standard grid systems at both the local- and global-scale. There are many benefits to grid systems. The many benefits of global grid systems include universality (fitting into existing alternative Earth models such as geodetic and astronomical models); authority (being supported by a permanent and credible agency, standard or law); succinctness (using as few codes as possible to ensure location); definitiveness (providing reference to one and only one unique location); exhaustiveness (applying to the whole Earth); being hierarchical (allowing references to multiple scales depending on the desired level of detail); intuitiveness (easy to understand and use); and tractability (able to be used computationally) (Clarke 2002). Not all commonly used grids share these properties, and consequently there exist multiple grid systems, in spite of national efforts at standardization around "National Grids" as in the United States (FGDC 2015) and the United Kingdom (Ordinance Survey 2015).

The Universal Transverse Mercator (UTM), Military Grid Reference System (MGRS), and the United States National Grid (USNG) and other popular global grid coordinate systems are modified versions of the Universal Transverse Mercator or Gauss-Kruger coordinate system. The basic division of the globe into 60 6-degree-wide UTM zones is common, although other systems use 3 degree zones such as China (Yang et al. 2000). Other referencing systems exist for satellite imagery such as the World Geographic Reference System (WGS), and sometimes underlie Web-based map projection systems as data structures, such as tessellated triangles and hexagons (e.g. Quaternary Triangular Mesh). National grid systems are standardized in many countries such as the Map Grid of Australia, the British Ordnance Survey National Grid, the Irish National Grid, the Qatar National Grid (QND95), and in the United States, the National Grid (Cavell 2005). Wikipedia uses a hybrid UTM/MGRS reference somewhat reflective of the USGS Civilian UTM system. In almost all cases, the Web also provides a forum for grid conversions, including geographic to projected and datum to datum transformations. Datums common to national grids are the International Terrestrial Reference Frame and the World Geodetic Datum of 1984 (Maling 1992). At a third level, there is a host of special purpose grids such as the Maidenhead Grid Square system (amateur radio), United States Public Land Survey system, Canadian Dominion Land Survey (cadastral), and the Universal Polar Stereographic, which completes the Military and Civilian versions of the UTM system for the two polar regions. Often local geospatial data servers, for example state and provincial data clearinghouses, use these projections for maps and data.

Lastly, there are highly localized projections for small areas. As an example, Egypt uses the Helmert 1909 ellipsoid and three mapping zones localized to the Purple, Red and Green zones defined colonially (see www.digitaleg.com). Perhaps the best examples of local projections are the 123 zones of the State Plane Coordinate system of the United States (Stem 1989). These are grouped by state and follow either the Transverse Mercator or Lambert conformal conic projections, with some cases using both, and one zone (Alaska Zone 1) using the oblique Mercator. The U. S. Geological Survey (www.usgs.gov) and the National Geodetic Survey (www.ngs.noaa.gov/TOOLS/spc.shtml) provide extensive Web support for conversion from other coordinate systems into the state plane system, which was extensively revised and metrified after adoption of North American Datum of 1983.

4.3 Web-Based Atlases and Map Projections

In this section we review issues of map projection selection for Web-based atlases. An atlas can be thought of as a bound collection of maps which are organized around a common theme and share a standard design. In most cases, an editor is responsible for the layout, choice of data to include, and the overall map design—including map projection. Two editors from the late 1500s share some historical significance with respect to map projections in the history of the atlas: Gerardus Mercator and Abraham Ortelius. Mercator's *Atlas sive Cosmographicæ Meditationes de Fabrica Mundi et Fabricati Figura*, published in 1595, is credited as the first publication to include 'atlas' in the title, and relied on the Mercator projection. Monmonier (2004) argues that Mercator's application of the term atlas had more to do with giving credit to the mythological figure Atlas than to a cartographic innovation of a book of maps. Ortelius'*Theatrum Orbis Terrarum*, published in 1570, is more consistent with the modern concept of an atlas but did not include "atlas" in the title, and utilized an oval-shaped projection with straight parallels and curved meridians.

During the infancy of the Web, online atlases were often no more than a collection of static paper maps that could be viewed and printed. However, the atlas has evolved from a purely print-based format to a Web-based interactive environment. For instance, since the late 1900s, the National Geographic Society has published their *World Atlas* in print version, but it is now also a Web-based world atlas titled *National Geographic's Atlas Explorer* (maps.nationalgeographic.com/maps/atlas-explorer.html). As technology increased, so have the interactivity levels of a Web-based environment (e.g., linking videos, animation, and querying data). As a result, more sophisticated Web-based atlases are now common, with options to select from an infinite number of map projections. Despite the atlas' evolution in the Web environment, one question remains challenging to cartographers: Which map projection should be selected to map the data?

4.3.1 A Review of Web-Based Atlases and Their Map Projections

What follows is a brief review of the projections that are used in Web atlases. The atlases included in this review are divided into three categories: World, specialty thematic, and individual countries. For each category, the review will first summarize the map projections that appear in atlases, examine the data sets that are being mapped, the associated projection properties, and then comment on their appropriateness for the data set mapped. All of the atlases reviewed here are freely available online and do not require any installation on a local computer.

4.3.2 Web-Based World Atlases

Table 4.1 lists five online world atlases (1) World Atlas, (2) Holt, Rinehart, and Winston Atlas, (3) Maps of World, (4) Swiss World Atlas, and (5) Grolier World Atlas. According to Table 4.1, when mapping the world, the following projections were selected by these atlases: Mercator, Miller, Plate Carrée, Mollweide, Eckert V, Robinson, and Aitoff. General reference maps were found in all of world atlases. These maps show political borders of individual countries or simple outline of the world's coastlines. With these maps, no specific projection property is required. Yet, these general reference maps include a mix of projection properties. General reference maps typically do not attempt to preserve angles, directions, distances, or areas correctly. The conformal property of the Mercator projection (Fig. 4.2) is not appropriate for maps showing the outline of the world's coastline (e.g., the WorldAtlas.com). There is gross distortion in the appearance of the landmasses in the upper latitudes and measuring angles is not the purpose of this general reference map. A compromise projection would be a better choice for this map purpose. It is interesting to note that the world time zones map presented in WorldAtlas is illustrated through the Mollweide projection (Fig. 4.3). This projection represents lines of longitude as curved lines that converge to points that denote the poles. Time zone maps are usually based on a cylindric projection with equally spaced meridians such as the Miller (Fig. 4.4) or Plate Carrée (Fig. 4.5) that suggest zones of equal time.

The Swiss World Atlas and Grolier World Atlas include more diverse thematic topics than the world atlases discussed previously. The Swiss World Atlas maps many thematic topics on the Plate Carrée projection. This cylindric equidistant projection shows all lines of longitude as equally spaced straight lines and at true scale (which is not a necessary projection property for thematic data). Lines of latitude are shown straight and of the same length as the equator which creates considerable distortion in the upper-latitude landmasses. Despite the distortion, this projection is a popular choice for Web mapping since its rectangular dimensions fit a computer screen well. The Grolier World Atlas maps its thematic data sets on the

Table 4.1 A listing of web-based atlases of world-wide data

World atlases of general thematic topics						
Atlas name	Geographic region mapped	Data set mapped	Projection name	Projection class	Projection property	Projection identified
World atlas (www.worldatlas.com)	World	Regional boundaries	Mollweide	Pseudocylindric	Equal area	Yes
	World	Major latitudes	Mollweide	Pseudocylindric	Equal area	Yes
	World	Time zones	Mollweide	Pseudocylindric	Equal area	Yes
	World	World coastline outline	Mercator	Cylindric	Conformal	Yes
	World	Political borders	Robinson	Pseudocylindric	Compromise	No
Holt, Rinehart, and Winston (go.hrw.com/atlas/norm_htm/world.htm)	World	Political borders	Robinson	Pseudocylindric	Compromise	No
Maps of world (www.mapsofworld.com/world-atlas/)	World	Political borders	All maps are available in one of five projections: Miller, Mercator, Robinson, Eckert V, or Aitoff	Miller & Mercator: cylindric Robinson & Eckert V: Pseudocylindric Aitoff: modified Azimuthal	Mercator: conformal Miller, Robinson, Eckert V, Aitoff: compromise	All projections are identified
	World	Physical landforms				
	World	Various thematic themes				
	World	Various historical themes				
	World	Satellite imagery				
	World	Various weather/environment/natural resources themes				
Swiss world atlas (schweizerweltatlas.ch/?lang = en)	World	Various thematic themes	Plate Carrée	Cylindric	Equidistant	No
Grolier world atlas (go.groiler.com/atlas?id = mtps018)	World	12 broad thematic themes	Robinson	Pseudocylindric	Compromise	No
	World	Time zones	Miller	Cylindric	Compromise	No

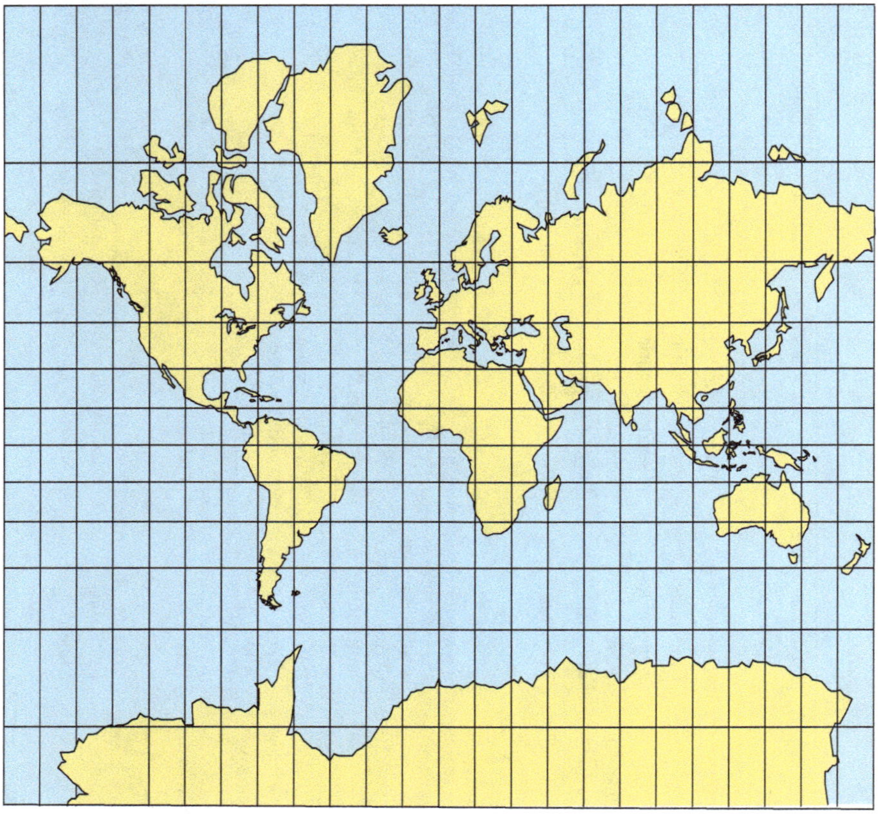

Fig. 4.2 The Mercator conformal cylindric projection

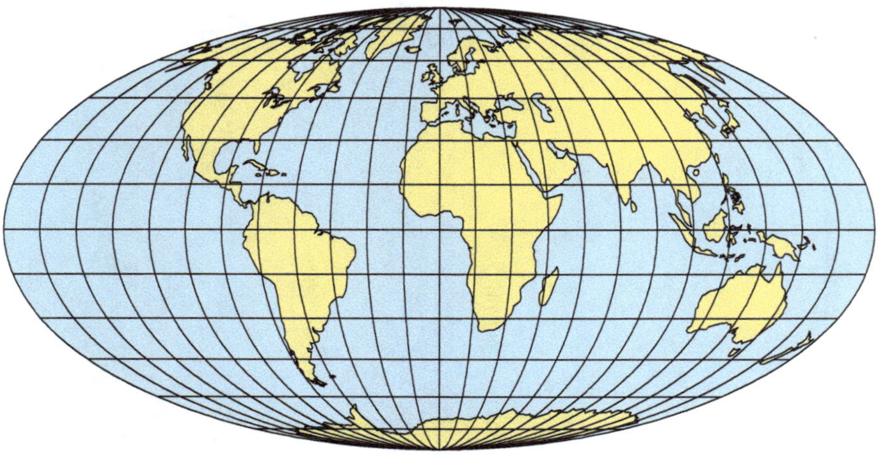

Fig. 4.3 The Mollweide equal area pseudocylindric projection

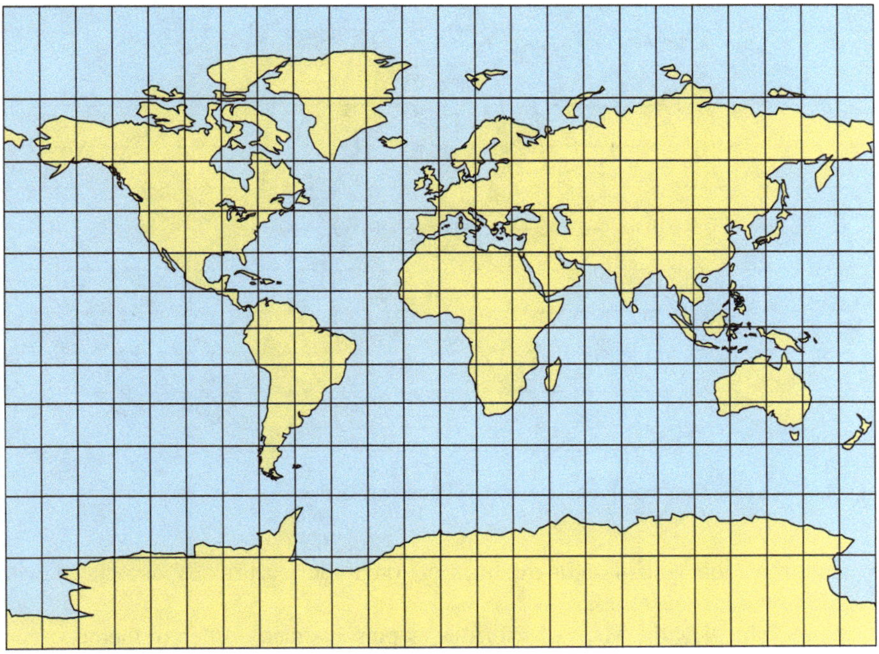

Fig. 4.4 The Miller compromise cylindric projection

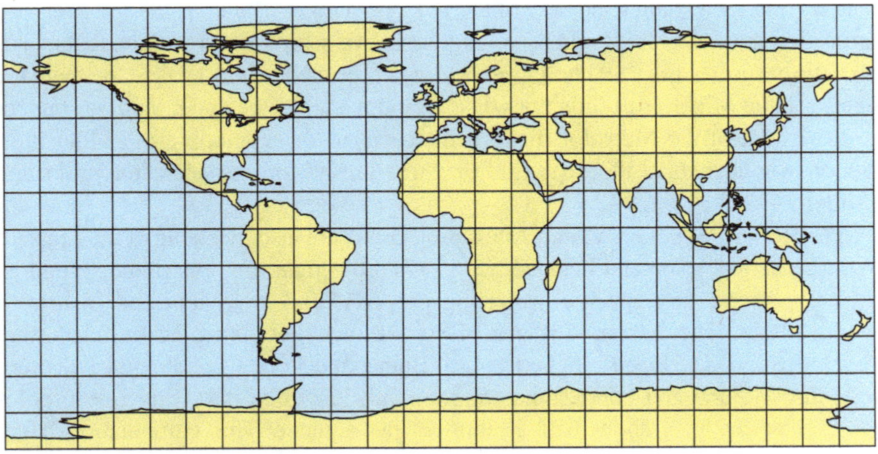

Fig. 4.5 The Plate Carrée equidistant cylindric projection

Robinson projection (Fig. 4.6). This projection was developed by Arthur Robinson at the request of Rand McNally. Since its development, the Robinson projection, although not equal area, is a popular choice for many thematic data sets since it gives a good visual representation of Earth's landmasses. Note that in this atlas, the

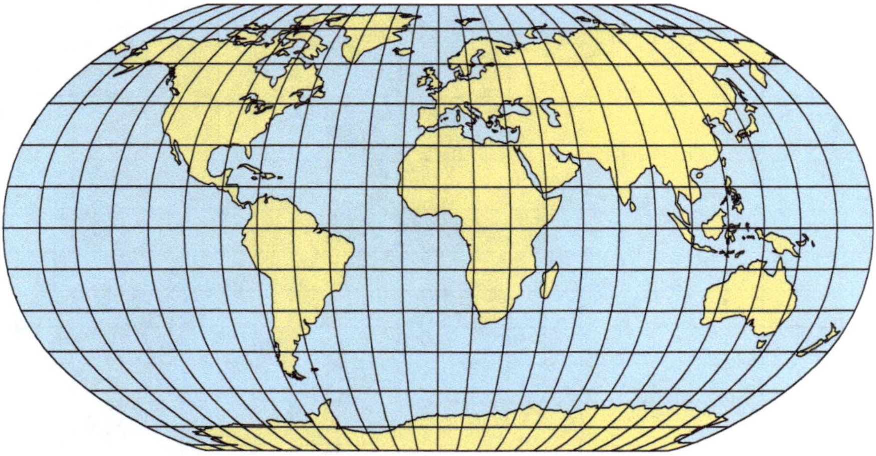

Fig. 4.6 The Robinson compromise pseudocylindric projection

Miller projection with its straight lines of longitude is correctly associated with mapping world time zones.

From Table 4.1, the Maps of World atlas presents many different thematic map topics. An interesting approach taken by Maps of World in mapping these topics is to illustrate each data set using all five of the following projections: Miller, Mercator, Eckert V, Robinson, and Aitoff. In other words, the user can select any data set and view them using any of the five projections. The utility of the Mercator projection and its ability to preserve angular relationships to map thematic data has already been discussed. Although appearing in printed world atlases for thematic data sets, the Miller projection does distort landmasses in the upper latitudes but not to the extent of the Mercator. In fact, the catalyst for developing the Miller projection was to reduce the distortion that was present on the Mercator projection (Miller 1942).

The Robinson, Eckert V, and Aitoff projections are also available in the Maps of World. The Robinson and Eckert V (Fig. 4.7) projections are both pseudocylindric, do not preserve any specific projection property, and have appeared in printed world atlases. The Eckert V is one of six pseudocylindric projections that were developed by Max Eckert (Eckert 1906). Both projections have straight lines representing latitude and curved lines representing longitude that converge to poles represented as lines about half as long as the equator. This combination better reflects the way in which these lines appear on a globe than, for example, cylindric projections. This is an appealing aspect of selecting pseudcylindrics for world maps. On the Robinson projection, the lines of longitude are drawn with elliptical arcs while the Eckert V uses sinusoidal curves (Snyder and Voxland 1989). For this reason, the lines of longitude curve to a greater extent on the Eckert V than on the Robinson. The Aitoff projection (Fig. 4.8) was presented by Aitoff (1899) and is also a compromise projection. However, due to the curved nature of the lines of

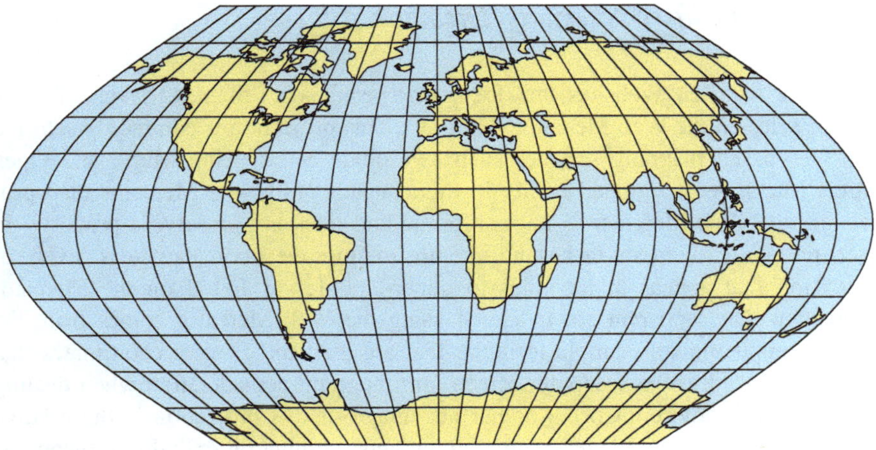

Fig. 4.7 The Eckert V compromise pseudocylindric projection

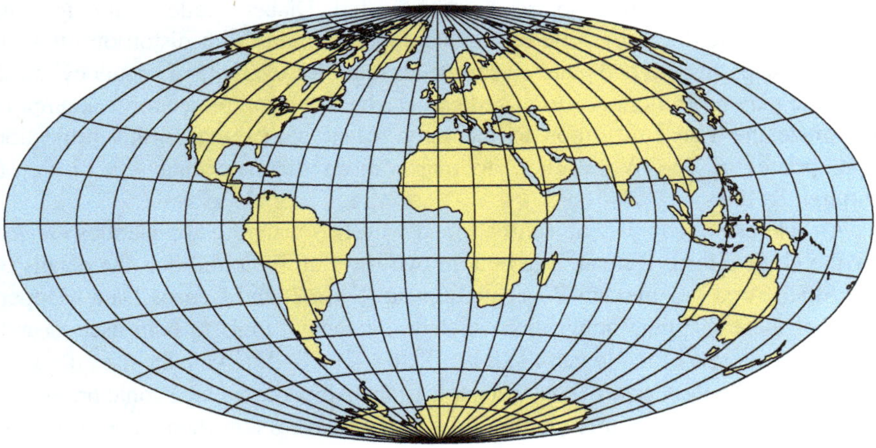

Fig. 4.8 The Aitoff compromise modified azimuthal projection

latitude, the projection is classified as a modified azimuthal. Although sharing some of the same visual characteristics as the Robinson and Eckert V, the Aitoff projection represents lines of longitude and latitude as complex curves (Snyder and Voxland 1989). In addition, the poles are characteristically represented as points which also mimics the way in which the poles appear on the Earth. However this characteristic results in a considerable compression of landmasses near the polar areas which can be visually undesirable. The Aitoff projection has also been selected for maps in printed world atlases. Collectively, due to their attempt at preserving the overall appearance of the Earth's landmasses and the representation of the lines of latitude and longitude, these three projections are suitable for a range of thematic and general reference world maps.

4.3.3 Web-Based Specialty Thematic Atlases

Table 4.2 lists ten Websites that focus on specific data themes. Seven of these atlases selected the Web Mercator (Fig. 4.9). This projection is commonly selected for Web-based mapping services such as Google Maps, Bing Maps, and Open Street Maps. As mentioned earlier in this chapter, unlike the Mercator map projection, the Web Mercator is not conformal and does not preserve a property. In fact, the Web Mercator makes no attempt to improve upon the gross levels of distortion that appear in the mid- to upper-latitudes as found on the Mercator projection. One apparent advantage of using the Web Mercator comes from the rather computationally simple formulas that are used to calculate coordinates that are then plotted to the computer screen (an important consideration when dealing with the Web environment). Another advantage of this projection is the relative programming ease that comes with panning and zooming on cylindric projections. Non-cylindric projections, such as the Robinson with its curved lines of longitude, need to have the longitude lines re-calculated every time panning takes place.

The Plate Carrée projection appears in the IDF Diabetes Atlas. Although this equidistant cylindric projection represents upper latitudes with distortion, its rectangular shape fits the computer screen better than, for example, a pseudocylindric or conic map projection. As discussed earlier, the Robinson projection's attempt to minimize the visual distortion to the Earth's landmasses makes this projection appropriate for The World Bank to map various thematic data sets displayed through its atlas.

An interesting comparison can be found with two mapping applications created by the U. S. Census Bureau. The bottom two rows of Table 4.2 lists the details of the Small Area Income and Poverty Estimate (SAIPE) and Census Data Mapper. Both of these mapping applications include interactive tools to help users search and map data sets of the lower 48 United States. The SAIPE Website uses the Web Mercator while the Census Data Mapper uses the Albers equal area conic projection (Fig. 4.10). Since the conterminous United States have a considerable east to west extent and is located along the mid-latitudes a conic projection has traditionally been favored. On conic projections in general, the locations of the standard parallels (lines of no distortion) can be specified to coincide over the geographic area of interest (e.g., the mid latitudes) which in turn reduces distortion for that geographic area. Contrary to the standard parallel placement options on conic projections, the equator is assigned as the single standard line on the cylindric Web-Mercator. For this reason, equatorial regions are mapped with lower distortion than landmasses appearing in the mid- and upper-latitudes.

Table 4.2 A listing of web-based atlases showing specialty thematic data

Atlases of specialty thematic data							
Atlas name	Geographic region mapped	Data set mapped	Projection name	Projection class	Projection property	Standard lines	Projection identified
IDF Diabetes atlas 6th edition (www.idf.org/diabetesatlas)	World	Diabetes	Plate Carrée	Cylindric	Equidistant	Equator	No
Ocean data viewer (data.unep-wcmc.org)	World	Ocean-related themes	Web Mercator	Cylindric	None	Equator	No
Global development atlas (atlas.esri.com/Atlas/NGO_Atlas.html)	World	5 Thematic themes	Web Mercator	Cylindric	None	Equator	No
UNESCO atlas of the world's languages in danger (www.unesco.org/culture/languages-atlas/)	World	6 Thematic themes	Web Mercator	Cylindric	None	Equator	No
Health map (healthmap.org/en/)	World	Health-related themes	Web Mercator	Cylindric	None	Equator	No
The world bank (www.app.collinsindicate.com/worldbankatlas-global/en-us)	World	12 Thematic themes on global development	Robinson	Pseudo-cylindric	Compromise	None	No
European environment agency (www.eea.europa.eu/data-and-maps/explore-interactive-maps)	Europe	17 Environmental themes	Web Mercator	Cylindric	None	Equator	No
Michelin route maps (www.viamichelin.com/Web/Maps/Map-Europe)	Europe	Transportation routes	Web Mercator	Cylindric	None	Equator	No
United States census bureau's small area income and poverty estimate (www.census.gov/did/www/saipe/index.html)	Lower 48 United States	Various poverty data	Web Mercator	Cylindric	None	Equator	No
United States census bureau's census data mapper (tigerweb.geo.census.gov/datamapper/map.html)	United States	Various demographic data	Albers	Conic	Equal area	Geographic location dependent	No

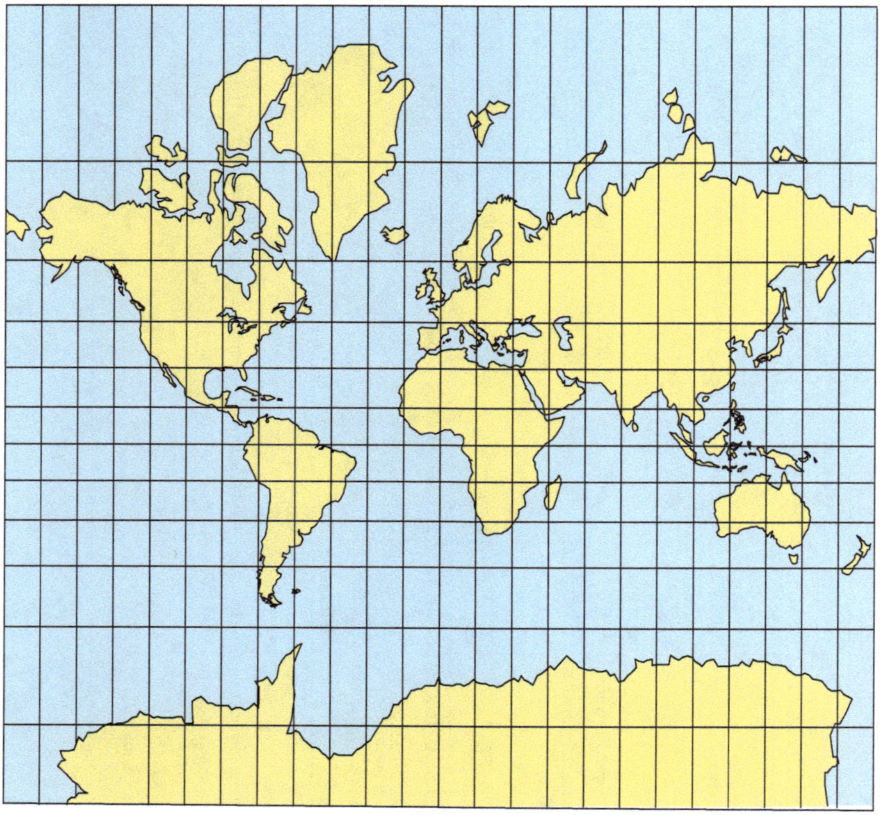

Fig. 4.9 The web Mercator cylindric projection

4.3.4 Web-Based Atlases of Individual Countries

Table 4.3 lists six atlases that focus on topics related to individual countries. Three atlases use the Web Mercator to map various themes. However, the United States National Atlas uses the Plate Carrée; the Atlas of Canada, the Lambert conformal conic (Fig. 4.11); and the Interactive Agricultural Ecological Atlas of Russia and Neighboring Countries, the Albers equal area conic map projection. Canada and Russia have a great east to west extent and by locating the standard parallels over the geographic areas of interest, distortion is reduced. The use of an equal area projection for Russia is logical since the maps show spatial distributions of various thematic topics. The choice of a conformal projection to map Canada does not follow traditional cartographic recommendations of selecting an equal area projection when the map purpose involves showing spatial distributions. However, choosing a projection property is not always the most important consideration. In some cases, the visual appearance of the landmasses or the graticule may take precedence. For example, Figs. 4.10 and 4.11 compare the landmasses and the way

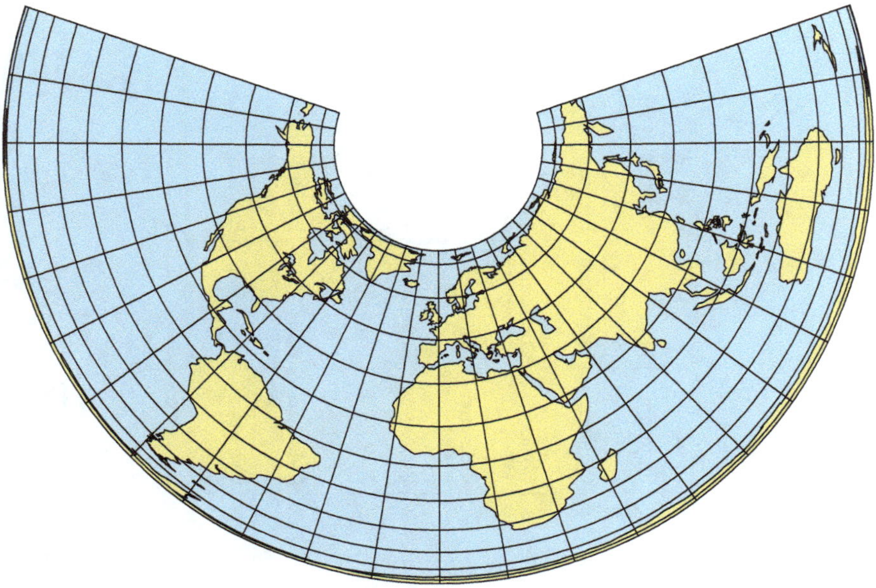

Fig. 4.10 The Albers equal area conic projection

in which the lines of longitude converge in the North Pole region. The lines of longitude converge to an arc that represents the North Pole on the Albers equal area projection (Fig. 4.10) which has the tendency to stretch the landmasses in this area east to west. Since the lines of longitude converge to the North Pole (represented as a point) on the Lambert conformal conic projection (Fig. 4.11) the landmasses are not subjected to this same stretching. The Atlas of Canada designers may have made a conscious decision to forego the traditional recommendation of using an equal area projection for their thematic maps. Instead, they deliberately selected a conformal conic projection for Canada as a result of the pole being represented as a point instead of a line. Showing the North Pole as a point produces landmasses in the upper regions of Canada with less east-west stretching and reflects the way in which the lines of longitude converge to the pole. Both of these characteristics create a visual appearance that is similar to what is shown on a globe.

4.3.5 Summary of Map Projections and Web-Based World Atlases

Based on the review of the Web-based atlases six observations can be made. First, it is apparent that there is disagreement between the data to be mapped and the selection of an appropriate map projection property to preserve. The Web-based atlases in this survey included general reference and thematic maps. According to

Table 4.3 A listing of web-based atlases showing thematic data of individual countries

Atlases of individual countries							
Atlas name	Geographic region mapped	Data set mapped	Projection name	Projection class	Projection property	Standard lines	Projection identified
United States national atlas (http://viewer.nationalmap.gov/viewer)	Lower 48 United States	13 Thematic themes	Plate Carrée	Cylindric	Equidistant	Equator	No
The atlas of Canada (http://atlas.nrcan.gc.ca/site/english/)	Canada	8 Thematic themes	Lambert	Conic	Conformal	49°N and 77°N	Yes
Atlas of living Australia (spatial.ala.org.au/)	Australia	Various thematic themes	Web Mercator	Cylindric	None	Equator	No
Interactive agricultural Ecological Atlas of Russia and neighboring countries (www.agroatlas.ru/en/content/climatic_maps/Pcp/Pcp/)	Russia and former member countries of soviet union	Crops, diseases, pests, weeds, and environment variables	Albers	Conic	Equal area	44°N and 68°N	Yes
Forest Atlas of Cameroon (www.wri.org/tools/atlas/map.php?maptheme=cameroon)	Cameroon	6 Thematic themes	Web Mercator	Cylindric	None	Equator	No
England and Wales Environmental Health Atlas (www.envhealthatlas.co.uk/homepage/gotoatlas.html)	England and Wales	21 Health-related themes	Web Mercator	Cylindric	None	Equator	No

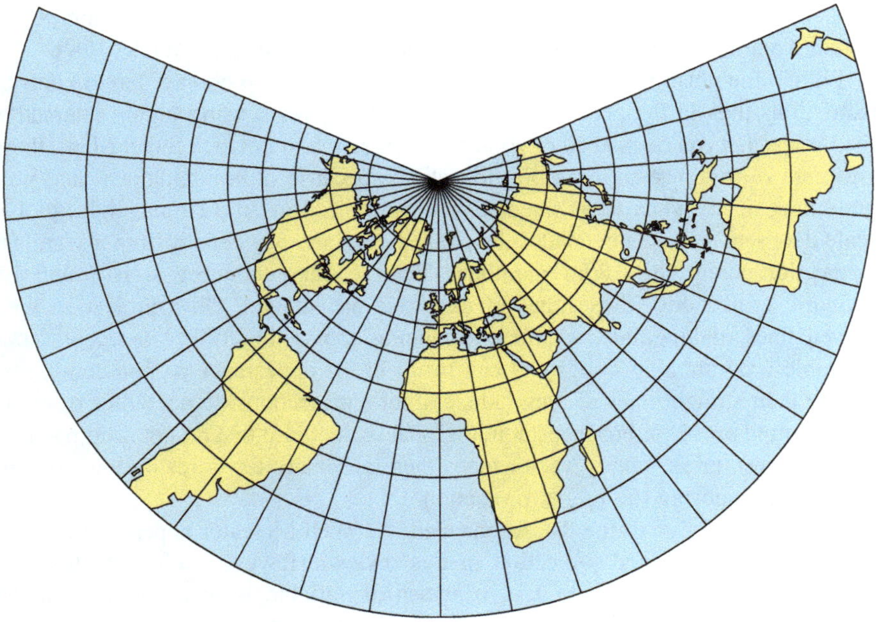

Fig. 4.11 The Lambert conformal conic projection

recommendations present in the cartographic literature, an equal area or compromise projection would be appropriate for these kinds of maps. Yet, those projection properties were not universally found. Second, from Tables 4.1 through 4.3 it is interesting to see that it is uncommon to disclose the projection name to the user. In fact, there was no instance where the Web Mercator was identified by name. Identifying the projection name may help the map reader understand how, for example, the projection's distortion pattern, impacts the spatial distribution represented on the map. At a minimum, providing the projection name would equip users with enough information that they could research the projection in question. Third, the selection of the Lambert conformal conic projection for the Atlas of Canada is contrary to recommendations to use an equal area property for thematic maps. However, this projection choice helped preserve the overall appearance of Canada's landmasses and demonstrates the flexibility to which a map projection can be tailored to meet the needs of a map purpose – provided that those making the selections are equipped with the appropriate knowledge.

Fourth, projections that have historically been used in the printed world atlas do not necessarily translate into the digital world atlas. For instance, unlike in the printed world, it is very easy to share data in the digital world (e.g., via map mash-ups—especially at the regional or smaller geographic area extents). The Web Mercator makes this sharing process easier by setting a common map projection which eliminates the need to convert between one projection and another. In another respect, sharing data is also made easier by removing the map projection altogether and storing vector data (such as Esri shapefiles) in "geographics" or latitude/longitude values based on the World Geodetic System (WGS) 1984.

Fifth, online mapping and the varied interactivity levels present different kinds of restrictions than in the printed world. For example, when using highly interactive tiled-maps that can be panned or zoomed (such as found in Open Street Map), there are more restrictions on the projection choice because of computational and programming needs when users interactively pan or zoom.. In a similar light, global scale data sets that are intended to be panned continuously benefit from a seamless connection along the 180th meridian. This seamless connection is found on cylindric projections whose lines of longitude are straight. On a projection with curved lines of longitude, seamlessly panning across this 180° line causes considerable programming issues and can result in the appearance of non-continuity. Due to their straight lines of longitude, cylindric projections do not suffer from the wrap-around problem. For maps at non-global scale, those that do not need panning (or zooming), or are purely vector maps there are many fewer restrictions to consider when choosing the proper projection.

Sixth, irrespective of the dataset mapped, the Web Mercator appeared to be the most frequently selected projection in this review. However, selecting this projection to map the distribution of a statistical variable is inappropriate. In the cartographic literature, the Mercator (or the visually similar Web Mercator) projection is not recommended for mapping thematic data. For instance, assume one were to symbolize a statistical variable (such as population) on a map using dots. When viewing the map, an important consideration is the interpretation of the visual density of the spatial pattern. The visual density results from the varied nature of the data (and dots) within each geographic region. If the map upon which the dots are plotted is equal area then the impression of the visual density is likely to be correct. Since neither the Web Mercator nor the Mercator are equal area both projections suffer from considerable exaggeration of landmass' shape and size in the upper-latitudes which makes neither projection suitable for representing distributions of a statistical variable. Unfortunately, the inertia of using the Web Mercator for Web mapping has largely been ingrained and its use will likely continue for the foreseeable future. In order for these Web-based atlases to better represent general reference maps or portray the data on thematic maps, projections need to be selected, on which the projection property preserved from the globe is matched to the map purpose.

4.4 Online Map Projection Software and Code Libraries

In this section we review a few of the more popular or practical and downloadable programs (or executables) and software code libraries. Most of these downloadable programs make their source code available under an open source license. In this section we focus on explaining the structure and utility of several commonly available map projection related software packages or online code libraries.

4.4.1 *Flex Projector*

Flex Projector (http://www.flexprojector.com/), is a cross-platform application for producing custom world map projections and has become a very good tool for practicing mapmakers and students of cartography (Jenny et al. 2008). The development of Flex Projector was inspired by the Robinson projection (Robinson 1974, pp. 147–148; Jenny et al. 2008). It allows users to modify the shape of a projection by altering the length of parallels and their distance from the equator (Fig. 4.12). According to Jenny et al. (2008), the aim of Flex Projector is to democratize the fabrication of world map projections. The program is downloadable as an executable for the Windows, Macintosh, and Linux platforms, and, in addition, the source code is available as open source at the Website noted above. From the map projection perspective, Flex Projector focusses on cylindric, pseudocylindric, and polyconic projections.

4.4.2 *Global Map Projector*

Global Map Projector (http://www.giss.nasa.gov/tools/gprojector/) is also known as G.Projector, is also a cross-platform application. The program is downloadable as an executable for the Windows, Mac OS X, and Linux platforms available at the Website noted above. It is a Java executable and requires the Java Runtime Environment to be installed on your computer. G.Projector converts an equirectangular map image into one of over 100 global or regional map projections (Fig. 4.13). It was developed and is being maintained by Robert B. Schmunk of the NASA Goddard Institute for Space Studies.

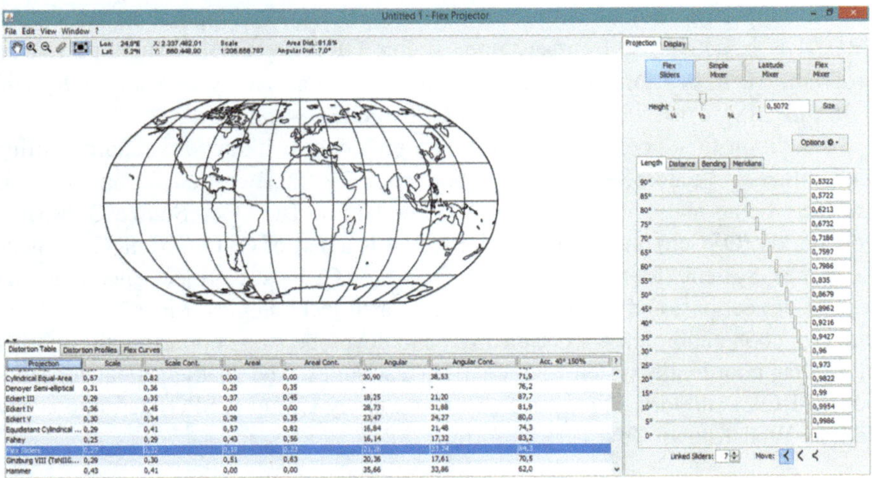

Fig. 4.12 FlexProjector. *Source* http://www.flexprojector.com/

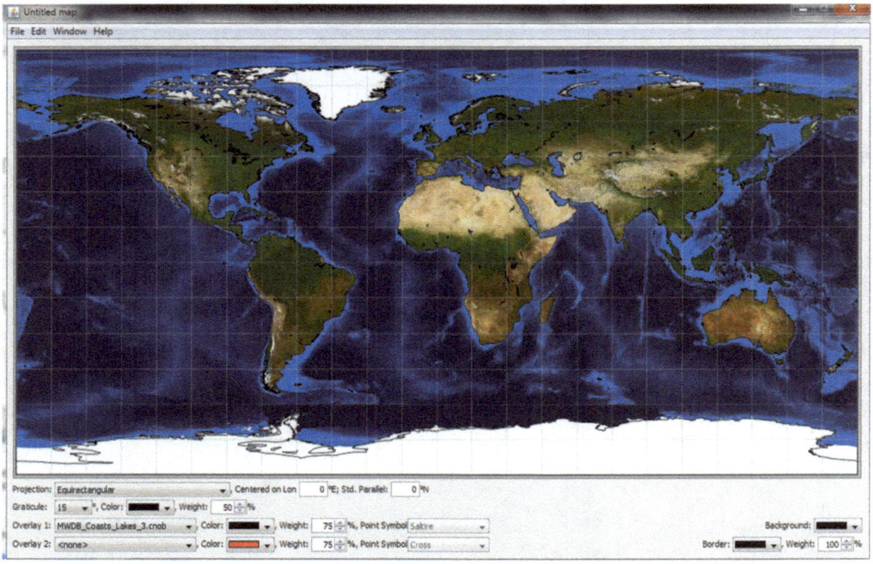

Fig. 4.13 G.Projector. *Source* http://www.giss.nasa.gov/tools/gprojector/

4.4.3 PROJ.4

Proj.4 (http://trac.osgeo.org/proj/) is a cartographic projections software code library originally written by Gerald Evenden, then of the US Geological Survey (USGS). It is commonly used in conjunction with the Geospatial Data Abstraction Library (GDAL). GDAL is a C++ library for raster and vector geospatial data formats that is released under an open source license (Warmerdam 2008). GDAL presents a data model for both raster and vector available to any calling application for all supported formats (http://www.gdal.org). GDAL supports its own API. PROJ.4 is in active use by many open source GIS/Geospatial technology related programs such as MapServer, Geographic Resources Analysis Support System (GRASS GIS), PostGIS, and Quantum GIS (QGIS).

In addition to supporting over 100 map projections, it supports datum shifting with either a 3-parameter transformation or a 7-parameter transformation. It is presently being maintained by Frank Warmerdam of The Open Source Geospatial Foundation (OSGeo) and Planet Labs. PROJ.4 is a part of OSGeo Meta Coordinate Reference System (MetaCRS) project. The MetaCRS is a project encompassing several projections, and coordinate system related technologies. MetaCRS seeks to improve cooperation among GIScientists and developers by supporting the sharing of similar coordinate system dictionaries, test suites, and mathematical formulations. PROJ.4 supports European Petroleum Survey Group (ESPG) definitions and various Well Known Text (WKT) formats.

4.4.4 JH Labs Java Library

Jerry Huxtable or JH Labs Java Library (http://www.jhlabs.com/java/maps/proj/index.html) is a Java-based implementation of a portion of the PROJ.4 map projection library. The code is open source and one goal for the library is to eventually incorporate all of the projections included in PROJ.4. In addition, on the Website JH Labs provides "The Globe Applet," which can be used to examine the supported projections. It displays several thematic layers and, importantly, a graticule and Tissot's indicatrix as selectable options. JH Labs also have a Java Map Editor application available for viewing and editing maps in various formats (http://www.jhlabs.com/maps/index.html). The editor uses Java 1.2 and Swing.

4.4.5 JavaScript Libraries

Java script libraries are becoming more popular, not only for map projections use, but also in the larger online mapping development/application area. JavaScript is a dynamic computer programming language, frequently used in conjunction with Web browsers (Flanagan and Ferguson 2006). Because JavaScript programming, especially advanced programming and associated complex handling of browser differences, can often be very difficult and time-consuming, JavaScript libraries have been developed (Mikkonen and Taivalsaari 2007). A JavaScript library is sometime referred to as a JavaScript framework. These libraries have been developed to deal with issues of advanced JavaScript programming that can be complicated and onerous to work with.

An example of a JavaScript library developed to aid in working with map projections on the Internet is the Geo Projections library (https://github.com/d3/d3-geo-projection/) extending d3 (Data-Driven Documents) and the d3.geo library, specifically. A JavaScript library for manipulating documents, d3 is based on the document object model (Bostock et al. 2011). d3 manipulates data using HTML, SVG and CSS, based on Web standards. The d3.geo library is used to create map projections within a Webpage through a simple function call. The following code example shows the specific d3.geo library needed to compute the plotting coordinates for the Mercator cylindric projection (Fig. 4.14).

```
var mercator = d3.geo.projection(function(λ, φ) {

return [

λ,
Math.log(Math.tan(π/4 + φ/2))

];
});
```

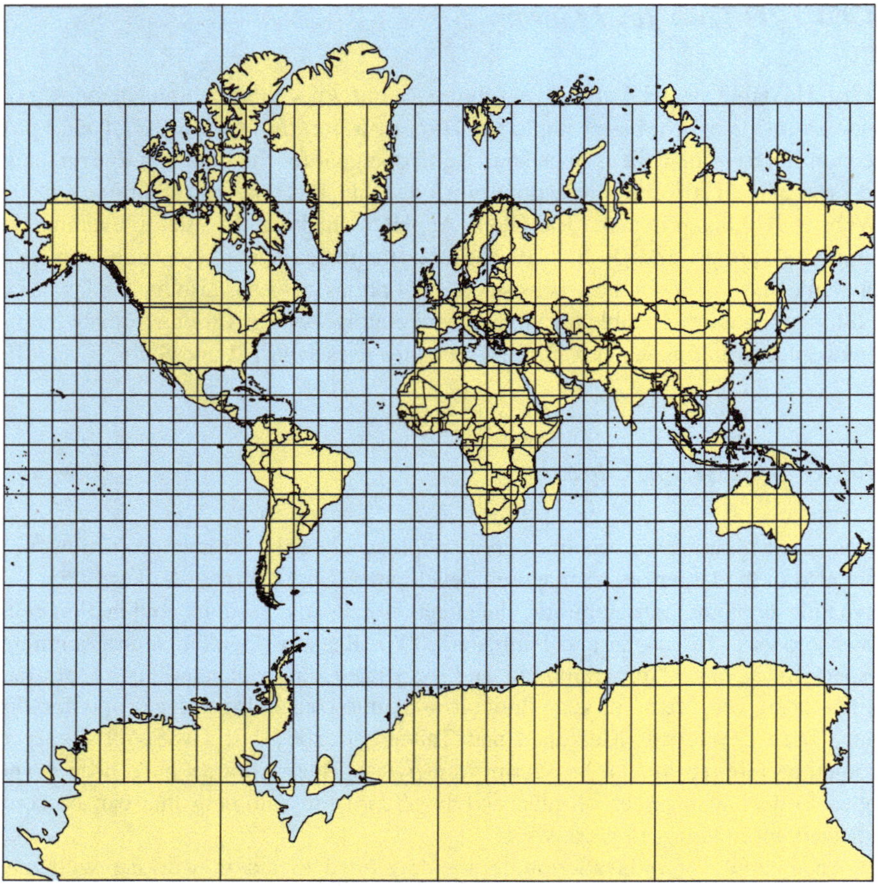

Fig. 4.14 The Mercator projection derived by using the d3.geo d3 JavaScript library (https://github.com/mbostock/d3/wiki/Geo-Projections)

4.4.6 Generic Mapping Tools (GMT)

Generic Mapping Tools is an open source assortment of approximately 80 core tools for controlling geographic and Cartesian data and producing PostScript illustrations, with supplements that add another 40 modules. GMT contains over 30 map projections and transformations (University of Hawaii 2014). It has long been popular with geophysicists and oceanographers working in the UNIX environment. GMT was written and is maintained by Paul Wessel (U. of Hawaii) and Walter H. F. Smith (NOAA) with support from a global set of volunteers, and is subsidized by the National Science Foundation (Wessel et al. 2013). The latest version is version 5, which added the option to call the code as an API. This API feature allows for rapid development of custom applications and increase speedup of scripted workflows. In addition to the standard download package, loaders for OS X and Windows are available (Wessel et al. 2013).

4.4.7 mapIMG

mapIMG is an open source software package that focuses on map projection/re-projection of small scale (coarse) raster data. It is primarily used for mapping raster datasets at continental- to global-scale (Finn et al. 2012). The source code, documentation, test datasets, and executable files (for Windows, Linux, and Solaris) are available at http://cegis.usgs.gov/projection/acc_proj_data.html. mapIMG is tailored to address issues unique to raster datasets such as areal accuracy (as a function of spatial resolution), distortions of cells (pixels or other gridded data like post space for elevation, as opposed to points as in a vector dataset), and the associated resampling that accompanies projection transformations (including alternative resampling methods) (Finn et al. 2012). mapIMG has been further developed as pRasterBlaster to work in a high performance parallel-processing based computing environment, where a separate core library of projection transformation and resampling functions is available for use by both pRasterBlaster and mapIMG (Finn et al. 2014a).

4.4.8 Dynamically Created Vector Maps

Another option for working with projections in spatial data is through dynamic creation and projection of vector data in maps. The intent of vector-based automation tools is to provide, in terms of map projections software, the capability to display multi-zoom datasets over maps. That is, the goal is to load and display data at a full range of map scales. Often Scalable Vector Graphics (SVG) are used to assist this effort (W3C 2010; http://polymaps.org/). SVG is usually encapsulated within a specialized compiled language or interpreted script. SVG is an eXtensible Mark-up Language (XML) based vector image format for two-dimensional graphics with support for interactivity and animation, with behaviors defined in XML text files (W3C 2003; Clarke 2005). This type of multi-scale mapping requires methods for automating generalization (map symbols and map content) to improve communication across scales. Guiding principles have been proposed for symbol selection and map content at related pre-defined scale ranges for maps (Cecconi and Galanda 2002; Brewer and Buttenfield 2007; Samsonov 2011).

More recently Jenny (2012), Gaffuri (2012) and Lienert et al. (2012) have investigated using variable scale with the related challenge of on-the-fly adjustment of symbology, blending geometries, and adapting projections for dynamic vector mapping, with Samsonov et al. (2013) proposing a new scale-adaptive approach to symbolization of vector data on multi-scale maps. From the perspective of this chapter, the adaptation of map projections for dynamic vector mapping is a key interest. As previously described in the section on Non-Web Mercator mapping options at the global-scale, Jenny's suggestion (2012) that there is a present requirement for adaptive composite map projections that can seamlessly switch between projections as a user pans and zooms around a map is very relevant. This is an area of active research with many recent developments occurring in a short amount of time (and more expected) and we encourage the reader interested in dynamic vector mapping to check the latest literature.

4.5 Map Projection Research/Education

In addition to the materials discussed above for map projection education, there is also an interesting body of literature related to challenges in teaching about map projections and the resulting misconceptions about projections and distortion. In considering the methods that we use to educate students about map projections, Downs and Liben (1991) examined cognitive challenges behind the concept of shadows and developable surfaces. They found that many college-age individuals do not have sufficient cognitive and spatial skills to understand shadow projections, thus would not be able to effectively understand map projections taught with this method. Olson (2006) suggests that we need to become better "visual detectives" and learn to more closely read the cues in the map graticule.

While Downs and Liben (1991) and Olson (2006) focused on how we learn about projections, Battersby and Kessler (2012) looked at how we apply our knowledge of projections in evaluating distortion. A surprising finding of this study is a substantial reliance on Mercator-based distortion cues (e.g., Greenland, Antarctica, and the general "polar regions") for explaining distortion in *any* of the projections included in the study. Anderson and Leinhardt (2002) also show that angular distortion is not well understood—even by experts. Battersby (2009) further supports this with demonstration that even when a projection is familiar most people have a difficult time adjusting for distortion in the map.

4.5.1 Tutorials and Decision Support Systems

Considering which map projection to choose for a map can be a challenging task. There are few guidelines to assist in the selection of an appropriate map projection. One guideline presented by Snyder (1987) begins the selection process by considering the geographic area of interest (world, hemisphere, or continent, ocean, or smaller region). Once the geographic area to be mapped is determined additional criteria are defined (e.g., projection property, location of geographic landmass, and the location to preserve scale) leading to a named projection. While this guideline is useful, the one making the selection needs to have a working knowledge of map projections to navigate their way through the guideline.

On the other hand, cartography and GIS texts discuss the specific projection parameters and numerical values that must be carefully specified to ensure that the geographic area of interest is mapped with distortion levels appropriate to the map purpose. Similarly, these texts explain that map projections have specific properties that should be matched to the map's data and purpose. One text (Slocum et al. 2009) discusses the utility of the projection properties (i.e., azimuthal, equal area, equidistant, conformal, and compromise). Directions between two locations on the Earth's surface can be preserved. Assume that a hurricane is brewing in the Central Atlantic and weather forecasters are interested in predicting the storm's direction to warn coastal communities of the impending danger. Azimuthal projections preserve the ability to measure directions from one point to all other points on a map. Since azimuthal projections are suitable for tracking they are often used for navigational purposes.

Additionally, Decision Support Systems (DSS) can serve as an aid to learning about projections. Finn et al. (2014b) discuss a DSS for small-scale map projections to help users identify appropriate map projections and to provide a tutorial detailing the logic guiding appropriate projection decisions for different purposes. This tutorial educates users on the derivation of each suggested projection and supplies a basic explanation of the individual projection design. The DSS is available at http://mcmcweb.er.usgs.gov/DSS/.

4.6 Conclusion

This chapter presented an overview of map projections and how they have been integrated into the Internet. The first section provided a general introduction to projections while the second section covered the map projections that are frequently used by Web maps and map services and addressed projections adopted for both global and local-scale mapping. Many global-scale maps use the Web Mercator map projection. While the cylindric nature of the Web Mercator facilitates the display of mapped data (e.g., panning) there are two major concerns related to the use of this projection. First, unlike the Mercator projection, the Web Mercator is not conformal. Second, like the Mercator, due to the considerable distortion found at the upper latitudes, the Web Mercator is not suitable when mapping at the global-scale. For non-global scale maps, the online map services typically use a variant of the Universal Transverse Mercator coordinate system. Many countries have developed their own national grid systems (e.g., Irish National Grid) and local-level administrative units have created their own coordinate system (e.g., United States State Plane Coordinate System) and distribute data formatted in these coordinate systems. The Web has many sites that are well equipped to handle complex conversions between these different grid and coordinate systems.

The third section presented a review of online atlases. This section examined online atlases of world, specialty thematic and individual countries. World atlases and atlases of individual countries use a greater variety of map projections and properties selected for included maps than the thematic atlases. The thematic atlases almost unanimously use the Web Mercator map projection.

The fourth section considered online map projection software. There is a growing number of Websites devoted to map projection software. Some of the Websites, such as Flex Projector, are interactive and allow users to interactively create a map projection. Other Websites, such as PROJ.4, are a set of useful code libraries that are freely available to the end user for integration into a number of Web-based and other mapping applications.

The fifth section discussed topics related to map projection research and education. There has been progress made in researching the map reader and their understanding of map projections. For instance, research has suggested that the map reader could use the graticule to better understand for example the projection property. In another finding, distortions in the upper latitudes common to the Mercator projection are cues for map. However, the number of research outcomes is still limited and there is much left to investigate regarding the map reader and the map projection.

Disclaimer Any use of trade, product, or firm names is for descriptive purposes only and does not imply endorsement by the U.S. Government

References

Aitoff D (1899) Projections des Cartes Géographiques. In: Atlas de Géographie Moderne Paris. Hachette, New York

American Cartographic Association, American Geographical Society, Association of American Geographers, Canadian Cartographic Association, National Council for Geographic Education, National Geographic Society, and Special Libraries Association, Geography and Map Division (1989) Resolution regarding the use of rectangular world maps. Am Cartographer 16(3):223

Anderson KC, Leinhardt G (2002) Maps as representations: expert novice comparison of projection understanding. Cogn Instr 20(3):283–321

Battersby SE (2009) The effect of global-scale map-projection knowledge on perceived land area. Cartographica 44(1):33–44

Battersby SE, Finn MP, Usery EL, Yamamoto KH (2014) Implications of web mercator and its use in online mapping. Cartographica 49(2):85–101. doi:10.3138/carto.49.2.2313

Battersby SE, Kessler FC (2012) Cues for interpreting distortion in map projections. J Geogr 111 (3):93–101

Bostock M, Ogievetsky V, Heer J (2011) D^3: data-driven documents. IEEE Trans Vis Comput Graph (Proc. InfoVis)

Brewer CA, Buttenfield BP (2007) Framing guidelines for multi-scale map design using databases at multiple resolutions. Cartography Geog Informa Sci 34(1):3–15

Bugayevskiy LM, Snyder JP (1995) Map projections: a reference manual. Taylor and Francis, London

Cavell JA (2005) USNG: getting it right the first time. The American Surveyor. June issue. http://www.amerisurv.com/PDF/TheAmericanSurveyor_USNationalGrid-Cavell_June2005.pdf. Accessed 10 Jan 2015

Cecconi A, Galanda M (2002) Adaptive Zooming in Web Cartography. Comput Graph Forum 21:787–799. doi:10.1111/1467-8659.00636

Clarke KC (2002) Criteria and measures for the comparison of global geocoding systems. In: Goodchild MF, Kimerling AJ (eds) Discrete global grids: a web book. http://www.ncgia.ucsb.edu/globalgrids-book/. Accessed 10 Jan 2015

Clarke P (2005) Dynamic web-mapping using scalable vector graphics (SVG). In: Proceedings ESRI user conference 2005. http://proceedings.esri.com/library/userconf/proc05/papers/pap1997.pdf. Accessed 10 Jan 2015

Downs RM, Liben LS (1991) The development of expertise in geography: a cognitive-developmental approach to geographic education. Ann Assoc Am Geogr 81 (2):304–327

Eckert M (1906) Neue Entwürfe für Erdkarten. Petermanns Mitteilungen 52(5):97–109

Eisenlohr F (1870) Ueber Flächenabbildung. Journal für rein und angewandte Mathematik 72 (2):299–307

Federal Geographic Data Committee [FGDC]. Educational resources for the United States National Grid. https://www.fgdc.gov/usng/educational-resources/index_html. Accessed 10 Jan 2015

Finn MP, Liu Y, Mattli DM, Behzad B, Yamamoto KH, Guan Q, Shook E, Padmanabhan A, Stramel M, Wang S (2014a) High-performance small-scale raster map projection transformation on cyberinfrastructure. Wang S, Goodchild MF (eds) Paper accepted for publication as a chapter in CyberGIS: fostering a new wave of geospatial discovery and innovation, Springer, Berlin

Finn MP, Steinwand DR, Trent JR, Buehler RA, Mattli D, Yamamoto KH (2012) A program for handling map projections of small scale geospatial raster data. Cartographic Perspect 71:53–67

Finn MP, Usery EL, Woodard L, Yamamoto KH (2014b) The logic of selecting an appropriate map projection in a decision support system (DSS). In: Lapaine M, Usery EL (eds) Chapter this volume: map projection choice. International Cartographic Association, Springer, Berlin

Flanagan D, Ferguson P (2006) JavaScript: the definitive guide, 5th edn. O'Reilly & Associates

Gaffuri J (2012) Toward web mapping with vector data. LNCS 7478: geographic information science. In: Proceedings of GIScience 2012. Springer, Berlin, pp 87–101

Hassler F (1825) On the mechanical organization of a large survey, and the particular application to the survey of the coast. Am Philos Soc Trans New Ser 2:358–408

Jenny B (2012) Adaptive composite map projections. IEEE Trans Visual Comput Graphics 18 (12):2757–2782

Jenny B, Patterson T, Hurni L (2008) Flex projector—interactive software for designing world map projections. Cartographic Perspect 59:12–27 and 68

Lienert C, Jenny B, Schnabel O, Hurni L (2012) Current trends in vector-based internet mapping —a technical review. In: Peterson MP (ed) Online maps with APIs and mapservices., Lecture notes in geoinformation and cartographySpringer, Berlin, pp 23–36

Maling DH (1992) Coordinate systems and map projections, 2nd edn. Pergamon Press, Elmsford NY

Mikkonen T, Taivalsaari A (2007) Using JavaScript as a real programming language. Technical Report TR-2007-168, Sun Microsystems Laboratories

Miller O (1942) Notes on cylindrical world map projections. Geogr Rev 32(3):424–430

Monmonier M (1995) Drawing the line: tales of maps and cartocontroversy. Holt, New York

Monmonier M (2004) Rhumb lines and map wars: a social history of the mercator projection. University of Chicago Press, Chicago IL

NGA (2014) NGA Advisory Notice on "Web Mercator". http://Earth-info.nima.mil/GandG/wgs84/web_mercator/. Accessed 10 Jan 2015

NGA Office of Geomatics (2014) Implementation practice Web Mercator projection. National Geospatial-Intelligence Agency (NGA) Standardization Document, version 1.0.0, 18 Feb 2014. http://Earth-info.nima.mil/GandG/wgs84/web_mercator/%28U%29%20NGA_SIG_0011_1.0.0_WEBMERC.pdf. Accessed 10 Jan 2015
Olson JM (2006) Map projections and the visual detective: how to tell if a map is equal-area, conformal, or neither. J Geogr 105:13–32
Ordinance Survey (2015) The National Grid. http://www.ordnancesurvey.co.uk/resources/maps-and-geographic-resources/the-national-grid.html. Accessed 10 Jan 2015
Robinson A (1974) A new map projection: its development and characteristics. Int Yearb Cartography 145–155
Robinson A (1990) Rectangular world maps—no! Profes Geogr 42(1):101–104
Samsonov T (2011) Multiscale hypsometric mapping. In: Advances in cartography and GIScience. Lecture notes in geoinformation and cartography (LNGC), vol 1. Springer, Berlin, Heidelberg, pp 497–520
Samsonov T, Podolsky A, Yurova N (2013) Multimapper—prototype system for designing multi-scale maps. In: Proceedings international cartographic conference 2013. http://icaci.org/files/documents/ICC_proceedings/ICC2013/_extendedAbstract/219_proceeding.pdf. Accessed 10 Jan 2015
Slocum T, McMaster R, Kessler F, Howard H (2009) Thematic cartography and geovisualization. Pearson Prentice Hall, Upper Saddle River, NJ
Snyder JP (1987) Map projections: a working manual (U.S. Geological Survey Professional Paper). U.S. Government Printing Office, 1395, Washington DC
Snyder JP (1993) Flattening the earth: 2000 years of map projections. University of Chicago Press, Chicago, IL
Snyder JP, Voxland P (1989) An album of map projections. US Geological Survey, Washington DC
Stem JE (1989) State plane coordinate system of 1983. NOAA Manual NOS-NG-5. U.S. Department of Commerce, Washington DC
University of Hawaii (2014) What is GMT? School of ocean and earth science and technology. http://gmt.soest.hawaii.edu/projects/gmt/wiki. Accessed 10 Jan 2015
Vujakovic P (2002) Whatever happened to the 'new cartography?' The world map and development mis-education. J Geogr High Educ 26(3):369–380
Warmerdam F (2008) The geospatial data abstraction library. In: Hall GB, Leahy MG (eds) Chapter 5 of open source approaches in spatial data handling. Springer, New York, pp 87–104
Wessel P, Smith WHF, Scharroo R, Luis JF, Wobbe F (2013) Generic mapping tools: improved version released. EOS Trans Am Geophys Union 94:409–410
W3C (World Wide Web Consortium) (2003) Extensible Markup Language (XML). http://www.w3.org/XML/. Accessed 10 Jan 2015
W3C (World Wide Web Consortium) (2010) Scalable Vector Graphics (SVG). http://www.w3.org/Graphics/SVG/. Accessed 10 Jan 2015
Yang Q, Snyder JP, Tobler WR (2000) Map projection transformation: principles and applications. Taylor and Francis, New York

Chapter 5
Cartograms as Map Projections

Waldo Tobler

Since the time of Ptolemy 2000 years ago the objective of map projections has been to produce maps of high metric fidelity, and this is still the prime objective today. Thus, the emphasis is on direction, distance, and area preservation. These properties are needed in order to serve a variety of purposes. This way of looking at maps considers them as a type of terrestrial graph paper. This is possible because, paraphrasing John Donne, we use "the net of Meridians and Parallels cast over the world". Different kinds of geographic graph papers are used for different kinds of problems. Thus there are the classic conformal, equidistant, and equal area projections. Particular solutions are provided by the conformal projections, such as the Mercator projection, or the stereographic. Other maps such as the azimuthal projections, including the gnomonic projection, show directions. Equal area projections are most often used for larger regions or countries and for the depiction of statistical information. There are also quite a large variety of lesser known properties for which special projections are used. Still other projections serve for general purpose maps.

Another type of map is referred to as a cartogram. It is the intent of this note to explain how these relate to conventional maps of the two dimensional surface of the Earth. But these cartograms are often not considered as belonging to the class of map projections. One difference between these and, say topographic maps, is that the phenomena depicted may change more rapidly in time, possibly often even hourly. This suggests that animation is a proper domain for cartograms. However not all history is quite this rapid and these maps can also be useful even if they are based on census information for which the change may be noted only every decade or so. In another respect cartograms have a property in common with the

W. Tobler (✉)
Department of Geography, University of California,
Santa Barbara, CA 93106-4060, USA
e-mail: tobler@geog.ucsb.edu

M. Lapaine and E.L. Usery (eds.), *Choosing a Map Projection*,
Lecture Notes in Geoinformation and Cartography,
DOI 10.1007/978-3-319-51835-0_5

Mediterranean portolan charts of the thirteen to fifteen hundreds in that they are based on empirical observations rather than strictly geometrical considerations.

5.1 Geographic Graphs

This notion of map projections as graph paper for spheres can be extended to serve additional, non-traditional uses. A considerable number of these are provided by the class of cartograms, taken to be a special type of map projection. These can be classified in at least two ways. One grouping might be by type of problem or purpose. One common use is simply to present a point of view. This can be as simple as the ego-enhancing "Here's a Representation of My Favorite Region, or City"; these are often intended to be humorous and colorful, and sometimes appear on post cards. Some of these can be referred to as 'Fisheye' maps (Rase 1997). One more serious recoded use is to aid pilots by enlarging the vicinity of an airport, with a kind of local bubble enlargement. Or they can depict the state of the world from an alternate point of view, as in "The Atlas of the Real World" (Fig. 5.1).

There is also the problem-solving point of view in which the cartogram, and its inverse, are a way of using an unusual coordinate system that renders a situation more understandable and simpler to manage (Bunge 1966). In many fields the choice of a proper coordinate system often clarifies a situation. An example of this is to render the Earth on a map in geomagnetic, rather than geographic, latitude and longitude coordinates in dealing with terrestrial magnetic problems. Today this might be done inside of a computer, switching between projections as needed, displaying only the final result. This problem-specific group would put the distorted Mercator projection into the category of a warping cartogram that assists in ocean navigation. Cartograms based on movement also seem possible but are rare.

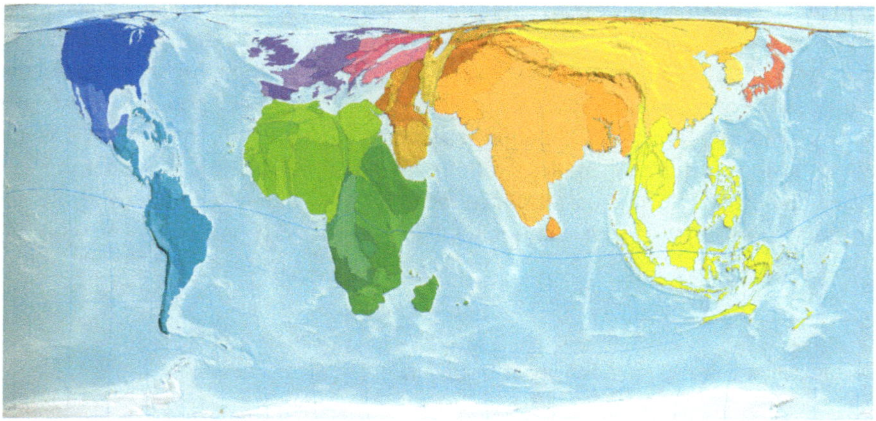

Fig. 5.1 Mathematically computed equal world population projection. Dorling et al. (2008)

Another approach is to group cartograms into the type of properties that they preserve. In such a classification one might find topological maps in which incidence and adjacency are preserved, but distance, area, and angles are not. The classic case is the well known London Underground map, a style now copied throughout the world. The usefulness of this depiction can hardly be questioned! Another category would be based on the adjustment of distances, using time or cost of travel as the metric instead of kilometers. Typically these cartograms resemble the azimuthal class of map projection in that they are based on a single center. Of course they are not geometric projections, but rather are based on empirical information. A third type of cartogram is related to area. Here the sizes of regions or countries are expanded, or shrunken, according to some numerical quantity, most often population magnitude, as in Fig. 5.1, but any non-negative measure can be used. This classification of cartograms into three general types seems to cover most examples.

5.2 Topological Transforms

All map projections attempt to preserve neighbor and adjacency properties, and mostly succeed except at the edges of the map. But if this is all that is preserved the maps may appear distorted. "Mental maps" wherein individuals are asked to draw maps of their neighborhoods, or of the world, often preserve these properties but latitude and longitude lines drawn on such maps appear as squiggly lines. The metric properties are not preserved, but sketch maps are still useful! I am certain that you have made or used them. Comparing the London Underground map from 1910 (www.ltmuseum.co.uk), metrically correct, with Beck's severely warped map of 1933 (and the modern derivatives in use today) clearly shows what can be done in this respect (Dobbin 2013). The current underground map has also been analyzed for its metric properties using Tissot's results by Jenny (2006). Such analyses are not often performed for cartograms but one can recognize and indicate the angular, area or distance properties if these are considered relevant. It might be of interest to minimize the distortion of these additional properties in the cartogram since there are always degrees of freedom in the choice of how to represent a specific cartogram—see the equations below. This also applies to the next category.

5.3 Distance Transforms

Constructing maps from measured distances is a well understood problem in surveying known as trilateration. It has been generalized in psychology and there goes under the name of multidimensional scaling. An example is children ranking how much they enjoy playing with other children. Well-liking is considered "closeness", that is, a small distance. In the computer programs relations such as this (the children's locations relative to each other) are converted to coordinates to yield a 'map' of the classroom. The same computer procedures can be used to make geographic maps of travel time or cost, or some other metric. As one example several studies have shown that the interaction between different linguistic groups in bilingual countries, such as Belgium and Canada, are not the same. They are quite 'apart' from each other, and this can be depicted in map form.

These maps seem to be of two types. In one case only isolated locations, for example, airports with travel cost (or time) taken from a schedule, are indicated and the intermediate places are set down between these. An example is Barrett's map of "The world based on airfares from London" (Barrett 1992). The intermediate places are really just arbitrarily inserted and thus misrepresented, and the map needs to be considered disconnected, and really as showing only discrete places. For display in these maps directions from a center are usually retained, i.e., azimuthal. This type of map can be extended, using multidimensional scaling, to represent time, or cost, etc., distance between two places, and then to all places (not only from one center), even using kilometer distances, but the distances will only be approximated, sometimes poorly. This can also be done with sociological distances, for example such as differences between languages, customs, gross domestic products, or any differentiating measure. An index of the degree of the fidelity of this representation is available as the 'stress', and this is an overall measure of fit; therefore it is somewhat of an improvement on Tissot's index which measures only local distortion.

The second type is usually implicitly based on continuous contours (isochrones; geographical circles in travel space) of travel time or cost from (or to) a location. Even in this case some interpolation is required, but an attempt is made to provide a spatially continuous representation. It could be based on any type of contour map; including for example, populations density, but this is more rare. Obviously in these cases there generally results in different distances in different directions—even though azimuths are preserved! This is quite unusual in conventional projections. Isochrones are often amoeba like in shape and even with disjoint pieces. An analogy can be made to pole centered azimuthal projections, with the irregular isochrones (or isotims) corresponding to parallels and the orthogonal trajectories (gradients) similar to meridians emanating from a pole. Directions and angles are clearly not preserved. And the maps generally relate to (or from) only one place, and at only one instance in time.

A related analytical use has been made by Hāgerstrand (1957) of logarithmically-scaled azimuthal maps of migration (Fig. 5.2) in order, as he puts it, to be able to count the symbols indicating the coming and going of people. Most of these individual movements are crowded about the origins and destinations. In other words, this is the resultant effect of the well known distance decay in human affairs. Obviously this logarithmic projection is a proper map to represent the phenomena; angular and a area distortions can be calculated. The local area is of course enlarged, and this leads directly to the next category.

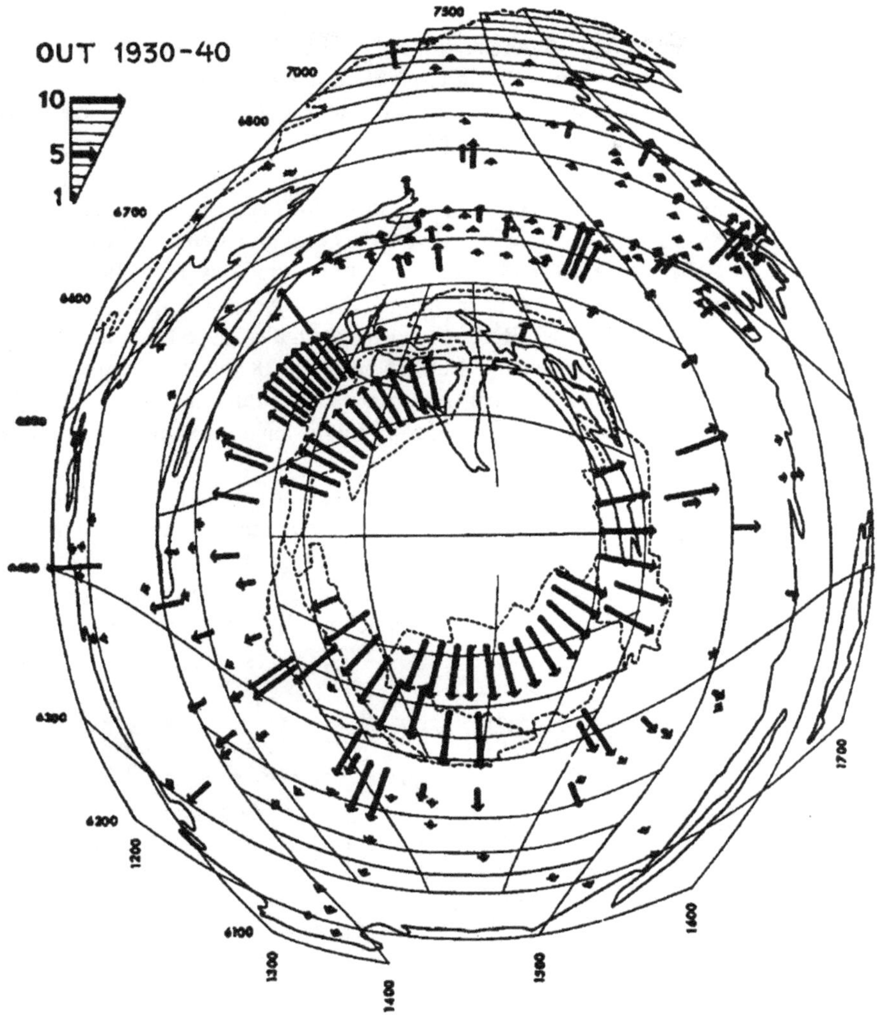

Fig. 5.2 Logarithmic map of Asby, Sweden, used to study migration. Hāgerstrand (1957)

5.4 Area Transforms

Area cartograms, with region sizes made proportional to some measure such as population or wealth, have recently become common now that a free fast computer algorithm has become available (Gastner and Newman 2004). Typical cartography texts include a smattering of such cartograms as examples. But these are generally not treated as map projections but rather as graphical or pictorial illustrations. And they often appear in newspapers, and even as exercises in lower school classes, as can be confirmed by entering the term 'cartogram' into a search engine on the internet. These maps offer a different view of the world and can be based on a variety of topics (Tobler 1996; Dorling et al. 2008; Hennig 2013).

The history of these cartograms goes back to the blossoming of statistical graphics in the mid eighteen hundreds (Tobler 1996), and this is reflected in the terminology applied to these maps. They have been referred to as anamorphoses (France), verzerte Karten (Germany), varivalent maps (Russia), and value-by area maps (Raisz 1934). Many of the last type of these were prepared by E. Raisz who, starting in the 1930s, presented rectangular depiction of regions and countries with sizes proportional to a variety of phenomena. They have also appeared in other publications as 'statistical maps'. Recently this 'rectangular' type of cartogram has been perfected by researchers in the Netherlands (Van Kreveld and Speckmann 2007) who have shown that they can be produced by computer (Fig. 5.3). The geographic graticule on such a map projection might have kinks in it and not be smooth, even though the populations are correct.

An additional use that has been made of area cartograms is for statistical purposes. When doing area sampling one likes to know that all subjects have an equally likely chance of being chosen. An area cartogram stretches (or shrinks) space and can warp geography so that areas are proportional to the target density (Fig. 5.4). After sampling in this domain—attacking the problem—the inverse is

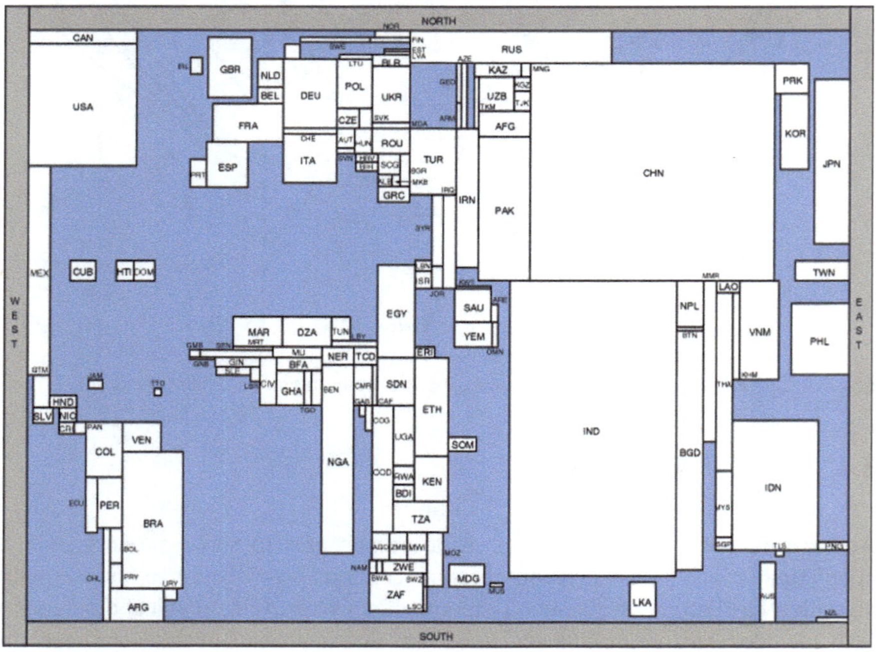

Fig. 5.3 Mathematically computed equal world population projection. Van Kreveld and Spaekcmann (2007)

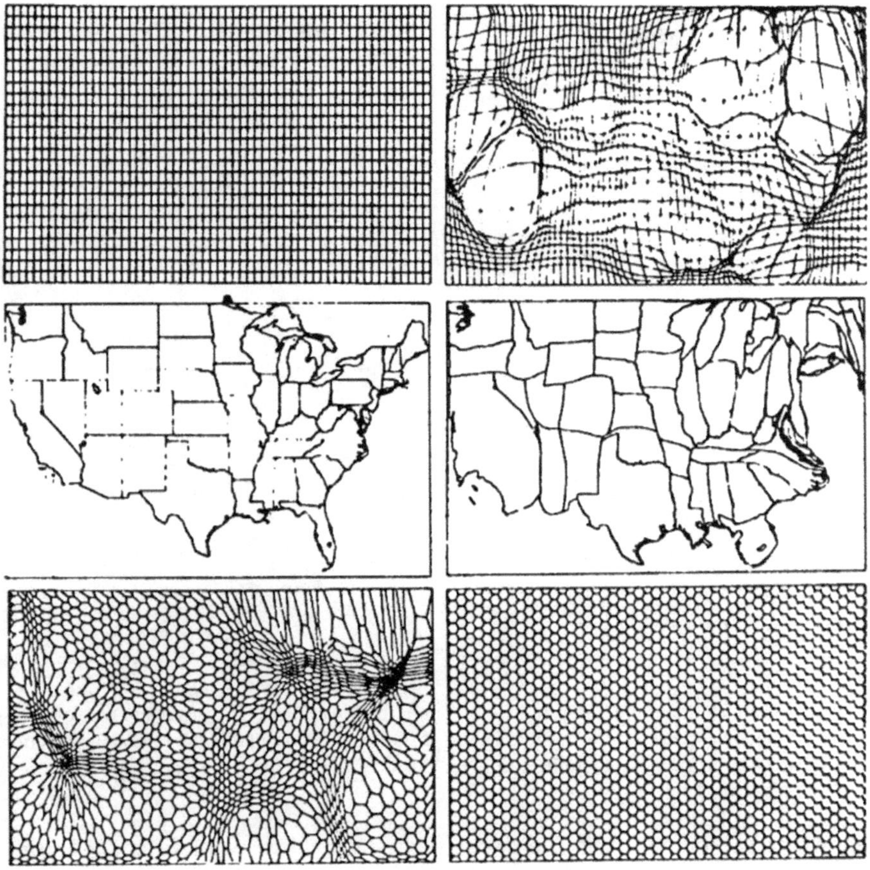

Fig. 5.4 The transform–invert–solve paradigm. Converting the USA to a population density space, overlaying a hexagonal pattern, and then taking the inverse to partition the country into approximately equal population regions. Tobler (1973)

used to apply the solution. This is quite similar to the application and warping of the world by the Mercator projection in connection with the gnomonic projection in ocean navigation. Thus the 'Transform–Solve–Invert' procedure is conceptually identical to what is done when using the Mercator projection.

5.5 Networks

When modern transportation situations are considered it is necessary to violate the usual continuous topology. From Los Angeles to New York now costs less than to many intermediate places. Thus the map may need to be turned inside out.

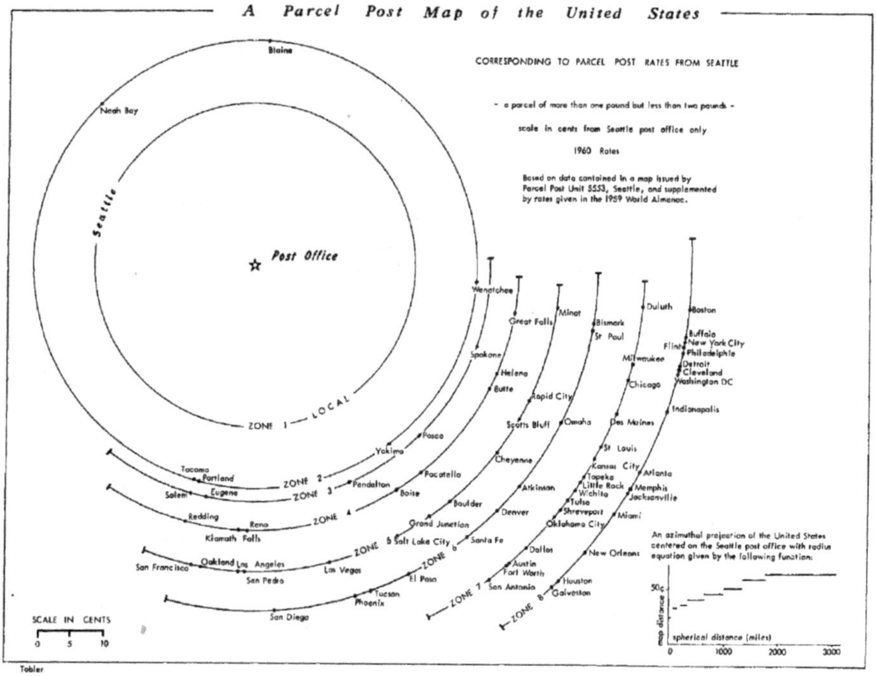

Fig. 5.5 The postal rates are a step function, seen in the lower right. This results in the collapsing of the continuous space into a discrete set of distances. In this particular rendition directions (azimuths) from the Seattle post office are maintained. Tobler (1961)

Many such conditions occur. Bunge has suggested balloons with the ends of strings glued on to constrain some places to be close together when balloon is inflated, with the other places bulging out. This is the spherical real world model to be mapped. In another example, parcel postage rates are a step function of distance, as are many other transportation costs. Thus places are lumped together. If one draws postal costs as a function of geographic distance it looks like several places coincide (Fig. 5.5).

In the time of the Roman Empire travel by water was more rapid than travel by land. Thus Britain was closer to Rome than was Vienna. Even central Spain was further than Britain. The resulting isochrones are rendered graphically in the Roman History project (orbis.stanford.edu) and depict this, but with inverted positions.

The geographic details in these situations are often best represented as a network. Positioning of the network graphically in a two dimensional diagram must attempt to maintain the measurements and this often distorts the conventional geography. The placement of the isolated locations is possible, most often using a form of multi-dimensional scaling, but if the geographic coordinates (latitude and longitude) of these locations are known then interpolation between them can be said to be 'difficult'. In a classification of the types of situations that occur one must recognize inversions (geographically far places are closer than conventionally near places), interruptions (nearby places are not together), and many other violations of the traditional assumptions. This is the case even if distances are considered from only one location. But how badly are these maps distorted? Can this disruption be measured? By Tissot's indicatrix or by the stress? Is it possible to detect (that is, calculate) how much a high speed road or railroad system distorts a country? How warped has it become due to an expressway? Perhaps it has been turned almost inside out, with some places very close and others relatively further away! What did the Romans think (metrically) of the shape of the world when land travel was so difficult? These kind of relations are quickly visible on cartograms. But parts of the maps may overlap, which most (but not all) classic map projections prevent. Another difficulty is that geographic travel times (or costs, etc.) are not symmetric: from A to B is not the same as from B to A. Cartographers most now take into account these global realities and this often requires ingenious cartograms.

5.6 The Equations

Most books on map projections do not consider cartograms. This in spite of the fact that it can be shown, by writing out the equations (see below), that area cartograms are a generalization of equal area map projections; the Earth's surface area is the particular measure preserved on an equal area projection. Stretching by population is just a different choice of property to be preserved. And many of the distance oriented cartograms are equivalent to azimuthal projections. A reason that such maps are not frequently discussed in the projection literature might be that they are of more recent origin, and that they are generally produced by individuals rather than government sponsored national mapping agencies. They do appear in some atlases, along with thematic maps. In this respect they are somewhat like the retroazimuthal projections that have very restricted usage. It is certainly the case that Tissot's measures of angular, area, and distance properties can be calculated for these type of maps. Measures can also be devised that indicate whether or not the resulting maps actually match the design objective—i.e., fit what they are intended to show. This may mean going a bit beyond the classical indices. At the moment there do not appear to be evaluation standards for cartograms, at least there is no consensus on this point. The large variety of possible uses makes this difficult.

There are really no books, and few scholarly articles, that consider cartograms as their main subject. The exception is for cartograms that modify areas according to some measure. Here a few recent books can be cited, along with those already mentioned. In dealing with cartograms using metrogenic substitutions (cost or time, etc.) instead of kilometers and based on a single center, the modifications can be evaluated as are azimuthal projecionts.

Appendix: Equations for an Area Cartogram

The equal area condition for a map projection in spherical (φ, λ) and plane rectangular coordinates (x, y) is:

$$\frac{\partial\varphi}{\partial x}\frac{\partial\lambda}{\partial y} - \frac{\partial\lambda}{\partial x}\frac{\partial\varphi}{\partial y} = R^2 \cos\varphi.$$

The condition equation for a cartogram is:

$$\frac{\partial\varphi}{\partial x}\frac{\partial\lambda}{\partial y} - \frac{\partial\lambda}{\partial x}\frac{\partial\varphi}{\partial y} = R^2 D(\varphi, \lambda) \cos\varphi,$$

where $D(\varphi, \lambda)$ is the density distribution on the Earth, considered spherical.

Clearly, when the density distribution is constant (unity), the cartogram becomes an equal area projection. There are many solutions to both of the foregoing partial derivative equations. In each case the one condition does not suffice to yield the two equations necessary [$x = f(\varphi, \lambda)$, $y = g(\varphi, \lambda)$] to completely define a map projection. Thus some other criteria is applied. The obvious second condition is to require that the angular distortion be minimized, making the image more recognizable. But other conditions, for example rectangular shape or symmetry about the equator, are often used for equal terrestrial area maps. The rectangular Value-by-Area maps of Erwin Raisz come to mind as alternatives.

The inverse of a cartogram of this type is to be found in the usual manner using the Jacobian defined by the partial derivatives, as described in books on advanced calculus, e.g., Kaplan (1952: 96–100). If the defining equations are not given explicitly then a two dimensional finite difference empirical iteration with interpolation needs to be used. A numerical example is given in Fig. 5.4.

References

Barrett F (1992) The World based on airfares from London. The Independent, London, 9 Feb, p 6

Bunge W (1966) Theoretical geography, 2nd edn. Lund University, Gleerup

Dobbin C (2013) London underground maps. London

Dorling D, Newman M, Barford A (2008) The atlas of the real world. Thames & Hudson, London. www.worldmapper.com

Gastner M, Newman M (2004) Diffusion-based method for producing density-equalizing maps. Proc Natl Acad Sci U S A 101(20):7499–7504

Hāgerstrand T (1957) Migration and area. In: Hannerberg D (ed) Migration in Sweden. University of Lund, Sweden

Hennig B (2013) Rediscovering the world: map transformations of human and physical space. Springer, Berlin

Jenny B (2006) Geometric distortion of schematic network maps. Bull Soc Cartographers 40:15–18

Kaplan W (1952) Advanced calculus. Addison-Wesley, Reading

Raisz E (1934) The rectangular statistical cartogram. Geogr Rev 24:292–296
Rase W (1997) Fischauge-Projektionen als kartographische Lupen (Fish-eye projections as cartographic lenses). Salzburger Geographische Materialien, Heft 25:115–122
Tobler W (1961) Map transformations of geographic space. PhD dissertation, Department of Geography, University of Washington, Seattle
Tobler W (1973) A continuous transform useful for disticting. Ann N Y Acad Sci 219:215–220
Tobler W (1996) Thirty five years of computer cartograms. Ann Assoc Am Geogr 94(1):58–73
Van Kreveld M, Speckmann B (2007) On rectangular cartograms. Comput Geom Theor Appl 37(3):175–187

Chapter 6
Anamorphoses as a Method of Visualization

Sabir M. Gusein-Zade and Vladimir S. Tikunov

6.1 Introduction

Development of a number of Earth sciences, connected with the spatial-temporal analysis, presupposes not only an improvement of methods of representation of spatially-coordinated phenomena, but also demonstration of their relations and connections with other phenomena, especially in cases when we are analyzing them as systems. Often it is necessary to examine changing over the space characteristics of several phenomena at once. It is more convenient to carry out such an analysis in the case when one of the characteristics is uniformly distributed over the territory and we are regarding all other characteristics against this one as the background. Of course such a situation is very rare and the idea to create it artificially arises. For this it is possible to transform the image of the phenomenon taken as the base from usual Euclidean metric of the space into a conditional thematic "space" of the uniformed phenomenon. Under the term "transformation" we understand a transition from the ordinary cartographic image, usually based on the topographic metric of the Earth surface, to another image, based on a metric connected with the phenomenon under consideration. Geographers express growing interest to such transformed images which are called anamorphoses. In other words anamorphoses can be defined as graphical images obtained from the traditional maps, the scale of which is not constant and varies depending on values of some indices, on which they are based.

S.M. Gusein-Zade
Faculty of Mechanics and Mathematics, M. V. Lomonosov Moscow State University, 119991 Moscow, Russia

V.S. Tikunov (✉)
Faculty of Geography, M. V. Lomonosov Moscow State University, 119991 Moscow, Russia
e-mail: tikunov@geogr.msu.su; vstikunov@yandex.ru

M. Lapaine and E.L. Usery (eds.), *Choosing a Map Projection*, Lecture Notes in Geoinformation and Cartography,
DOI 10.1007/978-3-319-51835-0_6

Fig. 6.1 Net polarized landscape on land and sea. Functional zones and ways of communication: ▶ **a** for a homogeneous plain inside a continent, **b** for a seaside. In the upper row of the legend for land, in the lower one for sea. *1* Urban historical-architectural reservations; *2* public service and utilitarian ways of communications; *3* permanent urban residences and manufacturing industry; *4* agriculture of high and medium intensity; *5* natural meadows, pastures, forest industry, hunting, suburban recreational parks; *6* natural reserves; *7* recreational residences and touristic roads (after Rodoman 1974)

In English speaking countries instead of anamorphoses terms including cartograms, transformed maps, pseudo-cartograms, topological cartograms and others are used. We prefer to use the term anamorphosis and to call an anamorphation (from the Greek word anamorphoo) the process of their creation. It seems that this term reflects the essence of the process, connected with the change of image proportions, more precisely. Besides that let us emphasize that this term is spread in a number of countries, first of all in the East Europe. In the Russian scientific language the word "anamorphosis" also has been used for a rather long time. Thus, as far back as well-known Russian linguist Dal' (1881) defined anamorphosis as a hideous but regularly deformed picture, which can be seen in a cut or curved mirror.

This definition coincides with one of the methods of creation of anamorphoses, which is still used today. However according to the contemporary concept of anamorphotic images the term "hideous" is hardly in its place. A transformation of cartographic images is produced for theoretical and practical purposes and serves as a tool of spatial analysis.

Anamorphated images differ from cartoids and from well-known mental maps (Gould and White 1974). Cartoids are abstract graphic images for compilation of which real spatial relations are not important, but there are shown some substantial characteristics: the main essence of phenomena, regularities in their allocation, in their development and in the reasons defining them. Examples of cartoids are: "the ideal continent", "typical relief forms", "inversion cartoid of the population system in Africa" (after Rogachev), reasons for rise of Moscow in the Russian state (after Saushkin and Rodoman), and polarized landscape (after Rodoman 1974; Fig. 6.1).

Mental images are graphic representations of ideas about spatial objects formed in human brains. They have been created by all of us when we drew schemes explaining, for example, how is it possible to find the desired place in a city. One can average such representations and obtain a collective mental image. A number of examples like characterization of different places of Los Angeles from the point of view of representatives of middle white, afro-american and hispanic-speaking population, ideas of Londoners about the North or the image of the world as it is seen from Van Hornsville village, USA (with such "provincial" centres as New York, London or Moscow) can be found in Gould and White (1974).

(a)

(b)

6.2 Compilation of Linear Anamorphoses

Among anamorphated images one can distinguish linear, area and volumetric ones. Moreover all of them can be animated. Linear anamorphoses often look like graph images. Changes of lengths of edges of them permit to change the distances between the regarded units (vertices) depending on values of characteristics taken as the base of the anamorphosis.

As an example of characterization of time changes let us give linear anamorphoses (Figs. 6.2 and 6.3) which show changes of mutual transport remoteness of regions of Russia (Malinovsky et al. 2002). On them, there are drawn symbolic straight (azimuth) lines from the centre of Moscow connecting it with each centre of a subject of Russian Federation (and also with Surgut since Khanty-Mansijsk (the administrative centre of the Khanty-Mansi Autonomous Region) does not have a rail-road station). Then the value of the “price distances” from Moscow (Fig. 6.2) and to it (Fig. 6.3) in 1985 and 2001 are put on these lines with the chosen. As a result one gets linear anamorphoses depicting not only the “price” remoteness, but also its change.

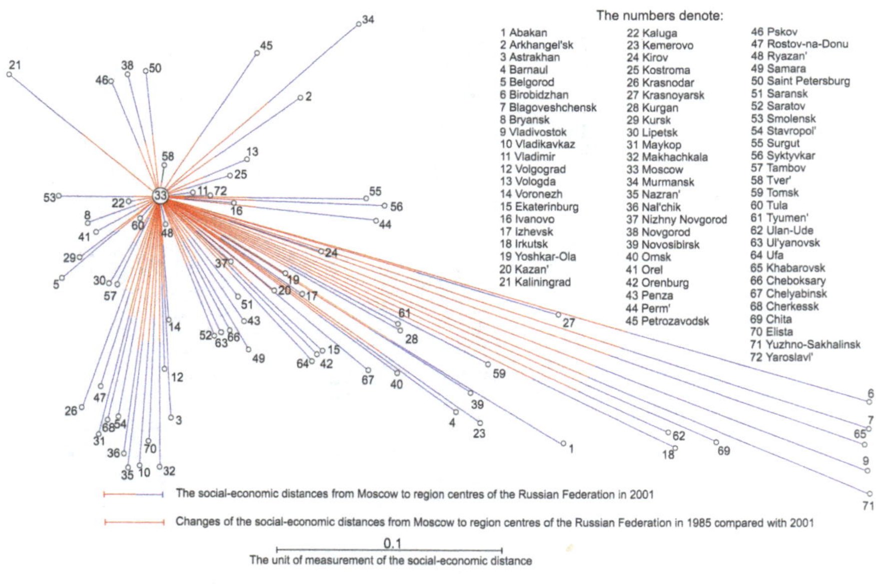

Fig. 6.2 Linear anamorphosis of change of the social-economic distance from Moscow to the centres of Russian Federation for the railroad passenger communications in 1985 and 2001

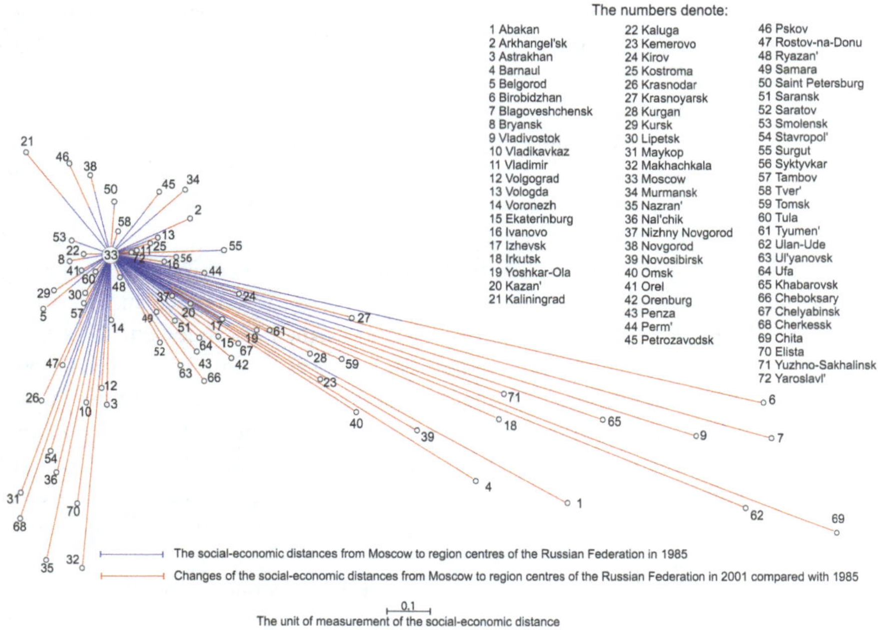

Fig. 6.3 Linear anamorphosis of change of the social-economic distance from the centres of Russian Federation to Moscow for the railroad passenger communications in 1985 and 2001

6.3 Compilation of Area Anamorphoses

Among anamorphoses the most widely distributed are area ones, which make uniform the density of an index (for example, the density of population, the density of the territorial distribution of incomes, the density of the territorial distribution of consumption of a product and so on). It means that in this case areas of images of territorial units become proportional to corresponding values of the index which is in the base of the anamorphosis. It is required that to the extent possible anamorphotic images preserve arrangement of territorial units with respect to one another and their shapes. Among compiled anamorphoses most often are found images, for which areas are proportional to numbers of residents of corresponding territories.

More rarely anamorphoses can be found in which gross revenue of population, gross yield of grain, gross national product and so on are shown. In the history of compilation and use of anamorphoses it is possible to find examples concerning very different fields. But most often they are used for representing various characteristics of population on territories, in electoral and medical geography, for representing quality of the environment, air pollution and so on (Levison and Haddon 1965; Forster 1966; Härö 1968; Ruston 1971; Malinovsky et al. 1977; Wonders 1980; Bochkareva 1981, 1983; Kadmon 1983; Pravda 1983; Selvin et al. 1984; Kelly and Neville 1985; Belov 1983; Uzan 1989). In spite of the variety of methods used earlier for compilation of anamorphosis, in this part of the paper we shall use only examples compiled with the algorithm, elaborated by the authors (Gusein-Zade and Tikunov 1993a).

First of all let us look at the images of the world. Let us give also a series of anamorphoses compiled by the authors on the base of data of the Department of Analysis of Economics, Social Information and Politics of the United Nations: actual and values of number of population of countries of the World for 1950–2050 (Fig. 6.4).

Countries are easily recognizable by their shapes (configuration). Of course the most attention is attracted by China and India. In Asia only Mongolia and Laos look more than unpretentious against the background of their encirclement. It is interesting that even such little "specks" on the initial map as Hong Kong and Singapore turn into considerable territories on the anamorphosis. Australia is "shrunk" rather heavily. In Africa, which is rather modest compared with Asia and even with Europe, first of all attention is attracted by the "ball-shaped" Nigeria. The most contrasts are inherent in Europe—compare the "tiny" Benelux and the countries of Scandinavia. Respectively uniformly are populated countries of America with the exception of Canada and Greenland.

Putting on an obtained anamorphosis characteristics connected with population, for example, the provision with food products, we shall get a more adequate impression about e.g. its deficiency. It will be so because it is related not to the territory, as it is on traditional maps, but more correctly—to the population which is needy in it. Now let us look at the second anamorphosis, compiled on the base of the data on the gross national product (Fig. 6.5).

Contrasts here are even sharper than on the previous one. First of all are striking three world "centres of welfare"—USA, Western Europe and Japan. Some countries of Europe are so heavily deformed that it is not easy to recognize them. However whereas in Europe the word deformation means expansibility, in Africa this process goes in the opposite direction. The majority of countries of Western and Central Africa are simply merged with each other. There are distinguished only the South African Republic and some oil-extracting countries. It is interesting to look at India and China—the leaders of the previous anamorphosis. Here they look more than modest. The relation between the areas of China and Hong Kong on the anamorphosis is curious. Australia acquires a more "worthy shape." In America all other countries look unpretentious against the background of the "moneybag"—the USA. Alaska has been included into consideration as a separate territory (in the sense that its own gross product has been attributed to it). It led to its "compression" into a narrow strip.

The almost regular rectangular shape of the islands of Puerto Rico and Trinidad is explained by their small sizes on the initial map. Therefore they were digitized very roughly, only with the use of four points each. On the anamorphosis their territories become more noticeable. The series of anamorphoses of the World can be easily continued (Figs. 6.6, 6.7 and 6.8).

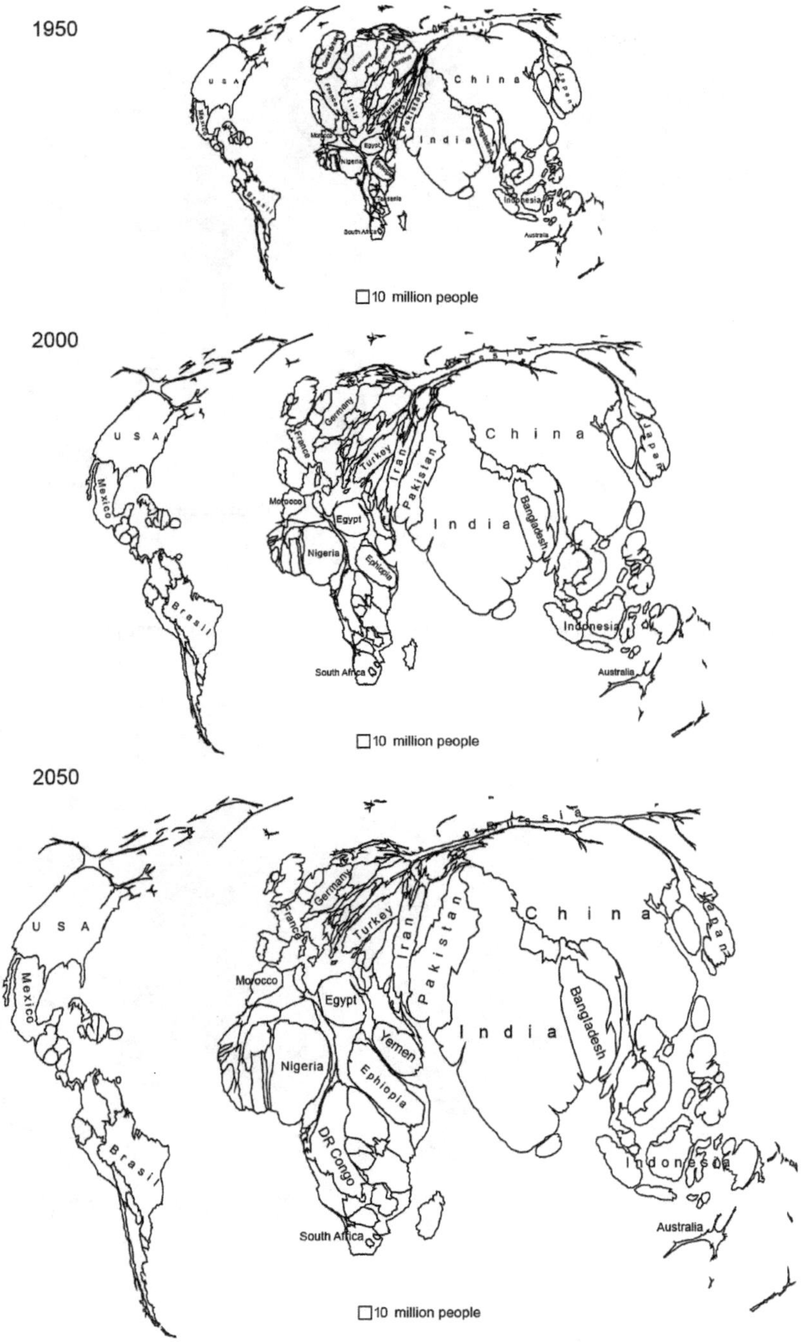

Fig. 6.4 Anamorphoses of the world compiled on the base of numbers of population of countries for 1950, 2000, and 2050

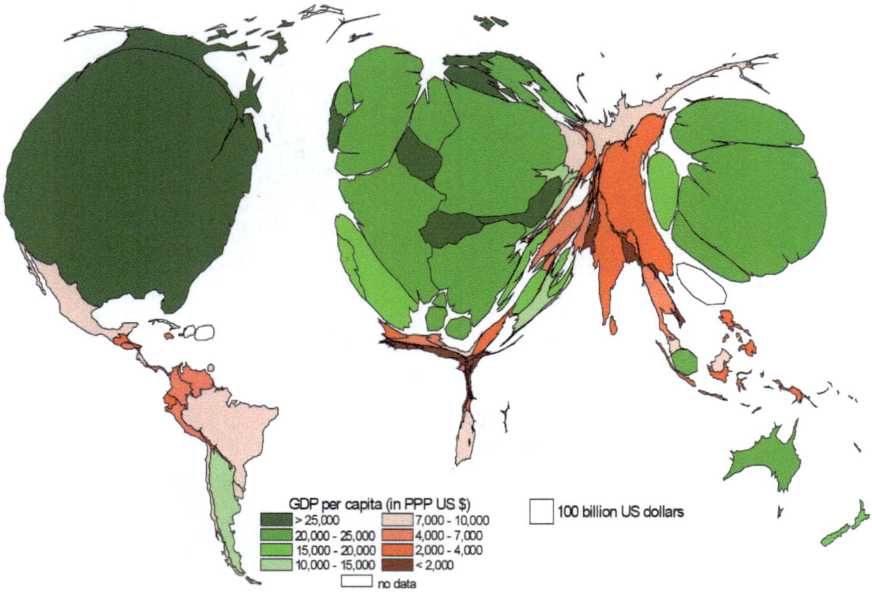

Fig. 6.5 Anamorphoses of the world compiled on the base of gross national product for 2003

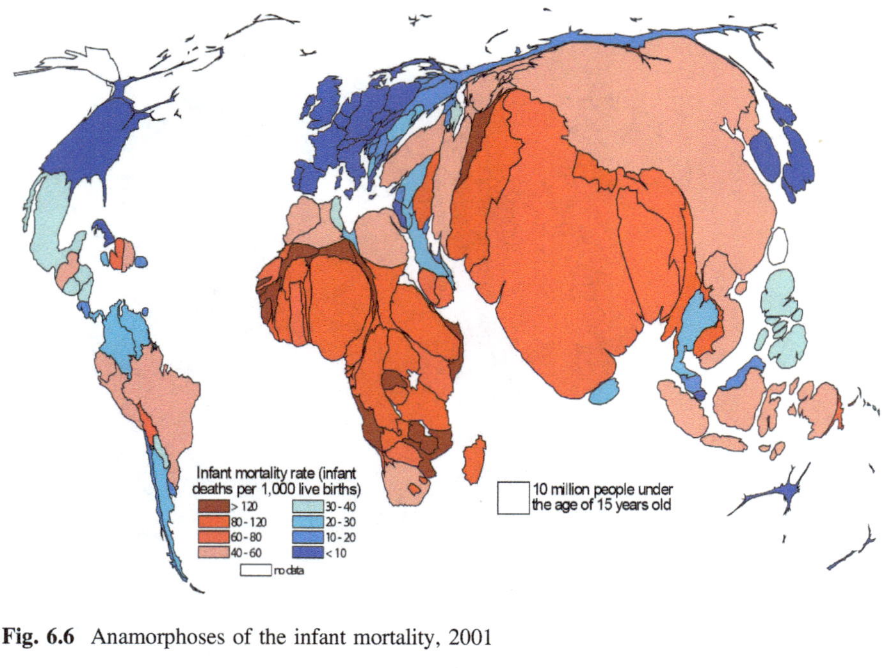

Fig. 6.6 Anamorphoses of the infant mortality, 2001

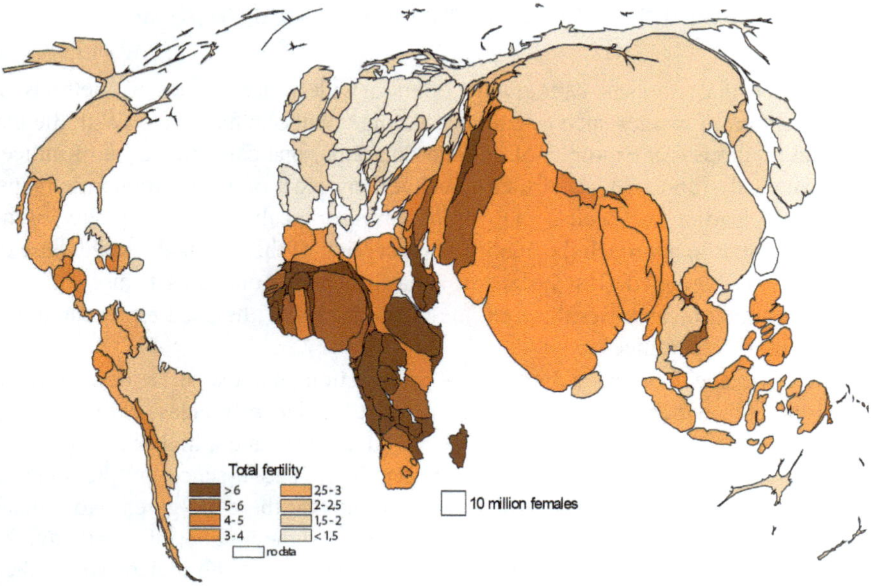

Fig. 6.7 Anamorphoses of the fertility, 1995–2000

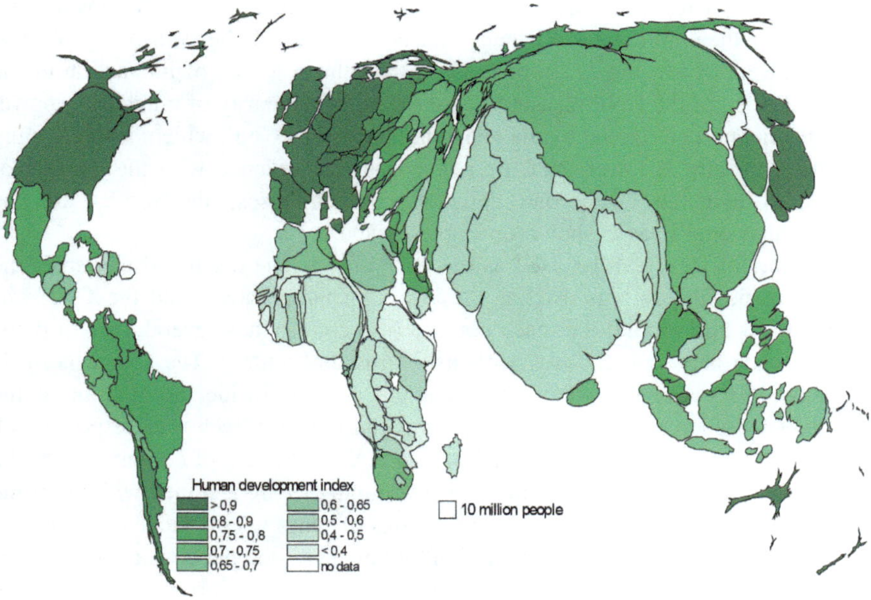

Fig. 6.8 Anamorphoses of the human development index, 1999

6.4 Compilation of Three-Dimensional Anamorphoses

It is known that classical cartography offers a widely used series of methods of representation of images such as colour, shading, symbolic marks etc. With the use of usual methods (for example, of column diagrams) one can show a lot of indices simultaneously (up to 30–40 of them), as one may meet on social economic maps. However, clearness of such a map drops catastrophically and the quality of its perception start to approach the quality of perception of the data table on the base of which it was compiled. Just because of that, on maps, one tries to use different representational means. Besides that, the effectiveness of the used representational method is of importance.

There were elaborated a method of compilation and use of a new style of representation, which, in principle, is somewhat similar to the classical method of column diagrams, but possesses a number of advantages and a higher effectiveness of representation of an index used for mapping. The essence of the method (Bogomolov et al. 2002) is the following. Assume that there are given two indices required to be mapped, for example, the real GNP and the number of population for countries of the World. In the example under consideration the indices were taken for the year 1996. Note that the real GNP is the GNP measured in international dollars on the base of parity of purchasing power of currencies. The international dollar has the same purchasing power with respect to the GNP as the US dollar inside United States. The real GNP of developing countries is usually considerably bigger than their nominal GNP and reflects the level of the well-being of their population more adequately. In contrast to that, the potential of interaction of the real GNP (i.e., of the GNP measured on the base of the parity of purchasing power of currencies) is determined by the value of the nominal GNP, which, in developing countries, usually is lower than the real one. In accordance with the number of population and the GNP, one compiles anamorphoses; the results are the three-dimensional images shown on Figs. 6.9 and 6.10.

Any one of them can be used as the base one. In the discussed examples the number of population was used as the base of anamorphoses and the GNP was shown on its background. For each country (if a country has several disjoint parts, then for each part) one chooses a point used as its "centre". The choice is made manually; the necessary condition is that the point is inside the contour of the country (if a country has several disjoint parts, then for each part respectively). The choice of centres is made in the Cartesian (X; Y) system of coordinates on the anamorphated image. All points of the state boundaries are assigned with the value of the Z-coordinate equal to zero and all centres of states get the value of Z equal to the value of the index under consideration (in our case—the per capita real GNP). If now using the obtained set of points one constructs the corresponding surface (automatically, using the method of Delaunay triangulation), one obtains a set of polyhedral pyramids with bases coinciding with the country territories and with the numbers of faces equal to the number of vertices used for digitizing the boundaries of the countries.

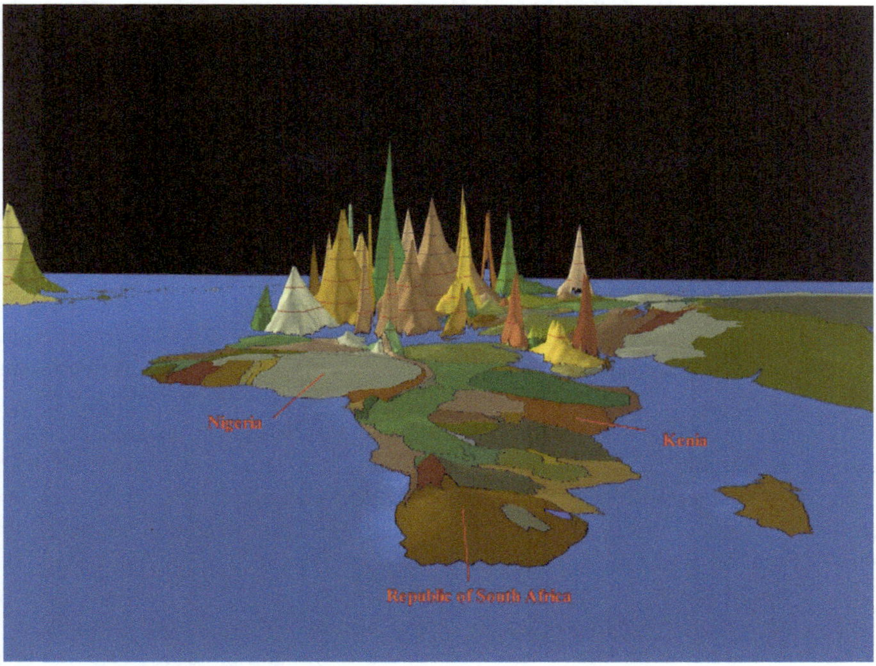

Fig. 6.9 Example of three-dimensional anamorphoses of real GNP

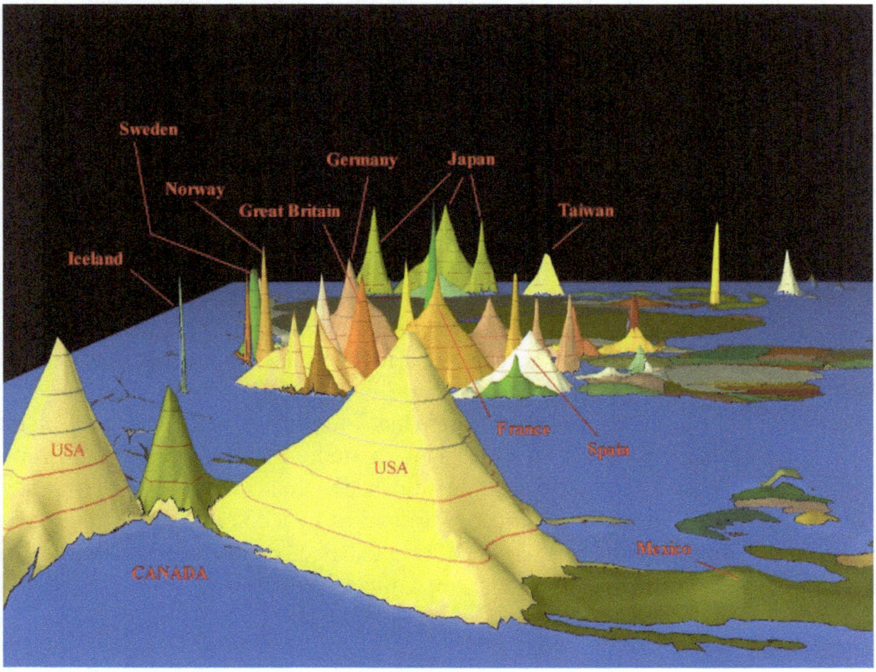

Fig. 6.10 Example of three-dimensional anamorphoses of real GNP

The anamorphosis of the World based on the population possesses the following property: the area of each country on it is proportional to its population. The height of the constructed pyramid is proportional to the per capita GNP. The volume of a pyramid is equal to V = SH/3, where S is the area of the base, H is the height of the pyramid. Therefore the volume of the constructed pyramid is proportional to the GNP of the country.

Let us note that it is somewhat difficult to compare GNPs of two countries by sight, since the outline of each pyramid is unique and all of them have rather irregular shapes. As a positive feature of this image, for sure, one should consider the showiness of the resulting image. It is more easy to make comparisons and to detect anomalies. One depicts two indices, in fact without using methods generally accepted in cartography, keeping them as a reserve. All the information about the two characteristics of objects are contained in the image.

It is also possible to construct column diagrams based on an anamorphosis, however the result is read and apprehended not so well since some countries cover other ones, countries with low GNP surrounded by neighbors with high GNP are simply not seen. The zero level is not well seen. Pyramid diagrams are free from this defect. They show the zero level for each contour of a country. One has a good field of vision, practically each country can be seen.

The technology of creation of such images is rather simple. The construction of the image was based on the political map of the World and the data on the number of population and real GNP for 1996. The initial map of the World was transformed into an anamorphosis using the algorithm of Gusein-Zade and Tikunov (1993a, b). The resulting data (files with the coordinates of points and with the topology of objects) were converted into the geographic information system (GIS) ArcView 3.0 format for subsequent export into the raster format *.JPG. There were exported two files of equal size rasters: one black and white (1) with contours of country boundaries and the other one coloured (2) with a unique colour shading for each country. The black and white raster was automatically converted into the vector format and polygons were attributed with central points with the help of the program Raster2Vector (Able Software). The result was saved in a file of *.SDL format and then converted into a *.CON file.

The latter one was the starting-point for the program DEMI, which constructs surfaces for a given set of irregularly arranged points. The result was saved in a file of *.DEM format, which represents a regular net of points with determined heights. With the help of a converter the file was saved in the *.BMP format (with grades of gray: more white for higher levels and more black for lower ones). After that (on the base of the obtained BMP file) the package 3D-Studio MAX 2.5 produced, from a regular net of size 300 × 300, the three-dimensional surface with the pyramidal block-diagrams. After that this surface was covered by a coloured raster image: for making it easier to use the image. Since for a detailed consideration of the surface one should look on it from different sides, there were created several images of different regions of the World.

6.5 Animations

As in the case of area anamorphoses, three-dimensional anamorphoses can be animated. Some general remarks first. With the start of the era of personal computers, which permit to reproduce full scale videos, one equates a possibility to show an object in dynamics to increasing the dimension of the initial static image by one level. Dynamic 2D-images (for examples cartograms where values of an index are shown by colour) are, as a matter of fact, three-dimensional.

Dynamic data for a series of indices are often met in the socioeconomic cartography. Construction of dynamic 2D images for each of the indices became a rather usual event. Analysis of one cartofilm is also not complicated. It is also possible to construct a cartofilm for each of the indices. However, it is somewhat difficult to find correlations between them. Thus we have arrived to the essence of one of the problems: depicting several dynamic indices simultaneously without constructing additional dynamic maps of correlations. To consider the problem, let us use three indices: population, per capita GNP, and average length of human life. All the indices were taken for each country of the World for the last 30 years. Showing the dynamics of all three indices on one model using methods of classical cartography is rather complicated. However, a solution can be the following one.

1. The population of the World: for each year one constructs an anamorphosis of the world population. Each of the images is two dimensional. If one shows the images subsequently, one after another, we see "inflations" and "compressions" of different countries. Since the images are raster ones, for each country let us choose a point which is not displaced for all the period and is situated approximately at the centre of the country and let us determine its XY-coordinates. One apprehends even small shape changes well enough and therefore the dynamic of the phenomenon can be seen very well.
2. Expected length of human life: its dynamics may be represented by colour. Let us choose a multi-level colour scale (say, 100 levels of colour) so that each level of the index is denoted by one colour). According to this, the dynamics of the length of human life in each country is reflected by a sufficiently continuous change of its colour. This is also well apprehended.
3. Per capita GNP. For that one uses the method of 3D-pyramidal block-diagrams. The essence of the method is as follows. Assume a 2D map of countries (or of other territorial units). For each country one chooses (by sight) a centre (X; Y) and the corresponding value of the Z-coordinate is the per capita value of GNP for this country in a given year. All vertices of contours of countries have their own XY-coordinates and let the value of Z for them be equal to 0. If, for the obtained XYZ-mass, we construct the surface in 3D space and, for better orientation, cover it by the initial 2D map, we shall see that each country became "a mountain", the base of which has the shape of the country and the height is proportional to the value of the index under consideration (per capita GNP in

our case). The initial two-dimensional map may be both the classical political map of the World and an anamorphosis, static or dynamic. A necessary condition is that XY-coordinates of the "mountain summit" for each country should not change in time.

Taking into account all discussed above, there were constructed dynamic pyramidal block-diagrams for GNP, moreover their "summits" were kept (on the image) at the same places and their heights changed in time. Shapes and areas of bases of these "mountains" changes depending on the country population (a dynamic anamorphosis). At the same time each "mountain" changes its colour in complete accordance with the expected length of human life. To make such images metric, values of the indices can be simply written over the summits of the "mountains". It is also possible to draw the level lines. "Dimension" of this image is very high. Indeed, a static non-coloured image would be a visualization of a certain surface. In time there are changing heights of the "mountains", shapes of their bases, and their colours.

6.6 Conclusion

Completing the chapter let us fix most essential reasons for practical use and perspective of anamorphoses. First of all, and this is not the most principal, they are convincing as illustrations, which permit one to imagine visually some non-evident facts and even to see some hidden geographical regularities. For example, theoretical principles of W. Christaller and A. Lösch about regularities of arrangements of hierarchical systems of centres get their confirmation only in regions with uniform distribution of population. It is possible that search of regularities in arrangements of populated places in the plane with the uniformed phenomenon could make the Christaller-Lösch principle much more often observable or even universal. It has good prospects to use anamorphoses for optimization of arrangement of nets of education, medical facilities and other service centres which, in general should be distributed uniformly on the artificially uniformed demographic plane. In the second place, anamorphoses make the relations between phenomena more visual if they are analyzed on the background of an image based on characteristics connected with them. Thirdly, it is reasonable to use anamorphoses for prediction of development of diffusive processes which take place in a non-uniform environment. If one transforms the resistance to the development of the diffusion into the uniform one, then most probably the diffusion will develop concentrically from the initial point. Thanks to that it is possible to forecast its development in time and to represent it in a graphical form. To get the picture of the spreading of the diffusion in the non-uniform environment, one should restore the image into the initial form. Anamorphoses can be used for the study of the diffusion of pollutions in the atmosphere and hydrosphere and also for a number of other problems.

The same can be said about compilation of maps of transport accessibility, which are created on the background of uniform practicability.

Thus compilation of anamorphated images in a number of cases is reasonable for modeling the structure, interconnections and dynamics of geographical phenomena. We hope that anamorphated images will attract attention of geographers of different interests and will become not eccentric illustrations, but tools of real geographical analysis. This field can appear to be a non-upturned scientific virgin land, which will give a rich harvest after a skillful processing.

Acknowlegements The study was supported by the grant of the Russian Science Foundation [Project No. 15-17-30009].

References

Belov AL (1983) Political geography of Canada: main contemporary problems. All-Union Geogr Soc Izv 115(4):363–368 (in Russian)

Bochkareva TV (1981) Tipology of largest urban agglomerations of the USA by the quality of environment and principal factors of its formimg. Acad Sci USSR, Izvestiya, Ser Geogr 5:74–85 (in Russian)

Bochkareva TV (1983) Influence of the ecological factor on changes of the dynamics and of the distribution of population in largest urban aglomerations of the USA. All-Union Geogr Soc, Izv 115(3):259–266 (in Russian)

Bogomolov N, Rylskiy I, Tikunov V (2002) Creation of the anamorphoses-based 3D-pyramidal block diagrams. In: Richardson DE, Oosterom P van (eds) Advances in spatial data handling: 10th international symposium on spatial data handling, Springer, Berlin, Heidelberg, New York, Barcelona, Hongkong, London, Mailand, Paris, Tokyo, pp 465–473

Dal' V (1881) Explanatory dictionary of the living Great Russian language. St. Petersburg, Moscow, 779 pp (in Russian)

Forster F (1966) Use a demographic base map for the presentation of a real data in epidemiology. Br J Prev Soc Med 20(4):165–171

Gould P, White R (1974) Mental maps. Penguin Books Inc., New York, Baltimore, 204 pp

Gusein-Zade SM, Tikunov VS (1993a) A new technique for constructing continuous cartograms. Cartography Geogr Inf Syst 20(3):167–173

Gusein-Zade SM, Tikunov VS (1993b) The transformed image in geographical analysis. Mapp Sci Remote Sens 30(4):306–319

Härö A (1968) Area cartogram of the SMSA population of the United States. Ann Assoc Am Geogr 58(3):452–460

Kadmon N (1983) Photographic, polyfocal and polar-diagrammatic mapping of atmospheric pollution. Cartographic J 20(2):121–126

Kelly JL, Neville RJW (1985) A population cartogram of New Zealand. NZ J Geogr 79:7–11

Levison M, Haddon W (1965) The area adjusted map: an epidemiological device. Publ Health Rep 80(1):55–59

Malinovsky DV, Tikunov VS, Trejvich AI (2002) Cartographic evaluation of changing mutual transport remoteness between Russian regions in 1985–2001 (at the example of railway tariff distances). In: Proceedings of the international conference GIS for sustainable development of territories. InterCarto 8, Helsinki, St.Petersburg, pp 180–186 (in Russian)

Malinovsky DV, Tikunov VS, Trejvich AI (1977) Perspective Canada: a compendium of social statistics, vol 2. Ottawa, p 335

Pravda J (1983) Metodicko-vyjadrovacie probl'emy tvorby tematick'ych map. Bratislava (in Slovak)
Rodoman BB (1974) Polarization of the landscape as means of preservation of the biosphere and recreational resources. In: Resources, environment, population. Nauka Publishers, Moscow, pp 150–162 (in Russian)
Ruston G (1971) Map transformations of point patterns: central place patterns in areas of variable population density. Pap Proc Reg Sci Assn 28:111–129
Selvin S, Merrill D, Sacks S, Wong L, Bedell L, Schulman J (1984) Transformations of maps to investigate clusters of disease. Univ. of California, Lawrence Berkeley Laboratory 33 pp
Uzan LJ (1989) Images d'espaces. Acta Geogr 79:12–19 (in French)
Wonders LJ (1980) The junior atlas of Alberta: introducing new mapping techniques to young students. Canadian Geogr 24(3):306–311

Chapter 7
Map Projections in Planetary Cartography

Henrik Hargitai, Jue Wang, Philip J. Stooke, Irina Karachevtseva, Akos Kereszturi and Mátyás Gede

7.1 Introduction

7.1.1 *Historical Background*

The first maps of the Moon showed the Moon in Orthographic projection, as seen by the eye, north up. Riccioli (in 1651) segmented his lunar map into 8 "octants" (Whitaker 2003: 63). Hevelius (1647) used a 360 degree system of azimuths around the limb of the Moon, with the pole of its coordinate system at the sub-Earth point. Tobias Mayer introduced the terrestrial latitude/longitude coordinate system with

H. Hargitai (✉)
NASA Ames Research Center, MS 239-20, Moffett Field,
CA 94035, USA
e-mail: hhargitai@gmail.com

J. Wang
Department of Earth & Planetary Sciences, Washington University
in St. Louis, 1 Brookings Dr., St. Louis, MO 63130, USA

P.J. Stooke
Department of Geography, The University of Western Ontario,
London, ON, Canada

I. Karachevtseva
MIIGAiK Extraterrestrial Laboratory (MExLab), Moscow State University of Geodesy
and Cartography (MIIGAiK), Moscow, Russian Federation

A. Kereszturi
Research Center for Astronomy and Earth Sciences, Konkoly
Thege Miklós Astronomical Institute, Budapest, Hungary

M. Gede
Department of Cartography and Geoinformatics, Eötvös Loránd University,
Budapest, Hungary

M. Lapaine and E.L. Usery (eds.), *Choosing a Map Projection*,
Lecture Notes in Geoinformation and Cartography,
DOI 10.1007/978-3-319-51835-0_7

meridians and equator (Greeley and Batson 1990: 147). The orientation was changed to south up by Schröter, showing the view as seen through an astronomic telescope (Mayer and Schröter 1791). This system remained in use until 1961. Lunar directions were stated with reference to their position in the sky as viewed from the surface of the Earth. Mare Orientale, the "Eastern Sea," is on the left limb of the Moon as seen in the Northern Hemisphere, that is, toward the east horizon. The orientation of Lunar maps was changed in 1961, when the "astronautical convention" was adopted in anticipation of human spaceflight. The direction from which the Sun rises on the Moon was henceforth called east, as it is on the Earth. This also reflected the fact that with the coming of Space Age, observations of the Moon were not relied on astronomic telescopes, but on spacecraft imagery instead. Map orientation was changed back to north up as in the very early decades of Lunar observations (Wilhelms 1987: 11).

Size and shape information, directions of the poles of rotation and definition of prime meridians of the planets and satellites have been published by the International Astronomical Union (IAU) Working Group on Cartographic Coordinates and Rotational Elements (WGCCRE) since 1979, with the addition of the same data for small bodies: minor planets (asteroids) and their satellites, and comets since 2003 and dwarf planets since 2009. This database is revised every three years (Archinal et al. 2011). The working group recommends a specific cartographic coordinate system for each body, although alternative coordinate systems may exist for various (e.g. dynamical) purposes. The mission of the working group is "to make recommendations that define and relate the coordinate systems of Solar System bodies to their rotational elements to support making cartographic products" of such bodies in a standardized way. High precision coordinates are essential for spacecraft operations, high-resolution mapping, and gravity field determination (Archinal et al. 2011).

7.1.2 Basic Terminology

Before the space age, unique prefixes like seleno- (Lunar) or areo- (Martian) were in use for each planetary body (e.g., Selenography, Areography) and these disciplines mainly dealt with the description of surface features, similar to what geography did at that time for Earth. Today most commonly the prefix geo- is used for any solid body considered under the umbrella term *planetary science*. This reflects the comparative planetological approach according to which the principles of terrestrial geology can be transferred to any other planets (Wilhelms 1990: 209). Unique prefixes or suffixes, however, are still used in some terms referring to coordinates, control points and global figures (with endings like -graphic, -centric, -detic, -id etc.). Coordinate systems are usually referred to as planetographic or planetocentric coordinates; however, in some cases the following prefixes are used: areo- (Martian), seleno- (Lunar), helio- (Solar), zeno- (Jovian) and geo-(Terrestrial).

In the following sections we present an overview of various cartographic aspects of the maps of solid-surface planets, gas giants, satellites, and small bodies with special emphasis on the projections of the end products. Finally, we discuss map-projected planetary images.

7.2 Planets

7.2.1 Rotation

The rotation of a planet may be direct (or prograde, or eastward or positive) or retrograde, depending on the direction of rotation (the Earth and the Sun rotate directly) (Archinal et al. 2011). Direct rotation is counterclockwise when viewed from above the north pole (JPL 2009).

7.2.2 Coordinate Systems

Planetary coordinate systems are defined relative to their rotational axis (that defines the latitudes) and prime meridian (Archinal et al. 2011). In planetary cartography two coordinate systems are used that are fixed to the body: the planetographic coordinate system (IAU 1971) and the planetocentric coordinate system. In both cases, the origin is the center of mass of the body. The coordinates of a planetocentric or spherical coordinate system are defined by a vector from the center of mass of the body to a particular point, while the planetographic coordinates are defined by a vector perpendicular to a reference surface (JPL 2009) (Fig. 7.1). The planetographic and planetocentric vectors are identical if the reference surface is a sphere (JPL 2009; Roatsch et al. 2008). On an oblate spheroid typical of rapidly rotating planetary objects such as Earth and Mars, both vectors are identical at the poles and equator, but the planetographic latitude at a given point in intermediate latitudes is larger than the planetocentric value. On Mars, for example, the difference between the latitudes amounts to approximately 0.3° (20 km) at 45° and vanishes at the equator and poles (Rosiek et al. 2005). A list of currently IAU-accepted coordinate systems for the planets and satellites is found at http://planetarynames.wr.usgs.gov/TargetCoordinates.

In terrestrial geodesy and in a spherical or planetocentric coordinate system, longitudes are always positive toward the east as defined by the "right-hand rule" (Archinal et al. 2011). An external observer would see planetocentric longitudes to decrease with time for directly rotating bodies (JPL 2009). This system is independent of the definition of the reference ellipsoid (Duxbury et al. 2002).

In a planetographic system, the range of longitudes extends from 0° to 360°. The planetographic longitude of the central (e.g., sub-Earth) meridian, as observed from

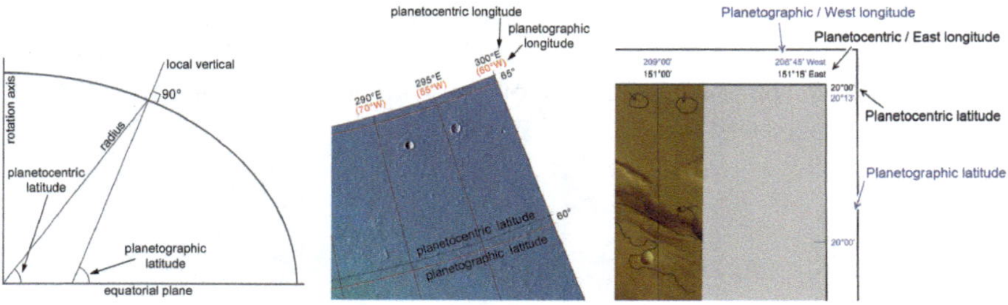

Fig. 7.1 Planetocentric and planetographic coordinates. *Left:* Geometric distinction between planetocentric and planetographic latitudes. The degree of polar flattening of this cross-section is greatly exaggerated. *Center:* Detail from the 1:5 M MOLA-based topographic map of the MC3 Arcadia Quadrangle of Mars that was made to show the official nomenclature and displays both sets of coordinates (USGS n.d.). This map has been replaced by a THEMIS-based map in 2014 that only displays planetocentric East coordinates. *Right:* Dual coordinates shown on the ortophoto-mosaic map of Albor Tholus, compiled at the Technical University of Berlin (Alberz et al. 2004a, b)

a direction fixed with respect to an inertial system, will increase with time (Archinal et al. 2011). This is the case in observing a planet's rotation from the Earth that explains why this system was chosen traditionally for mapping planetary bodies through telescopes from Earth.

Thus, west longitudes that are measured positively to the west are used for bodies with direct rotation, and east longitudes measured positively to the east when the rotation is retrograde, except for the Earth, Moon, Sun which have longitudes run both east and west 180°, or east 360° for historical reasons (Archinal et al. 2011).

Traditionally the planetographic system with west-positive ("west-'ographic") coordinates has been used for the mapping of planetary bodies. An areographic system was used for maps of Mars produced from the 1970s through the late 1990s (Duxbury et al. 2002). However, several recent missions adopted the planetocentric, east-positive ("east-'ocentric") system (Archinal et al. 2011). An areocentric system with east-positive longitudes was introduced with the Mars Global Surveyor (MGS) mission's MOLA (Mars Orbiter Laser Altimeter) team whose product, the MOLA grid, serves as a standard of geodetic control (Duxbury et al. 2002). This system was adapted in subsequent missions, including the US Mars Odyssey, European Space Agency's Mars Express (Gehrke et al. 2003), and US Mars Reconnaissance Orbiter missions. The MESSENGER mission adopted the planetocentric system for Mercury (Seidelmann et al. 2007). The Lunar Reconnaissance Orbiter mission recommended the planetocentric coordinates and east-positive longitude from 0° to 360° range for Lunar mapping, breaking the Lunar traditions (LGCWG 2008).

The IAU Working Group on Cartographic Coordinates and Rotational Elements allows the use of either planetographic or planetocentric system for a given body (JPL 2009). For Mars, both traditionally used planetographic latitudes with west-positive longitude (Inge and Batson 1992) (control network: MDIM (Mosaiced Digital Image Model) 1.0 or MDIM 2.0), and planetocentric latitudes with east-positive longitude (MDIM 2.1) co-existed in the 2000s, although the planetographic system has been dropped in most newer maps. The Mars Transverse Mercator (MTM) Map Series in which geologic and controlled photomosaic maps are published since 1984, have been updated using the planetocentric coordinates as the primary grid with a secondary grid showing the planetographic coordinates (Rosiek et al. 2003) (Fig. 7.1, Center), and HRSC maps use a similar dual system (Fig. 7.1, Right).

Most commercially available GIS systems (without special plug-ins) don't support the 0–360° longitude range and the west-positive longitude system. Mapping with the −180 to +180° east-positive coordinate system is recommended even if the final product will be labelled differently.

7.2.2.1 Prime Meridian/Fixed Reference Feature

A dynamically defined coordinate system does not suit the needs of mapping of surface features, therefore body-fixed, mass-centered coordinate systems are recommended for rocky bodies. The prime meridian of most rocky planets is defined arbitrary, usually by references to a suitable, prominent or easily observable surface feature such as a circular crater whose center point can be easily determined. This fixed reference feature may or may not be located on the prime meridian. As long as the original definition is maintained to within the accuracy of previous determinations, smaller features may be chosen to define the origin for longitude more precisely (Archinal et al. 2011).

The prime meridian of Mars was first defined by the German astronomers Beer and Mädler in 1830–32, who used a small dark feature to determine the rotational period of Mars. Later observers adapted this reference point that was named Dawes' Forked Bay by Proctor in 1864 (Proctor 1873) and Baie du Méridien by Flammarion (1890) after the bay-like appearance of the feature on Dawes' 1864 Mars drawings (Flammarion 1888 [1978]). The name was later latinized to Sinus Meridiani, by Antoniadi (1930) ('Meridian Bay', similar to Sinus Medii ['Middle Bay'] on the Moon).

After the Mariner-9 mission, it was redefined and fixed to the center of the small crater Airy-0 (de Vaucouleurs et al. 1973) by Merton Davies (MSSS 2001), a small crater inside the larger Airy crater, whose name commemorates GB Airy, who established Greenwich as the location of the prime meridian on Earth. Both craters are shown in Fig. 7.2.

For Mercury, the center of Hun Kal crater was used to define the 20°W longitude and it served as a reference to locate the prime meridian. Hun Kal means the 'number 20' in Mayan. This crater is shown in Fig. 7.3.

The rotation period of Venus was first determined from radar data in 1964, and the zero meridian was set to run through one of the first radar-bright features to be located (Carpenter 1966), later called Alpha (today's Alpha Regio). This feature

(a) (b)

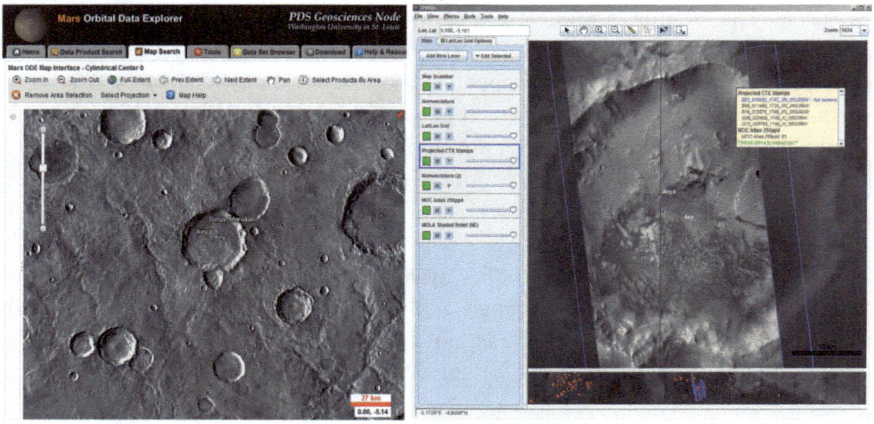

Fig. 7.2 Airy-0 and Airy craterson Mars. Screen copies from two GIS systems: **a** the NASA's Planetary Data System (PDS) Orbital Data Explorer (ODE). The basemap is Mars Odyssey THEMIS (Thermal Emission Imaging System) day IR (infrared) global mosaic generated by the Arizona State University THEMIS team (NASA's PDS Geosciences Node). **b** JMARS (Christensen et al. 2009) showing closeup of Airy crater in CTX image B03_010855_1742_XN_05S359W on MOC photomosaic background (NASA, JPL, Malin Space Science System)

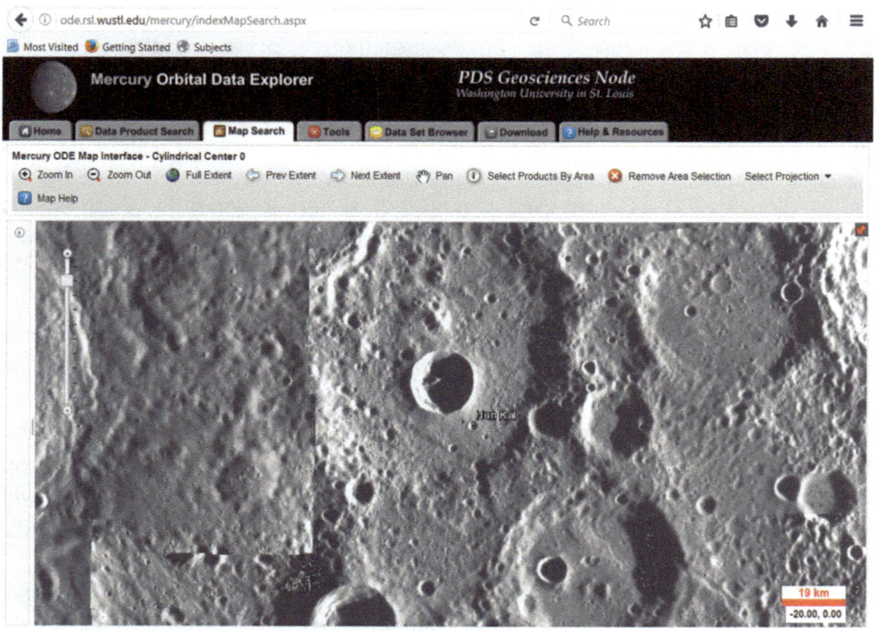

Fig. 7.3 Hun Kal crater on Mercury. The basemap was created using MESSENGER (Mercury Surface, Space Environment, Geochemistry and Ranging) orbital images including MDIS (Mercury Dual Imaging System) Narrow Angle Camera (NAC) and Wide Angle Camera (WAC) images. The map was generated by NASA's MESSENGER mission team and cartographic experts from the U. S. Geological Survey (USGS) (NASA's PDS Geosciences Node)

faced Earth at the inferior conjunction in 1964 and approximately in later years, since the rotation is nearly but not exactly resonant (IAU 1971: 128).

Because of uncertainties of the rotational period, a new definition based on a prominent surface feature was created in 1985. D. Campbell and Y. Tjuflin selected six craters that are common to both Russian Venera and American Arecibo datasets and selected one, named Eve (Davies and Rogers 1991). It was later replaced by another crater on the same longitude, so it passes the central peak of Ariadne crater at latitude 43.8°N (Hirsch 1994) (Fig. 7.4).

Unlike most other planets, Venus has a naturally defined particular meridian that has a special importance: this meridian corresponds to the sub-Earth longitude at the inferior conjunction (cf. Marov et al. 1973). When every 583 days Venus is positioned between the Earth and the Sun, the hemisphere of Venus centered at 320°E longitude turns toward the Earth (Burba 1996). This peculiarity can be visualized when selecting the cartographic central meridian of Venus or its hemispheres: 320° and 140°E longitudes are in fact used as the central meridians for the Venus relief map compiled in Lambert Azimuthal Equal Area projection (Lazarev and Rodionova 2011).

Pluto forms a double planet with its largest moon, Charon. They are tidally locked in a synchronous orbit, in other words, they face each other continuously. No other planet (or known dwarf planet) in the Solar System is tidally locked to its moon. The prime meridian of Pluto is crossing the sub-Charon point. The definition of Pluto's coordinate system is especially complicated (Zangari 2015).

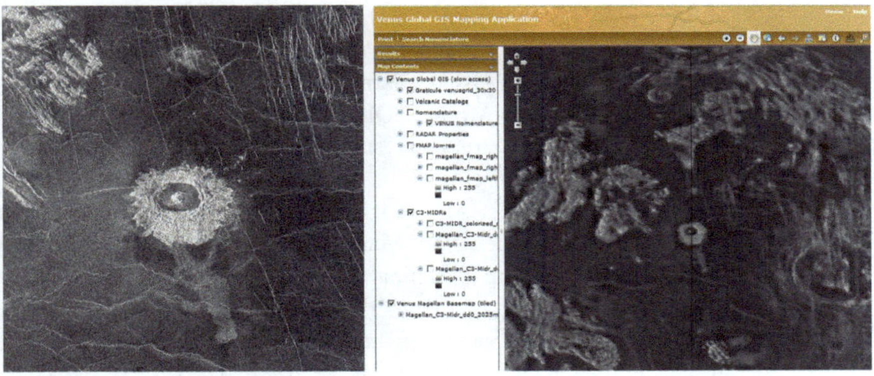

Fig. 7.4 *Left* Ariadne crater in Venus Magellan Synthetic aperture radar (SAR) FMAP left-look (full-resolution left-looking map) image mosaic created by the USGS Astrogeology Research Program (from Map-a-Planet of NASA's PDS Imaging Node). *Right* A broader geological context is shown in the display of the Venus Global GIS Mapping Application of UGSG in Equidistant Cylindrical projection

7.2.2.2 Latitude Systems

The North Pole is the pole of rotation that lies on the north side of the invariable plane of the Solar System which is close to but not the same as the ecliptic (JPL 2009). The direction of the North Pole at a given epoch is specified by the value of its right ascension α_0 and declination δ_0 (Archinal et al. 2011). North latitudes are designated as positive.

7.2.3 Topographic Reference Surfaces

Traditionally a spherical or ellipsoidal shape is used for mapping. Earth (geoid) and Mars (areoid) have datums that are rotational ellipsoids for which the radius at the equator is larger than the polar semi-axis (Archinal et al. 2011). The Mars reference body was redefined several times by the improved data on the planet's shape (Davies et al. 1992, 1995). The current version is defined by IAU as the "Mars IAU 2000" ellipsoid (Seidelmann et al. 2002). On Mars, the zero elevation level is defined by the mean planetary radius. The zero elevation of previous models of the Mars topography (Wu 1991), coincides with the average atmospheric pressure (6.1 mbar) measured by Mariner 9 probe (Zuber and Smith 1998; Kliore et al. 1972). MOLA topography is believed to be a significant improvement over previous models and was related to the 6.1 mbar atmospheric pressure surface of Mars by Zuber and Smith (1998).

Standard models of the reference surface are digital terrain models (DTMs). A DTM defines body radius or geometric height above the body reference surface as a function of cartographic latitude and longitude (JPL 2009). Laser altimetry based accurate topographic data is available for Mars (MOLA), Mercury (MLA) (Neumann et al. 2011) and Moon (LOLA: Lunar Orbiter Laser Altimeter) (Smith et al. 2010), while radar altimeters surveyed the topography of Venus (Muhleman 1961; Campbell et al. 1976).

Horizontal and vertical control bases for Mars topographic mapping are provided by the 231 m/pixel MDIMs generated from Mariner 9 and Viking Orbiter imagery and 500 m/pixel MGS MOLA grid (Archinal et al. 2004; Rosiek et al. 2005). The MOLA ground point spacing is about 300 m along track and 1/64 degree across track. This causes a wide spacing in the equatorial zone, which results in interpolation artifacts in the data gaps (Gwinner et al. 2010). The MOLA radii have accuracy $\sim$10 m vertically and $\sim$100 m horizontally (Neumann et al. 2001). A Martian areoid is generated from the original MOLA points and can be downloaded from the PDS Geosciences Node at a resolution of 32, 64 and 128 pixel/degree. MOLA data are considered the best Mars global control up to date. The MOLA global topographic map was generated by the MOLA science team as shown in Fig. 7.5.

In addition to the MOLA-based DTMs, other high resolution DTMs produced from the Mars Express High-Resolution Stereo Camera (HRSC) stereo images and

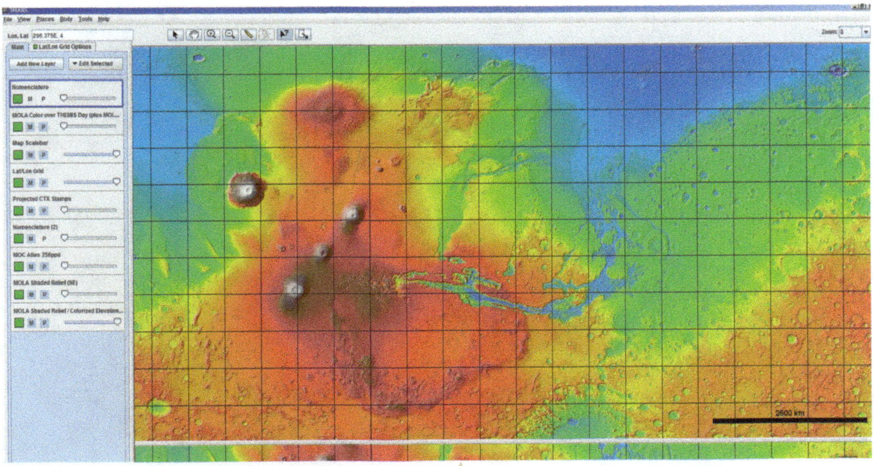

Fig. 7.5 Mars global surveyor MOLA global topographic map displayed in JMARS (MOLA Science Team) (Smith et al. 1999; Christensen et al. 2009)

the Mars Reconnaissance Orbiter HiRISE stereos (McEwen et al. 2007) are widely used. The HiRISE stereo pair is triangulated and controlled to MOLA elevation values, when possible (Eliason et al. 2006). HiRISE DTMs have a spatial resolution around 1–2 m. HRSC DTMs have a spatial resolution around 200 m, depending on the version and algorithm of the produced dataset. The HRSC DTM has a more accurate representation of areas with strong relief, whereas the MOLA DTM is often more reliable on level and very smooth terrain as well as on ice surfaces, because stereo image correlation may fail completely without visible albedo or morphologic features (Gwinner et al. 2010).

7.2.4 Projections

For small scale (global) maps Batson (1990) recommends the Lambert Azimuthal Equal-Area projection that retains the area of features, regardless of location, but it distorts the shapes and craters appear as ellipses at the edge of the map. This phenomenon of apparent foreshortening is well known in orthographic projections or by direct observations of the Moon. This effect gives the user a spherical feeling of the map. This projection is used in two-hemisphere views of the three series of the "Multilingual Maps of the Terrestrial Planets and Their Moons" (Dresden series: Shingareva et al. (2005), Budapest series: Hargitai and Bérczi (2006); Children's map series: Hargitai et al. (2015), Fig. 7.6—all three series are supported by ICA Commission on Planetary Cartography).

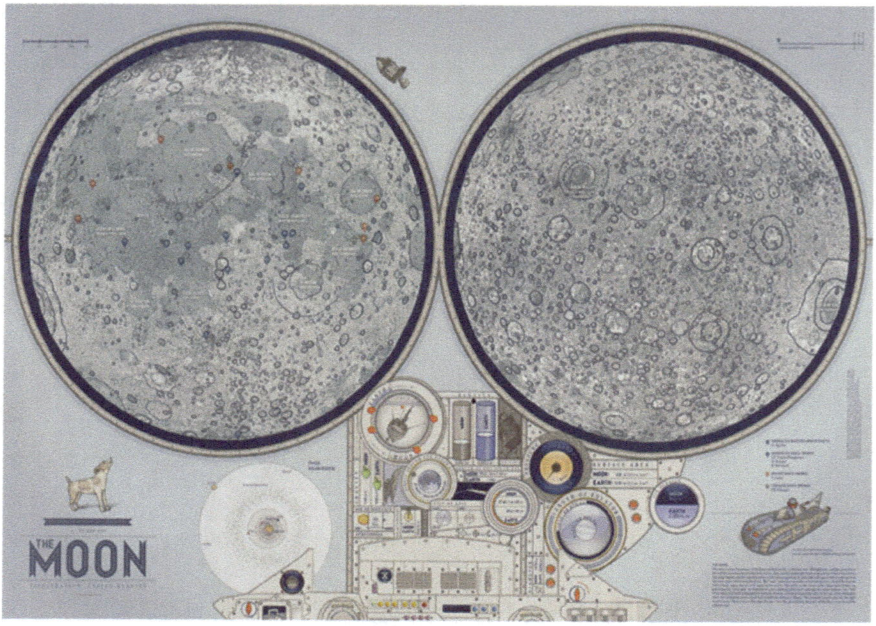

Fig. 7.6 Map of the Moon, designed by L. Herbszt (Hargitai et al. 2015). Lambert Azimuthal Equal-Area projection

Global small-scale maps, especially thematic maps, typically use the Mercator (e.g., Wilson 1912) or Equirectangular projection. Polar views may be shown in Azimuthal Equidistant or Polar Stereographic projection (Fig. 7.7).

The *Orthographic projection* provides a visually attractive global view, but the entire surface of a body can only be displayed in three views, and limb portions are heavily distorted. For global figures, the *Robinson projection* is increasingly popular. A recent example is the new global Geologic map of Mars (Tanaka et al. 2014).

For systematic mapping, the planet may be divided into projection zones: the Mercator projection is used in equatorial latitudes (e.g., 0–30°); the Lambert Conformal Conic in mid-latitudes (e.g., 30–65°) and the Polar Stereographic in polar latitudes (Batson 1990) (e.g., 65–90°) (examples of latitudinal ranges are from MOLA 1:5 M atlas).

Digital, interactive applications typically display maps in a simple cylindrical (equirectangular) "database-" or on-the-fly projections, or the sinusoidal projection (an equal-area projection, used for tiled data products), equally sampled in either planetographic or planetocentric latitudes. In these interactive digital GIS applications the users may modify or re-center the projection according to their needs.

The NASA's PDS Imaging Node has developed an online tool called "Map-a-Planet" to extract science-ready, map-projected images from global mosaics (Akins et al. 2014). This tool can display user-defined portions of 16-planetary

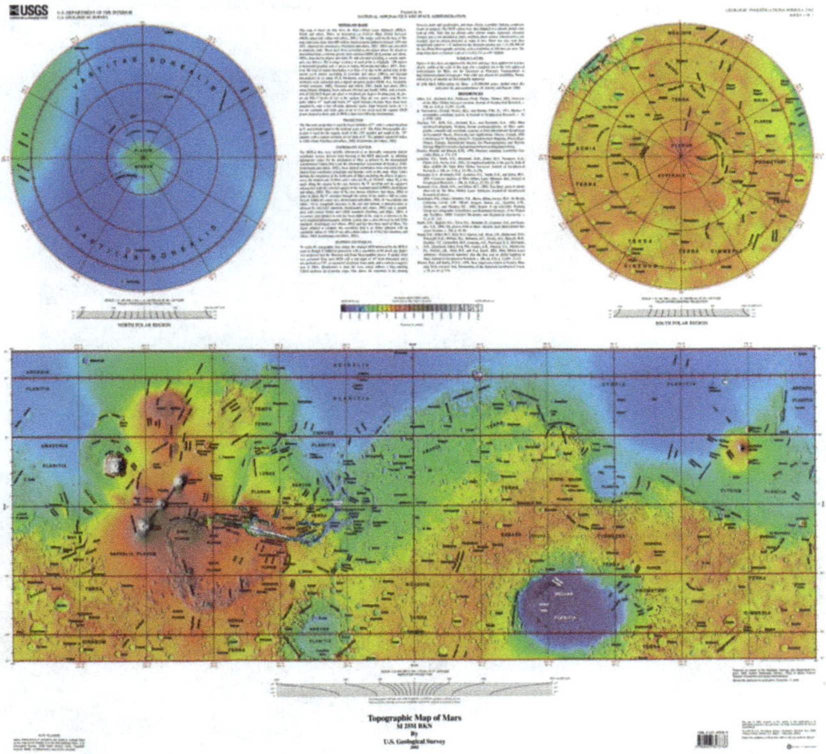

Fig. 7.7 MGS MOLA topographic map in print. Mercator projection 0 ± 57°, polar stereographic ±57–90° (USGS 2003). Cf. the digital version in Figs. 7.5 and 7.9

bodies in several thematic layers or image bands. Map outputs can be selected from four projections. In the Sinusoidal Equal-area projection Map-a-Planet always uses a central meridian at the center of the image map, in order to minimize map distortion. In the Simple Cylindrical projection features on a planet, such as round impact craters, become flattened at the higher latitudes. In the Mercator projection polar regions are extremely distorted but angles and shapes within any small area are essentially true. The Polar stereographic projection can be pre-ordered for a later download. Examples of different projections of a cratered surface, the Schrödinger Basin (75.0°S 132.4°E) region on the Moon are show in Fig. 7.8 (planets have similar viewing options). Map-a-planet 2 (https://astrogeology.usgs.gov/tools/map-a-planet-2) has been in operation since 2015 in order to replace the old version of Map-a-planet (Akins et al. 2014).

JMARS is a cross-platform GIS application developed by the Mars Space Flight Facility (Christensen et al. 2009). When started, data is displayed in Equirectangular Projection centered and projected at 0°E, 0°N. Users can re-project the maps with the center of the viewing window as the new center-point (Fig. 7.9).

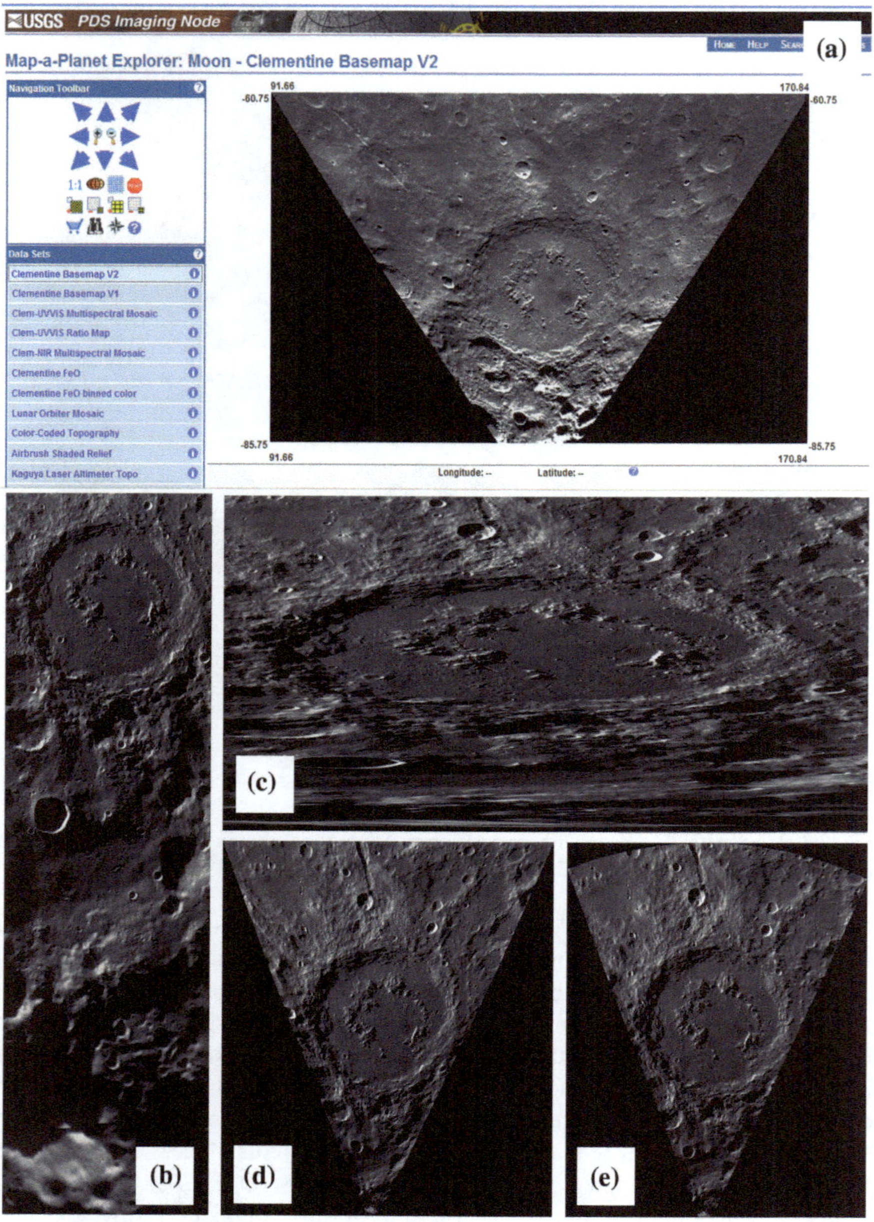

Fig. 7.8 **a** Map-a-planet: Schrödinger Basin in Clementine Basemap in sinusoidal equal-area projection (NASA's PDS Imaging Node, captured in 2013.). **b–d** The same crater in Lunar Reconnaissance Orbiter Camera Wide Angle Camera mosaic, produced in Map a Planet 2, 2017. **b** Mercator Projection, **c** Simple Cylindrical Projection, **d** Sinusoidal Equal-Area Projection, **e** Polar Stereographic Projection

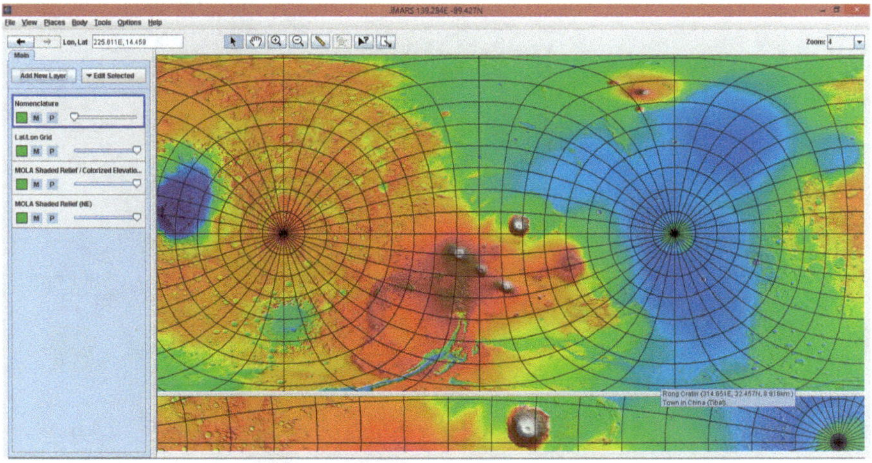

Fig. 7.9 Reprojected image of that in Fig. 7.5 in JMARS, re-centered at the southern pole

Finally, Esri's ArcGIS Online platform offers planetary maps and map layers (global background maps, quad chart gridlines etc.) from several Web Map Servers maintained by universities or research institutions. These can be directly imported into ArcMap in any projection.

Printed (or static, pdf) maps, however, traditionally use *conformal projections* (Mercator, Transversal Mercator, Lambert Conformal Conic, Polar Stereographic) which have no distortion of the shape and are independent of the actual—planetocentric or planetographic—coordinate system (Duxbury et al. 2002). Conformal projections are favored by astrogeologists because shape information is essential in interpreting surface features by photogeologic methods (Batson 1990). One particularly important landform in this respect is impact craters that are circular in shape. As it is put by Veverka (1985) when speaking of the choose of conformal projections: "most of us like to see round craters round". Distortion-free shape is especially important for precise measurements, like those made in the statistical analysis of crater size-frequency distributions where incorrect diameter measurements due to map distortions may lead to younger or older extracted surface ages. Distortions of diameters and areas within different map projections cause considerable errors during such measurements. In order to address this problem, an ArcGIS software module, called CraterTools, was developed by Kneissl et al. (2011) in which impact crater images are internally projected to a stereographic map projection with the crater's central-point set as the projection center, where the circle is defined without any distortion of its shape. The diameter is then measured using a sinusoidal map projection with a center longitude set to the crater's central-point, which does not show any distortion. *Equal-area projections* are useful in evaluating distribution, density and area of surface features (Batson 1990). The 1:200000 topographic image map series based on Mars Express HRSC imagery used equal-area map projections: sinusoidal projection between 85°N and 85°S,

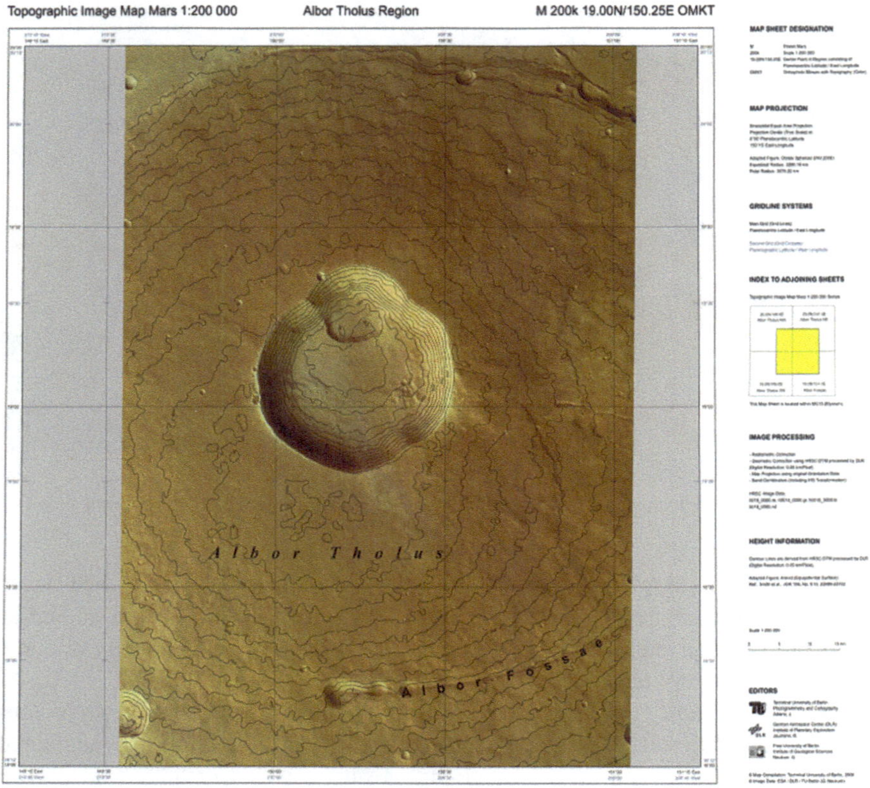

Fig. 7.10 HRSC 1:200000 topographic image map M 200k 19.0N/150.3E OMKT of Albor Tholus, Mars (Alberz et al. 2004a, Sinusoidal equal area projection

and Lambert Azimuthal projection for the polar areas (Alberz et al. 2004a, b) (Fig. 7.10).

Globes: traditional globe map prints usually consist of 12 gores (sometimes 9 on very small globes or 18 on very large globes) completed with two polar caps. The gores are in Cassini projection (transverse form of Simple Cylindrical projection) or in another similar projection such as Polyconic with central longitudes matching the central longitudes of the gores, while the projection of polar caps (usually stretching to the 80° or 70° latitude) is Azimuthal Equidistant (Wagner 1962). A globe map example is shown in Fig. 7.11.

Virtual globes: Google Earth's digitized globe uses Simple Cylindrical (Plate Carrée) projection for its imagery base including all add-on planetary maps. These maps are further reprojected into a General Perspective projection for viewing (Di Palma 2009). The General Perspective projection is similar to the orthographic projection, but its point of perspective is located at a scalable and finite distance. In Google 2D Maps (Earth, Mars and the Moon), the Mercator projection is used from the equator to approximately ±85° (y coordinates of the poles are infinite in the Mercator projection so the mapping limits are set to have the whole map

Fig. 7.11 Räths Erdmondglobus. VEB Räthgloben-Verlag, Leipzig, 1976; Courtesy of the Department of Cartography and Geoinformatics, Eötvös Loránd University

square-shaped). Due to technical reasons, longitudes may run over ±180° when defining features across the 180° longitude. Although the projection can be customized in Google Maps, using a custom projections requires new raster tile sets as the built-in map types are provided only in the Mercator projection. Another limitation is that custom projections are assumed to be rectilinear (i.e. x is a function of the longitude and y is a function of the latitude only). Defining non-rectilinear (e.g. azimuthal) projection is possible but the map interface will behave incorrectly (Google 2012).

VRML/X3D globes of the Virtual Globes Museum (VGM) (Márton 2008) use maps in three different projections to improve visualization and optimize texture sizes: two azimuthal equidistant maps for the polar regions and a Plate Carrée map between the ±50° latitudes (Gede 2009).

Regardless of the target virtual globe applications, globe maps have to be compiled keeping the distortions in mind. As for instance, when stretching a Plate Carrée map to the virtual globe, distances along the latitudes are reduced by cos φ times—above 60° the reduced width is less than half of the original which makes especially symbols and labels very distorted. In order to prevent these errors Hargitai and Gede (2009) recommended compiling maps of the higher latitudes in azimuthal equidistant projection and a Plate Carrée map for the equatorial regions, similarly to maps used in the VGM. Labels beyond the 30° latitude on the Plate Carée map should be horizontally stretched by cos φ and diagonally placed labels should be avoided if possible.

7.3 Gas Giant-Specific Parameters

7.3.1 Longitude System

The rotation of those bodies that have no solid surface (gaseous giant planets) can be either defined by the apparent rotation of the equatorial (System I) and mid-latitude (System II) cloud tops (for Jupiter) or by the rotation of their internal magnetic fields (System III) as determined by their radio signals during fly-by or orbital missions (Newburn and Gulkis 1973)

7.3.2 Size and Shape

The radii and axes of the large gaseous planets are determined at the one-bar-pressure surface. (Archinal et al. 2011).

Stellar occultations are often used to analyze the atmospheric structure of giant planets, and these measurements also provide information on the shape of the planetary body (Lecacheux et al. 1973), while analysis of probes orbiting around giant planets as well as the orbital changes of those probes that pass by them are used to analyze their field of gravity, and the inferred oblateness of their shape (Null 1976). Beside these methods, the shape determinations are completed by theoretical calculations using rotational speed of interior (observed by changes in the radio wavelength) (Anderson 1975) and correlating the oblateness with models of internal structure and hydrodynamic behavior of the fluid interiors (Helled et al. 2010).

7.4 Satellite-Specific Parameters

7.4.1 Longitude System

The axis of rotation of most satellites is normal to the mean orbital plane of the satellite. The rotation rate of most satellites is equal to their mean orbital period (i.e. they have a synchronous rotation due to tidal locking). Since on the tidally locked satellites there is a specific point located at a particular longitude defined by physical parameters, the positioning of the prime meridian on these bodies is not arbitrary. It is determined by the mean sub-planetary point (i.e. the center point of the hemisphere that always faces the parent body). Nevertheless, in most cases there is typically an impact crater at an arbitrary longitude that defines the longitude system, for example the Cilix crater on Europa that defines the 182° meridian (Archinal et al. 2011). On Io there are no impact craters observed yet and its surface is being actively resurfaced and thus modified by volcanism, so the 0° meridian is not fixed to a surface feature, but has been defined using the astronomical definition: the prime meridian of Io is the sub-Jupiter longitude at the first superior conjunction after 1950.0 (Veverka 1985).

7.4.2 Reference Surfaces

Although calculations predict that the hydrostatic shapes of several mid-sized and large satellites are triaxial ellipsoids, spherical reference surfaces are used due to complicated computation of triaxial ellipsoids and lack of agreement on basic definitions like longitude and latitude (Archinal et al. 2011). A spherical reference surface has the advantage that planetographic and planetocentric latitudes are numerically equal (Roatsch et al. 2008).

The shape determination series of satellite images are also used, especially from such missions where the probes orbited around a giant planet: Galileo around Jupiter (Davies et al. 1997) and Cassini around Saturn (Davies and Katayama 1984). In the cases of small or poorly imaged/visited satellites basic parameters like radius might contain substantial uncertainty.

Laser altimetry based accurate LOLA topographic data are available for the Moon (Smith et al. 2010), while radar altimeters surveyed the topography of Titan locally, by Cassini Radar (Radebaugh et al. 2007).

For most of the Solar System bodies no accurate topographic data are available, and shapes are determined using images of limb shapes (Thomas 1989), terminator positions, with methods of shape from shading (Lohse et al. 2006) and stereo image pair analysis. Numerical shape models are generated using these data (Simonelli et al. 1993).

7.4.3 Projections

Kaguya terrain camera images of the Moon are map-projected similarly, in simple cylindrical projection. Global mosaic maps of the Saturnian satellites from the Cassini mission are prepared in simple cylindrical projection. These serve both scientific interpretation and future mission planning (Gehrke et al. 2006).

7.5 Small/Irregular Body Specific Parameters

7.5.1 Rotation

The North Pole may change into South Pole for some of the comets with very large precession over a few decades (Archinal et al. 2011). In the case of some asteroids, rotation around several (usually two) axes might be present (Harris 1994).

7.5.2 Coordinates

In some cases, two coordinates may not be enough to uniquely identify a surface point, because a line from the center of the object may intersect the surface more than once. Examples are Eros, Kleopatra, Itokawa and possibly Toutatis and Ida (Archinal et al. 2011).

7.5.2.1 Longitude

For irregular bodies, the prime meridian may be chosen to align with its longest axis (e.g., 433 Eros). In the case of irregular bodies, it is desirable to relate each of its coordinate axes to a prominent landmark feature, rather than a single feature that defines a prime meridian, because their rotational axes and rates may be poorly defined or may vary in time. Currently, the 0° meridian is defined in a variety of ways for small bodies, including a bright albedo feature (Ceres); the direction of the long axis (Pallas); an arbitrary selected point based on light curve information (Lutetia); an arbitrary selected date on which the direction of the long axis pointed toward the Earth (Davida); the mean sub-body meridian (Pluto, Charon); impact craters (Eros, Ida) and other circular features. It is recommended that small bodies have an east-positive system that run 0–360° (Archinal et al. 2011).

7.5.3 Reference Surfaces

For convenience, sometimes spherical reference surfaces are used for mapping purposes (Archinal et al. 2011). In a few cases shapes could be determined by radar observation when the given asteroid has close flyby the Earth (Ostro et al. 1999). In such cases the coverage is poor if the rotation of the body is slow and the resolution is inhomogeneous because its distance from the Earth changed during the observation. The most detailed shape models for small and irregular bodies are available for Phobos (Willner et al. 2010) and Eros (Thomas et al. 2002) that have been imaged from close orbits.

7.5.4 Projections

Orthographic projections are convenient (Archinal et al. 2011) but they are redundant because six views are needed for a good surface coverage (Stooke 1992). The earliest maps of irregular bodies were sketches on unmodified cylindrical and azimuthal projections on a sphere (e.g., the map of Phobos in 1974 and Amalthea in 1981), containing large distortions. Modified cylindrical projections were developed by John Snyder and Lev Bugaevsky (Bugaevsky 1987). Unlike Snyder, Bugaevsky used planetocentric coordinate system that was implemented for Soviet maps of Phobos during planning research experiments of Phobos-1,-2 missions (Bugaevsky et al. 1992) and later it was used in the Atlas of Terrestrial Planets (Marov et al. 1992). The modified Bugaevskiy projection represents a real shape of the irregular celestial body (the parallels are sinusoidal) and is based on analytical method of calculations (Bugaevsky 1999; Fleis 2004). The modified Bugaevskiy projection is also used for the most recent mapping of Phobos (Fig. 7.12), based on Mars Express SRC (Super-Resolution Channel) orthoimages (Karachevtseva et al. 2015) and accessible online (Nyrtsov et al. 2012; http://geocnt.geonet.ru/en/3_axial).

Stooke (1992) developed a so-called morphographic azimuthal projection in which the radius constant of the spherical projection is replaced with a local radius. The coordinate grid is modified to follow topographic features, being pushed out from the centre of the map by larger radii and pulled towards the centre by smaller radii, enhancing global shape visualization and allowing the outer boundary of a map of a hemisphere to duplicate the shape of the cross-section of an object. Traditional cylindrical and conic projections designed for use with spherical objects were also adapted to irregular objects by Nyrtsov and Stooke (2002). An equal-area mapping technique is described by Berthoud (2005). The equal area projection retains the true relationship of areas and is useful in estimating the surface

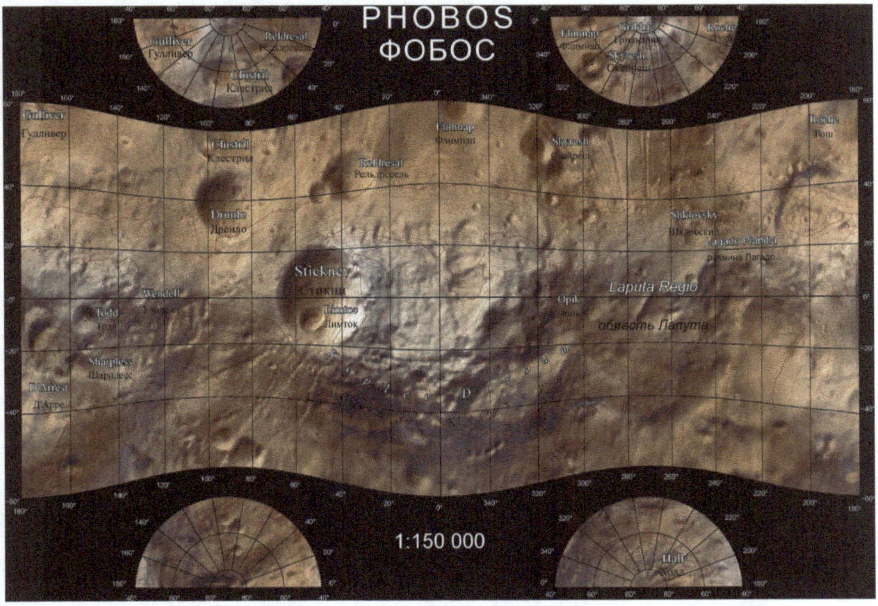

Fig. 7.12 Topographic map from new Phobos atlas (MIIGAiK, 2015), based on new Mars Express SRC DTM (Karachevtseva et al. 2015): equatorial part: modified Bugaevsky (normal conformal cylindrical) projection for three-axial ellipsoid. Polar areas: Azimuthal equidistant along meridian projection for three-axial ellipsoid

distribution, density and area of geologic features and regions (Berthoud 2005). Clark et al. (2008) described the Constant Scale Natural Boundary (CSNB) map projection used for small, irregular body mapping. This projection allows relationships between noses, saddles, and poles to be observed without areal distortion.

7.6 Data Archive and Processing

Individual images sent back from space missions are stored in NASA's PDS in unprocessed (Experiment Data Record—EDR) raw data format. Some mission teams provide map-projected images, but many other images had to be processed by the individual researcher or mapper (Hare et al. 2014). The raw PDS images can be processed and map-projected to be cartography- and science-ready. Where available, images may be projected onto a DTM, and not a perfect sphere (Fig. 7.13).

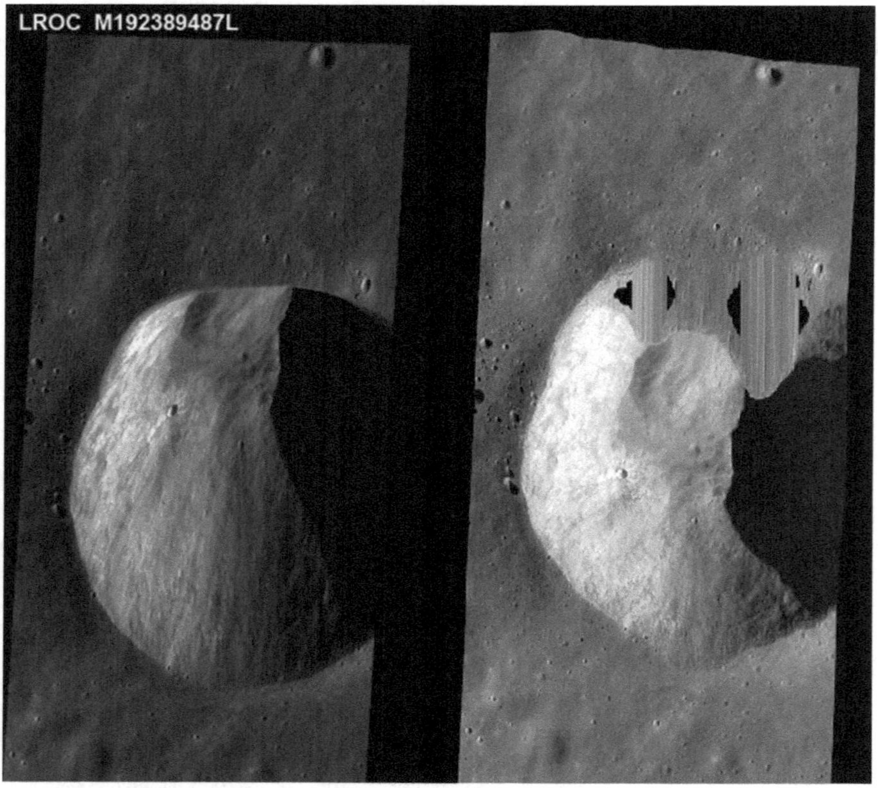

Fig. 7.13 The view of a lunar crater in an unprocessed image (*left*) and a geometrically and radiometrically corrected image (*right*) that is projected onto a Lunar elevation model. Note the image fill of the crater wall that was not visible due to perspective viewing geometry

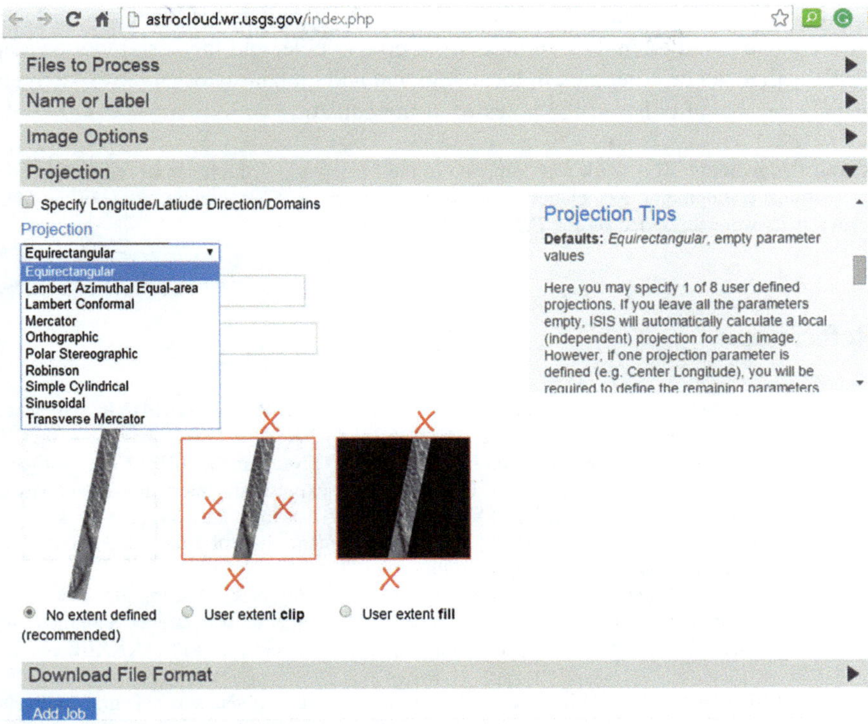

Fig. 7.14 Job submission screen of POW, showing projection options and tips. (Captured 8/29/2015)

The most common image processing tool is ISIS (Integrated Software for Imagers and Spectrometers) developed by the Astrogeology Program of the United States Geological Survey for NASA. ISIS can be used for the radiometric and geometric/cartographic processing of the PDS raw images (Anderson et al. 2002). The projection modules of ISIS3 are used in the considerably more user-friendly Map Projection on the Web Service (POW) (Hare et al. 2013, 2014). POW supports ten different output projections (Fig. 7.14). Calibrated and projected images are ready to be imported to any GIS application where the display- or export-projection can be selected or defined by the user. Geospatial Data Abstraction Library (GDAL) can also be used for conversion from ISIS format to other, common image formats.

7.7 Conclusions

Controlled photomosaics and data-derived thematic planetary maps are a primary tool of mission planning and discovery for the professional planetary scientists (Christensen et al. 2009). Geological planetary maps interpret these data and serve as a basis for subsequent studies. Features and characteristics of planetary surfaces are frequently studied in their fullness from pole to pole, and this task requires map displays that can be compared in local, regional and global scales and irrespectively of location, being either at the poles or at the equator.

Acknowledgements This work was compiled in the framework of the terms of reference of the International Cartographic Association Commission on Planetary Cartography. Some of the Mars related parts were supported by the COST TD1308 project.

References

Akins SW, Hare TM, Sucharski RM, Shute J, Gaddis L, Gaither T, Richie J (2014) Map-a-planet 2 mosaic projection web service. In: 45th LPSC, abstract #2047

Alberz J, Gehrke S, Wählisch M, Lehmann H, Schumacher T, Neukum G, HRSC Co-Investigator Team (2004a) Digital cartography with HRSC on Mars express. In: International archives of photogrammetry and remote sensing (IAPRS), vol XXV, Istanbul, Part B4, pp 869–874

Alberz J, Jaumann R, Neukum G (2004b) Topographic image map of Mars 1:200.000 Albor Tholus region. Technical University of Berlin

Anderson JD (1975) Planetary geodesy. Rev Geophys Space Phys 13:274, 275, 292, 293

Anderson JA, Cook DA, Thompson KT (2002) Rapid geometric software development for scientific analysis and cartographic processing of planetary images. In: LPSC XXXIII, abstract #1853

Antoniadi EM (1930) La Planète Mars: étude basée sur les resultats obtenus avec la grande lunette de l'observatoire de Meudon et expose analytique de l'ensemble des travaux éxécutés sur cet astre depuis 1659. Libraire scientifique Hermann et cie, Paris

Archinal BA, Lee EM, Kirk RL, Duxbury TC, Sucharski RM, Cook DA, Barrett JM (2004) A new mars digital image model (MDIM 2.1) control network. In: XXth ISPRS Congress, Istanbul, Turkey, July 2004

Archinal BA, A'Hearn MF, Bowell E, Conrad A, Consolmagno G, Courtin R, Fukushima T, Hestroffer D, Hilton J, Krasinsky G, Neumann G, Oberst J, Seidelmann P, Stooke P, Tholen D, Thomas P, Williams I (2011) Report of the IAU working group on cartographic coordinates and rotational elements: 2009. Celest Mech Dyn Astr 109:101–135

Batson RM (1990) Cartography. In: Greeley R, Batson RM (eds) Planetary mapping. Cambridge University Press, Cambridge

Berthoud MG (2005) An equal-area map projection for irregular objects. Icarus 175(2):382–389

Bugaevsky LM (1987) On the problem of elaborating of isometric coordinates and equiangular cylinclric projection of the tri-axial ellipsoid. Izvestiya VUZov, ser Geod i Aerofot 4:79–90 (in Russian)

Bugaevsky LM (1999) The theory of map projections of the regular surfaces. M.: Zlatoust 144 (in Russian)

Bugaevsky LM, Krasnopevtseva BV, Shingareva KB (1992) Phobos map and phobos globe. Adv Space Res 12(9):17–21. doi:10.1016/0273-1177(92)90314-N

Burba GA (1996) Cartographic aspects of Venus global geologic mapping at 1:10,000,000 scale. Vernadskiy-Brown Micro 24 abstracts, Moscow, 11

Campbell DB, Dyce RB, Pettengill GH (1976) New radar image of Venus. Science 193:1123–1124

Carpenter RL (1966) Study of Venus by CW radar—1964 results. Astron J 71(2):142–152

Christensen PR, Engle E, Anwar S, Dickenshied S, Noss D, Gorelick N, Weiss-Malik M (2009) JMARS—a planetary GIS. American Geophysical Union, Fall Meeting 2009, abstract #IN22A-06. http://adsabs.harvard.edu/abs/2009AGUFMIN22A.06C

Clark PE, Clark CS, Stooke P (2008) Using boundary-based mapping projections for morphological classification of small bodies. In: LPSC XXXIX abstract #1371
Davies ME, Katayama FY (1984) The control network of Iapetus. Icarus 59:199–204
Davies ME, Rogers PG (1991) The preliminary geodetic control network of Venus. N-3437-JPL. A Rand Note
Davies ME, Abalakin VK, Brahic A, Bursa M, Chovitz BH, Lieske JH, Seidelmann PK, Sinclair AT, Tjuflin YS (1992) Report of the IAU/IAG/COSPAR working group on cartographic coordinates and rotational elements of the planets and satellites: 1991. Celest Mech Dyn Astron 53(4):377–397. doi:10.1007/BF00051818
Davies ME, Abalakin VK, Bursa M, Lieske JH, Morando B, Morrison D, Seidelman PK, Sinclair AT, Yallop B, Tjuflin YS (1995) Report of the IAU/IAG/COSPAR working group on cartographic coordinates and rotational elements of the planets and satellites: 1994. Celest Mech Dyn Astron 63(2):127–148. doi:10.1007/BF00693410
Davies ME, Colvin TR, Thomas P, Veverka J, Belton MJS, Oberst J, Zeitler W, Neukum G (1997) Control networks of the Galilean satellites: solutions for size and shape. In: LPSC XXVIII, abstract #1097
de Vaucouleurs G, Davies ME, Sturms FM Jr (1973) Mariner 9 areographic coordinate system. J Geophys Res 78:4395–4404
Di Palma V (2009) Zoom: google earth and global intimacy. In: Di Palma V, Periton D, Lathouri M (eds) Intimate metropolis: urban subjects in the Modern City. Routledge, pp 239–270
Duxbury TC, Kirk RL, Archinal BA, Neumann GA (2002) Mars geodesy/cartography working group recommendations on Mars cartographic constants and coordinate systems, ISPRS, vol 34, part 4. Geospatial Theory, Processing and Applications, Ottawa
Eliason E et al (2006) Software interface specification for HiRISE reduced data record products, mars reconnaissance orbiter, JPL Document Number D-32006, MRO-M-HIRISE-3-RDR-V1.0, NASA Planetary Data System
Flammarion C (1890) Mappemonde Géographique de la Planete Mars. In: Astronomie Populaire. Paris p 484
Flammarion C (1888 [1978]) Les fleuves de la planète Mars. Astronomie 457–462 (Reprinted in L'Astronomie 92, p 49)
Fleis ME (2004) Non-spherical coordinate systems transformations of the celestial bodies. Institute of Geography of the Russian Academy of Sciences, Russia (in Russian)
Gede M (2009) The projection aspects of digitising globes. ICC 2009, Chile, Santiago
Gehrke S, Wählisch M, Lehmann H, Schumacher T, Albertz J (2003) Cartography with HRSC on mars express—the new series "topographic image map mars 1:200,000". Publikationen der DGPF 12:451–458
Gehrke S, Wählisch M, Lehmann H, Albertz J, Roatsch T (2006) Generation of digital topographic maps of planetary bodies. In: International archives of photogrammetry and remote sensing (IAPRS), vol XXVI, Goa, Part B4
Google 2012 Map Types. https://developers.google.com/maps/documentation/javascript/maptypes#MapCoordinates
Greeley R, Batson RM (eds) (1990) Planetary mapping. Cambridge planetary science series, vol 6. Cambridge, New York, xiii+296 pp
Gwinner K, Scholten F, Preusker F, Elgne S, Roatsch T, Spiegel M, Schmidt R, Oberst J, Jaumann R, Heipke C (2010) Topography of Mars from global mapping by HRSC high-resolution digital terrain models and orthoimages: characteristics and performance. Earth Planet Sci Lett 294(3–4):506–519
Hare TM, Akins SW, Sucharski RM, Bailen MS, Anderson JA (2013) Map projection web service for PDS images. In: 44th LPSC, abstract #2068
Hare TM, Akins SW, Sucharski RM, Bailen MS, Shute J, Anderson JA, Gaddis LR (2014) POW: update for the PDS map projection web service. In: 45th LPSC, abstract #2474

Hargitai H, Bérczi S (2006) Multilingual maps of the terrestrial planets and their moons: the east and central european edition. European Planetary Science Congress Berlin, Germany, 18–22 Sept, pp 515

Hargitai H, Gede M (2009) Multilingual virtual globes of Venus and Mars. ICC 2009, Chile, Santiago

Hargitai H, Gede M, Zimbelman J, Kőszeghy C, Sirály D, Marinangeli L, Barata T, López I, Szakács A, Dębniak K, Feuillet T (2015) Multilingual narrative planetary maps for children. In: Sluter R, Madureira Cruz C, Bernadete C, de Menezes L, Márcio P (eds) Cartography—maps connecting the world. In: 27th international cartographic conference 2015. Springer, 17–30

Harris AW (1994) Tumbling asteroids. Icarus 107:209

Helled R, Anderson JD, Schubert G (2010) Uranus and Neptune: shape and rotation. Icarus 210:446–454

Hevelius (Johann Höwelcke) (1647) Selenographia sive lunae descriptio. Typis Hunefelianis, Gedani

Hirsch D (1994) Atlas of Venus I-2523 sheet 3 of 4. NASA-USGS

IAU (1971) Proceedings of the fourteenth general assembly (de Jagger C, Jappel A (ed) D. Reidel), Dordrecht

Inge JL, Batson RM (1992) Indexes of maps of the planets and satellites, NASA Technical Memorandum 4395

JPL (2009) Cartographic standards. In: Planetary data system standards reference. Version 3.8. JPL D—7669, part 2, Jet Propulsion Laboratory, California Institute of Technology, Pasadena, California

Karachevtseva IP, Kokhanov AA, Rodionova JR, Konopikhin AA, Zubarev AE, Nadezhdina IE, Mitrokhina LA, Patratiy VD, Oberst J (2015) Development of a new Phobos atlas based on Mars Express image data. Planetary and Space Science, Vol 108. pp. 24–30 doi: 10.1016/j.pss. 2014.11.024

Kliore AJ, Cain DL, Fjeldbo G, Seidel BL, Sykes MJ, Rasool SI (1972) The atmosphere of Mars from mariner 9 radio occultation measurements. Icarus 17:484–516

Kneissl T, van Gasselt S, Neukum G (2011) Map-projection-independent crater size-frequency determination in GIS environments—new software tool for ArcGIS. Planet Space Sci 59: 1243–1254

Lazarev E, Rodionova J (2011) Venus mapping at small scale: source data processing and cartographic interpretation. In: Ruas A (ed) Proceedings of the 25th international cartographic conference. CO-160. ISBN: 978-1-907075-05-6

Lecacheux J, Combes M, Vapillon L (1973) The FI Scorpii occultation by Jupiter. I. The Jovian Diameter. Astron Astrophys 22:289

LGCWG: Lunar Geodesy and Cartography Working Group (2008) Recommendations for formatting large lunar datasets, May 18, draft

Lohse V, Heipke C, Kirk RL (2006) Derivation of planetary topography using multi-image shape-from-shading. Planet Space Sci 54(7):661–674

Márton M (ed) (2008) Virtual globes museum. http://vgm.elte.hu

Marov MY, Avduyevskiy VS, Borodin NF, Kerzhanovich VV, Lysov VP, Moshkin BY, Rozhdestvenskiy MK, Ryabov OL, Ekonoraov AP (1973) Preliminary results of measurements of the Venera 8 automatic spacecraft. NASA Technical Translation NASA TT F-14, 909

Marov MY, Krasnopevtseva BV, Shingareva KB (eds) (1992) Atlas of the terrestrial planets and their satellites. M. MIIGAiK, 208 p (in Russian)

Mayer T, Schröter JH (1791) Tob Mayeri Tabula Selenographica. In: Schröter JH: Selenotopographische Fragmente zur genauern Kenntniss der Mondfläche, ihrer erlittenen Veränderungen und Atmosphäre, sammt den dazu gehörigen Specialcharten und Zeichnungen. C.G. Fleckeisen, Lilienthal

McEwen AS, Eliason EM, Bergstrom JW, Bridges NT, Delamere WA, Grant JA, Gulick VC, Herkenhoff KE, Keszthelyi L, Kirk RL, Mellon MT, Squyres SW, Thomas N, Weitz CM (2007) MRO's high resolution imaging science experiment (HiRISE). J Geophys Res 112(E5). doi:10.1029/2005JE002605

MSSS: Malin Space Science Systems (2001) The martian prime meridian—longitude "zero" MGS MOC release No. MOC2-273, 31 Jan 2001. http://mars.jpl.nasa.gov/mgs/msss/camera/images/01_31_01_releases/airy0/
Muhleman DO (1961) Early results of the 1961 JPL Venus radar experiment. Astron J 66:292
NASA's PDS Geosciences Node at Department of Earth & Planetary Sciences in Washington University in St. Louis, Available online: http://geo.pds.nasa.gov/default.htm. Accessed 1 Sept 2015
Neumann GA, Rowlands DD, Lemoine FG, Smith DE, Zuber MT (2001) Crossover analysis of MOLA altimetric data. J Geophys Res 106:23723–23735
Neumann GA, Mazarico E, Smith DE, Zuber MT, Torrence MH, Barnouin OS, Solomon SC (2011) Laser altimetry of Mercury, Moon, and Mars. American Geophysical Union, Fall Meeting 2011, abstract #P41F-03
Newburn RL, Gulkis S (1973) A survey of the outer planets Jupiter, Saturn, Uranus, Neptune, Pluto, and their satellites. Space Sci Rev 14(2):179–271
Null GW (1976) Gravity field of Jupiter and its satellite from pioneer 10 and pioneer 11 tracking data. Astron J 81:1153–1161
Nyrtsov MV, Stooke PJ (2002) The mapping of irregularly-shaped bodies at planetary scale. In: Proceedings from international conference InterCarto 8. Helsinki–St. Petersburg, 28 May–1 June, pp 433–436
Nyrtsov MV, Fleis ME, Borisov MM (2012) Mapping of asteroid 433 Eros in cylindrical equidistant along the meridians and azimuthal projections triaxial ellipsoid. Izvestiya VUZov, ser. Geodesy Aerial Photogr 1:54–61 (in Russian)
Ostro SJ, Hudson RS, Rosema KD, Giorgini JD, Jurgens RF, Yeomans DK, Chodas PW, Winkler R, Rose R, Choate D, Cormier RA, Kelley D, Littlefair R, Benner LAM, Thomas ML, Slade MA (1999) Asteroid 4179 Toutatis: 1996 radar observations. Icarus 137:122–139
Proctor RA (1873) Half-hours with the telescope. G.P. Putnam's Sons, New York
Radebaugh J, Lorenz RD, Kirk RL, Lunine JI, Stofan ER, Lopes RMC, Wall SD, Cassini RADAR Team (2007) Mountains on Titan observed by Cassini Radar. Icarus 192(1):77–91
Roatsch T, Wälisch M, Giese B, Hoffmeister A, Matz K-D, Scholten F, Kuhn A, Wagner R, Neukum G, Helfenstein P, Porco C (2008) High-resolution Enceladus atlas derived from Cassini-ISS images. Planet Space Sci 56:109–116
Rosiek MR, Howington-Kraus E, Hare TM, Redding BL (2003) Mars transverse mercator (MTM) map series updated with plan-etocentric grid, lunar and planetary science, XXXIV, Lunar and Planetary Institute, Houston, unpaginated CD-ROM, abstract 1371
Rosiek MR, Kirk RL, Archinal BA, Howington-Kraus E, Hare T, Galuszka D, Redding B (2005) Utility of viking orbiter images and products for Mars mapping. Photogram Eng Remote Sens 71(10):1187–1195
Seidelmann PK, Abalakin VK, Bursa M, Davies ME, de Bergh C, Lieske JH, Oberst J, Simon JL, Standish EM, Stooke P, Thomas PC (2002) Report of the IAU/IAG working group on cartographic coordinates and rotational elements of the planets and satellites: 2000. Celest Mech Dyn Astron 82(1):83–111. doi:10.1023/A:1013939327465
Seidelmann PK, Archinal BA, A'Hearn MF, Conrad A, Consolmagno GJ, Hestroffer D, Hilton JL, Krasinsky GA, Neumann G, Oberst J, Stooke P, Tedesco E, Tholen DJ, Thomas PC, Williams IP (2007) Report of the IAU/IAG working group on cartographic coordinates and rotational elements: 2006, Celestial Mechanics and Dynamical Astronomy, vol 98, pp 155–180 [WGCCRE #10]
Shingareva KB, Zimbelman J, Buchroithner MF, Hargitai HI (2005) The realization of ICA commission projects on planetary cartography, Cartographica, vol 40, no. 4/Winter 2005. doi:10.3138/3660-4078-55X1-3808
Simonelli DP, Thomas PC, Carcich BT, Veverka J (1993) The generation and use of numerical shape models for irregular solar system objects. Icarus 103:49–61
Smith DE, Zuber MT, Solomon SC, Phillips RJ, Head JW, Garvin JB, Banerdt WB, Muhleman DO, Pettengill GH, Neumann GA, Lemoine FG, Abshire JB, Aharonson O,

Brown CD, Hauck SA, Ivanov AB, McGovern PJ, Zwally HJ, Duxbury TC (1999) The global topography of Mars and implications for surface evolution. Science 284:1495–1503
Smith DE, Zuber MT, Neumann GA, Lemoine FG, Mazarico E, Torrence MH, McGarry JF, Rowlands DD, Head JW, Duxbury TH, Aharonson O, Lucey PG, Robinson MS, Barnouin OS, Cavanaugh JF, Sun X, Liiva P, D-d Mao, Smith JC, Bartels AE (2010) Initial observations from the lunar orbiter laser altimeter (LOLA). Geophys Res Lett 37:L18204. doi:10.1029/2010GL043751
Stooke P (1992) Cartography of asteroids and comet nuclei from low resolution data. Asteroids, Comets, Meteors 1991. LPI, Houston, pp 583–586
Tanaka KL, Skinner JA Jr, Dohm JM, Irwin RP III, Kolb EJ, Fortezzo CM, Platz T, Michael GG, Hare TM (2014) Geologic map of Mars: U.S. Geological Survey Scientific Investigations Map 3292, scale 1:20,000,000, pamphlet 43 p. http://dx.doi.org/10.3133/sim3292
Thomas PC (1989) The shapes of small satellites. Icarus 77:248–274
Thomas PC, Joseph J, Carcich B, Veverka J, Clark BE, Bell JF, Byrd AW, Chomko R, Robinson M, Murchie S, Prockter L, Cheng A, Izenberg N, Malin M, Chapman C, McFadden LA, Kirk R, Gaffey M, Lucey PG (2002) Eros: shape, topography, and slope processes. Icarus 155:18–37
USGS (2003) Topographic map of Mars. Geologic Investigations Series i–2782
USGS (n.d.) Arcadia, MC-3. http://planetarynames.wr.usgs.gov/images/mc3_mola.pdf
Veverka J (1985) Planetary geology in the 1980s. NASA Office of Space Science and Applications, Washington DC
Wagner K (1962) Kartographische Netzentwürfe. Bibliografisches Institut Mannheim. pp 155–156
Whitaker EA (2003) Mapping and naming the moon: a history of lunar cartography and nomenclature. Cambridge University Press, pp 264
Wilhelms DE (1987) The geologic history of the Moon. USGS Professional Paper 1348. pp 11
Wilhelms DE (1990) Geologic mapping. In: Greeley R, Batson RM (eds) Planetary mapping. Cambridge University Press, New York, pp 208–260
Willner K, Oberst J, Hussmann H, Giese B, Hoffmann H, Matz KD, Roatsch T, Duxbury T (2010) Phobos control point network, rotation, and shape. Earth Planet Sci Lett 294:541–546
Wilson LJ (1912) A Mercator's projection of Mars. Popular Astron 20:65–65
Wu SSC (1991) Topographic maps of the polar, western, and eastern regions of Mars, Scale 1:15,000,000, U.S. Geological Survey Map I-2160
Zangari A (2015) A meta-analysis of coordinate systems and bibliography of their use on Pluto from Charon's discovery to the present day. Icarus 246:93–145
Zuber MT, Smith DE (1998) The relationship of MOLA northern hemisphere topography to the 6.1 mbar atmospheric pressure surface of Mars. In: 29th LPSC, 16–20 Mar 1998, Houston, TX, abstract #1724

Chapter 8
Combining World Map Projections

Bernhard Jenny and Bojan Šavrič

Abstract Cartographers have developed various techniques for deriving new projections from existing projections. The goal of these techniques is to substitute a disadvantageous trait of one of the source projections with the second source projection. This chapter discusses creating new projections by the juxtaposition and blending of two existing projections. It also presents a new approach for selectively combining projection characteristics. The emphasis in this chapter is on projections for world maps, as the described techniques are most useful for this scale.

8.1 Introduction

There are various techniques for creating new map projections from existing projections. These techniques are applied to substitute a disadvantageous trait of one of the source projections with the second source projection. This chapter discusses juxtaposition and blending, the two most commonly used techniques for combining two existing projections. A recent technique for combining selected projection characteristics is also presented. The techniques are useful for creating new projections for world maps.

B. Jenny (✉)
School of Science, Geospatial Science, RMIT University, Melbourne, Australia
e-mail: bernhard.jenny@rmit.edu.au

B. Jenny · B. Šavrič
College of Earth, Ocean and Atmospheric Sciences, Oregon State University, Corvallis, OR, USA

B. Šavrič
Esri Inc., Redlands, CA, USA

M. Lapaine and E.L. Usery (eds.), *Choosing a Map Projection*,
Lecture Notes in Geoinformation and Cartography,
DOI 10.1007/978-3-319-51835-0_8

8.2 Compositing Projections Along Parallels

The first technique creates a new projection by the juxtaposition of two existing projections. This fusion technique is almost exclusively applied to create pseudocylindrical projections by aligning two pseudocylindrical source projections along two straight parallels. It is generally limited to projections with regularly spaced meridians, otherwise the longitudes would not align along the fusion parallel. Compositing two cylindrical projections with this technique is possible, but used rarely. While compositing azimuthal projections is a theoretical possibility that serves little purpose, the technique cannot generally be applied to conic projections (Fenna 2007).

The Goode homolosine is the best-known example of a combined projection (Goode 1925). It combines the Sanson sinusoidal and the Mollweide projection (Fig. 8.1). The projection is commonly used in interrupted form and is equal-area. The Goode homolosine, like most other combined projections, shows a discontinuity in the graticule where the sinusoidal and Mollweide projections join (Gede 2011). This crease, which may be visually disturbing, is a typical trait of combined projections.

Goode chose to combine the sinusoidal and the Mollweide projections along the parallel at 40° 44′ 12″N and S, since this is the latitude of equal length for both projections. The sinusoidal projection retains the scale along all parallels, and can therefore be combined with a variety of pseudocylindrical projections along the parallel of the second projection where true scale is retained (Fenna 2007, p. 174).

When none of the source projections is the sinusoidal, or when an arbitrary parallel is chosen to combine the two source projections, one of the two projections must be scaled. The scaling brings the parallels along which projections are fused to the same length. For example, Érdi-Krausz connected a transformed sinusoidal projection with the Mollweide projection (Érdi-Krausz 1968). To bring the fusion parallels to the same length, he scaled one of the source projections (Gede 2011).

McBryde (1978) has created combined projections called P3, S2, S3, and Q3. The S3, perhaps the best known of this group, is a fusion along 55° 51′N and S of the McBryde-Thomas flat-polar and the sinusoidal projection.

A composition of two equal-area projections can still be equal-area as in the Goode homolosine. This can be achieved when the x coordinate is multiplied by a

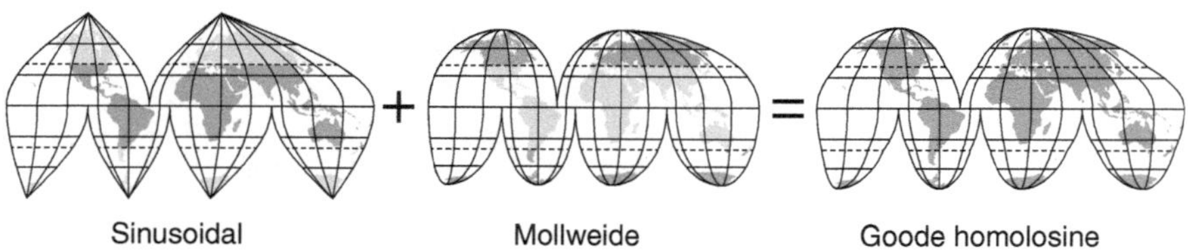

Fig. 8.1 The Goode homolosine combines the sinusoidal and the Mollweide projections. It is commonly used in interrupted form

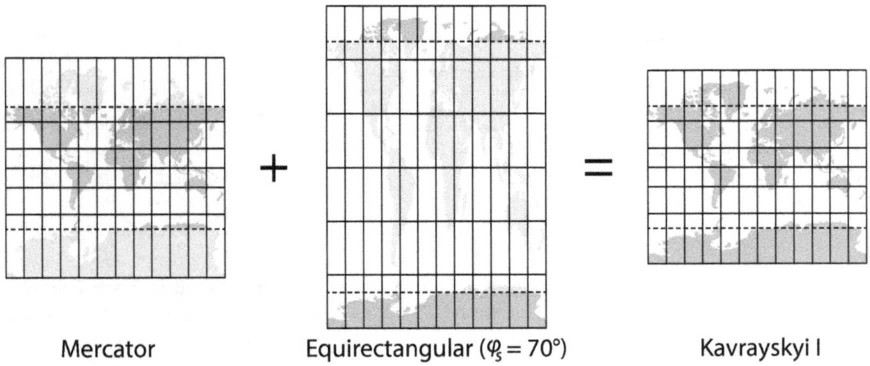

Fig. 8.2 The cylindrical Kavrayskyi I combines the Mercator and the equirectangular projections

scale factor that brings the length of the fusion parallels to the same length, and the y coordinate is divided by this same factor. This results in stretching and compressing one of the two source projections, but this transformation does not alter its equal-area property.

A rare example of a composite of two cylindrical projections is the Kavrayskyi I projection (Fig. 8.2). It combines the Mercator projection with the equirectangular projection along 70°N and S, thereby reducing the gross areal distortion of the Mercator projection. The equation for the x coordinate is $x = \lambda$ for the unary sphere. The vertical coordinate is computed with the equations below

$$y = \begin{cases} \frac{\varphi - \varphi_s}{\cos \varphi_s} + d_y & \text{for} \quad \varphi > \varphi_s \\ \frac{\varphi + \varphi_s}{\cos \varphi_s} - d_y & \text{for} \quad \varphi < -\varphi_s \\ \ln \tan(45^\circ + \frac{\varphi}{2}) & \text{otherwise} \end{cases}$$

where: $d_y = \ln \tan(45^\circ + \frac{\varphi_s}{2})$ and $\varphi_s = 70^\circ$.

8.3 Projection Blending

An alternative approach to combining two projections along parallels is computing the mean of two source projections. The spherical coordinates, longitude and latitude, are first converted with both projections, resulting in two pairs of Cartesian coordinates, x_1, y_1 and x_2, y_2. The blended coordinates are then computed with an arithmetic mean: $x = w \cdot x_1 + (1-w) \cdot x_2$ and $y = w \cdot y_1 + (1-w) \cdot y_2$, with the weight $0 \leq w \leq 1$.

Any pair of projections can theoretically be blended, but a self-intersecting graticule may result if the projection geometries differ considerably. For example,

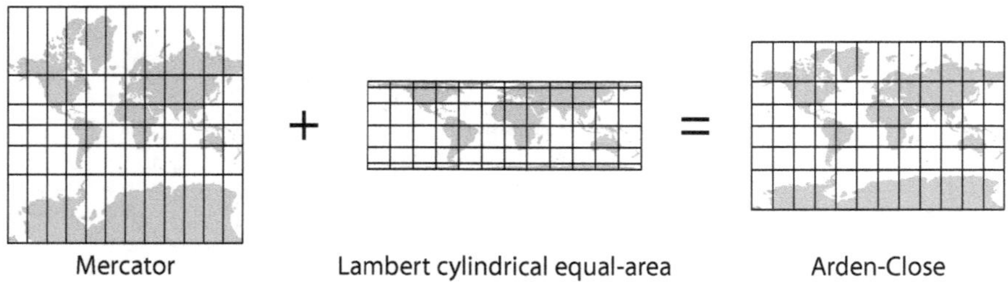

Fig. 8.3 The cylindrical Arden-Close is a blending of the Mercator and the Lambert cylindrical equal-area projection. The Mercator and the Arden-Close projections are truncated at 85°N and S

blending an azimuthal and a pseudocylindrical projection would result in a curvy and folded graticule (Jenny 2012).

The Arden-Close projection (Fig. 8.3) is a cylindrical projection (straight parallels and straight meridians), created by blending the Mercator and the Lambert cylindrical equal-area projections (Arden-Close 1947). It is a rare example of a cylindrical projection created by blending two cylindrical source projections. This combination compensates the gross enlargement of polar areas by the Mercator projection with the vertical compression of the Lambert cylindrical projection. However, since the Mercator projection has an infinite vertical extension, the blended Arden-Close projection also places the poles at infinite distance from the equator, which requires the graticule to be truncated along parallels close to the poles. The projection is not frequently used for this reason.

A series of more useful pseudocylindrical projections have been created in the past by blending two source projections with an arithmetic mean. The majority of these blended projections combine a cylindrical projection, such as the equirectangular projection, and a pseudocylindrical projection with meridians converging at pole points, such as the sinusoidal. For example, the Eckert V projection combines the sinusoidal and the plate carrée projections (Eckert 1906). Winkel generalized this approach by replacing the plate carrée with the equirectangular projection using any standard parallel (Winkel 1921). For his Winkel I projection (Fig. 8.4), he proposed standard parallels at 50° 28′N and S such that the total area of the map is at correct scale (Snyder 1993, p. 195). Putniņš proposed two projections called P1′ and P3′, which are both arithmetic means of the plate carrée and projections created

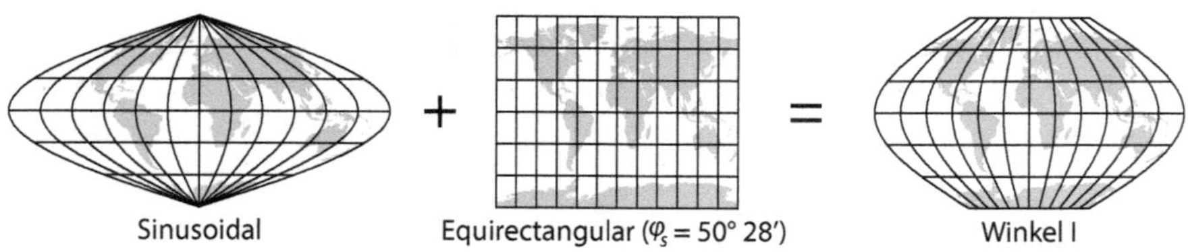

Fig. 8.4 The Winkel I is a blending of the sinusoidal and the equirectangular with standard parallels at 50° 28′N and S

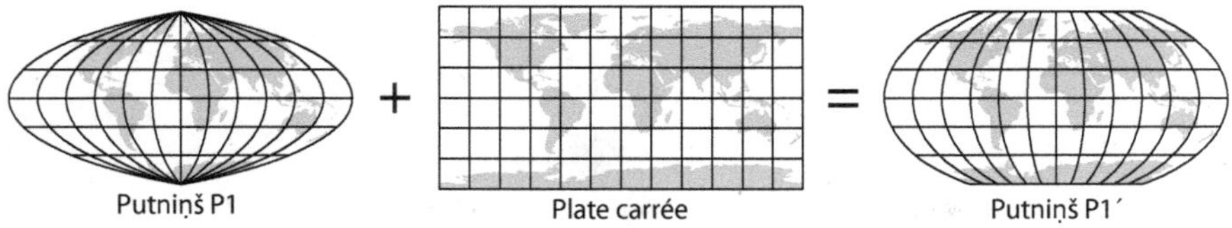

Fig. 8.5 The Putniņš P1′ is a blending of the plate carrée and the Putniņš P1 projections

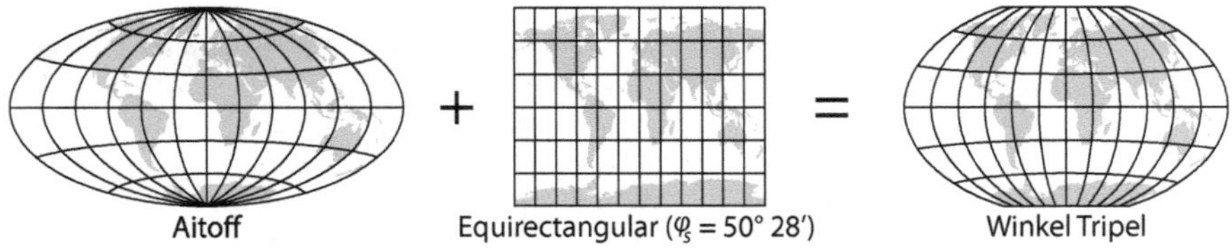

Fig. 8.6 The Winkel Tripel is a blending of the Aitoff and the equirectangular with standard parallels at 50° 28′N and S

by the same author (Putniņš 1934). Figure 8.5 shows the Putniņš P1′ projection (name with apostrophe) with a polar line as a result of the arithmetic mean between the Putniņš P1 (name without apostrophe) that has pointed poles and the cylindrical plate carrée projection.

The Winkel Tripel projection (Fig. 8.6) combines the Aitoff and the cylindrical equirectangular projection with standard parallels at 50° 28′N and S (Winkel 1921). Because the Aitoff projection does not have straight parallels, the Winkel Tripel also has bent parallels and is not a pseudocylindrical projection.

Unlike the compositing technique discussed before, blending two source projections does not create an equal-area projection, even if the two source projection are equal-area. If the equal-area property is to be retained, only the x or the y coordinate can be a blend, and the other coordinate must be mathematically derived from the equal area condition (Tobler 1973). An example of a well known equal-area blend is the Boggs eumorphic projection (Fig. 8.7), which is the arithmetic mean of the sinusoidal and the Mollweide projections only for the y coordinate (Boggs 1929). The equation for the x coordinate is adjusted such that the resulting projection does not distort area. Foucaut (1862), Hammer (1900), and

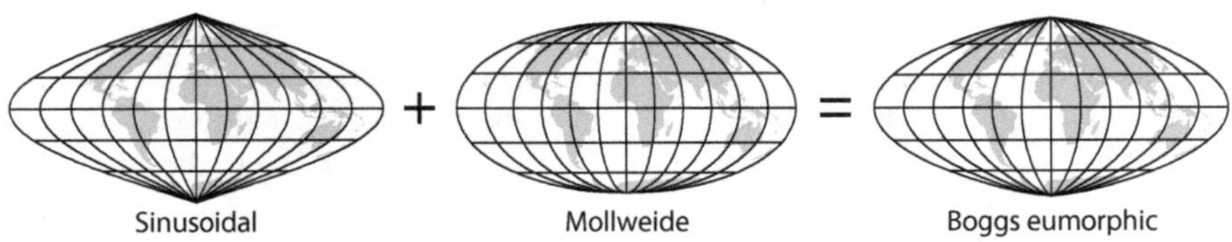

Fig. 8.7 The equal-area Boggs eumorphic projection blends the y coordinate of the sinusoidal and Mollweide projections

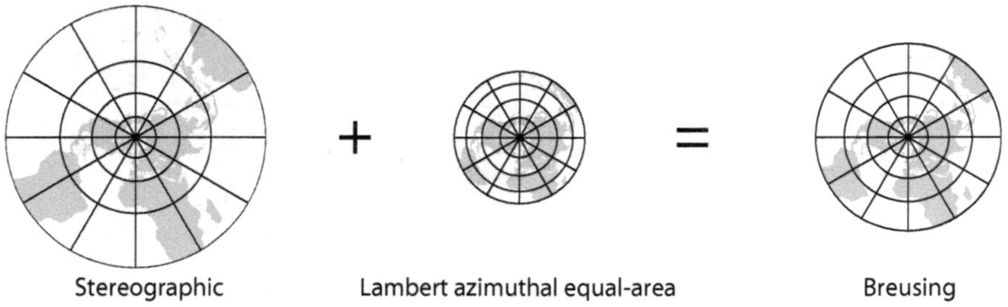

Fig. 8.8 The azimuthal Breusing projection uses a geometric means of the radii of the conformal stereographic and the equal-area azimuthal projections

Tobler (1973) have averaged the cylindrical equal-area and the sinusoidal projections to create other equal-area projections with similar techniques (after Snyder 1977).

The arithmetic blending technique, described above, uses constant weights w_1 and w_2 for the two projections. An extension to this approach is to vary weights with the location. A simple approach is to vary weights with the latitude, as described by Anderson and Tobler (2011): $w_1 = w(\varphi) \geq 0$, and $w_2 = 1 - w_1$, with one of the weights decreasing from one at the equator to zero at the poles.

Computing the geometric mean of two projections is an alternative to the arithmetic means. Tobler (1973) explores this method for creating pseudocylindrical projections. The projection by Breusing is a rare example of an azimuthal projection that blends two source projections involving a geometric mean (Snyder 1993, pp. 129–130). It combines the azimuthal equal-area projection and the conformal stereographic azimuthal projection (Fig. 8.8). The radius from the projection center is computed with a geometric mean of the radii of the two source projections.

8.4 Selective Combinations

When combining two source projections to create a new projection, the goal is to replace an unfavorable trait of one projection with the characteristics of another projection. Jenny and Patterson (2013) introduce an alternative approach to the blending and compositing techniques discussed above that aims at selectively combining the traits of two projections. This software-based technique first extracts the geometric characteristics of the two source projections and builds tables with according values, then lets the user selectively mix these tables, and finally converts the mixed tables back to a new projection. The tables encode (1) the horizontal length of parallels, (2) the vertical distance of parallels from the equator, (3) the distribution of meridians, and (4) the bending of parallels. The tables contain values for every 5° or 10° of increasing longitude and latitude. This selective combination

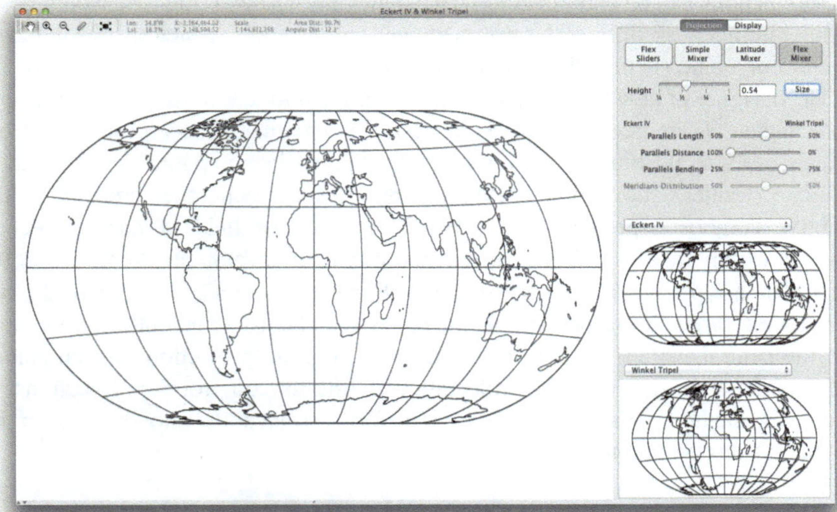

Fig. 8.9 Combination of selected projection traits of the Eckert IV and the Winkel Tripel projections with Flex Projector

technique is available in Flex Projector, a specialized software application for designing new world map projections (Jenny and Patterson 2007; Jenny et al. 2008). The user can adjust the geometry of a single projection, or combine two source projections using the methods described in this chapter.

The user can adjust weights using four sliders (Fig. 8.9, top right). For example, in Fig. 8.9, the Eckert IV and the Winkel Tripel (itself a blend of the Aitoff and the equirectangular) are combined. The length of the resulting parallels is a 50/50 blend of the two source projections. The parallels are vertically distributed as in the Eckert IV projection. Parallels are bent, but not as severly as in the Winkel Tripel projection, due to a 25/75 blend. With a height-to-width aspect set to 0.54, a compact graticule results with distinctly curved meridians near pole lines. The reader is referred to Jenny and Patterson (2013) and Jenny et al. (2010) for additional examples and more details on the algorithmic procedure.

8.5 Conclusion

The techniques for combining two source projections to create a new projection, outlined in this chapter, allow for the creation of a large variety of projections. The mentioned techniques can also be extended. For example, the Geocart software by Mapthematics can blend projection parameters, such as the latitude of standard parallels, between two source projections (Strebe 2010). Alternatively, more than two projections can be combined to form a new one. The extreme case would be an infinite number of differently parameterized projections, which is the concept behind polyconic (Fenna 2007) and polycylindric projections (Tobler 1986).

The designer of new map projections is not limited to the techniques discussed in this chapter. There are alternative methods for creating a new projection from scratch, deriving it from existing ones, or adjusting projection parameters to create a new one (Canters 2002; Snyder 1993). Some of these techniques are used in the adaptive composite projections for Web maps, a new field of map projection research (Jenny 2012). The goal of this research is to develop an alternative to the Web Mercator projection for small-scale Web maps, where maps automatically use an optimum projection depending on the map scale, the map's height-to-width ratio, and the central latitude of the displayed area. Multiple projections are combined with seamless transitions, using projection blending for compromise projections, or Wagner's method (Wagner 1949) to transform projections into each other by adjusting transformation parameters.

Acknowledgements The authors would like to thank Sebastian Hennig, TU Dresden, for his help with preparing the illustrations for this chapter, and the two anonymous reviewers for their valuable comments.

References

Anderson PB, Tobler WR (2011) Blended map projections are splendid projections. http://www.geog.ucsb.edu/~tobler/publications/-pdf_docs/inprog/BlendProj.pdf. Accessed 3 Aug 2011

Arden-Close CF (1947) Some curious map projections. In his: geographical by-ways: and some other geographical essays. Edward Arnold, London, pp 68–88

Boggs S (1929) A new equal-area projection for world maps. Geogr J 73(3):241–245

Canters F (2002) Small-scale map projection design. Taylor & Francis, London

Eckert M (1906) Neue Entwürfe für Weltkarten. Petermanns Mitteilungen 52(5):97–109

Érdi-Krausz G (1968) Combined equal-area projections for world maps. In: Hungarian cartographical studies, pp 44–49

Fenna D (2007) Cartographic science: a compendium of map projections, with derivations. CRC Press, Boca Raton

Foucaut HC de Prépetit (1862) Notice sur la construction de nouvelles mappemondes et de nouveaux atlas de géographie. Arras, France

Gede M (2011) Optimising the distortions of sinusoidal-elliptical composite projections. Ruas A (ed) Advances in cartography and GIScience. Volume 2: selection from ICC 2011, Paris. Lecture notes in geoinformation and cartography, vol 6. Springer, Berlin, pp 209–225. doi:10.1007/978-3-642-19214-2_14

Goode JP (1925) The homolosine projection: a new device for portraying the Earth's surface entire. Ann Assoc Am Geogr 15(3):119–125

Hammer E (1900) Unechtcylindrische and unechtkonische flächentreue Abbildungen. Petermanns Geogr Mitt 46:42–46

Jenny B (2012) Adaptive composite map projections. IEEE Trans Vis Comput Graphics (Proceedings Scientific Visualization/Information Visualization 2012) 18(12):2575–2582

Jenny B, Patterson T (2007) Flex projector. http://www.flexprojector.com. Accessed 12 Sept 2016

Jenny B, Patterson T (2013) Blending world map projections with Flex Projector. Cartogr Geogr Inf Sci 40(4):289–296

Jenny B, Patterson T, Hurni L (2008) Flex Projector—interactive software for designing world map projections. Cartogr Perspect 59:12–27

Jenny B, Patterson T, Hurni L (2010) Graphical design of world map projections. Int J Geogr Inf Sci 24(11):1687–1702

McBryde FW (1978) A new series of composite equal-area world maps projections. In: International Cartographic Association, 9th international conference on cartography, College Park, Maryland, Abstracts, pp 76–77

Putniņš RV (1934) Jaunas projekci jas pasaules kartēm, Geografiski Raksti, Folia Geographica 3 and 4, pp 180–209 [Latvian with extensive French résumé]

Snyder JP (1977) A comparison of pseudocylindrical map projections. Am Cartographer 4(1):59–81

Snyder JP (1993) Flattening the earth: two thousand years of map projections. University of Chicago Press, Chicago

Strebe D (2010) Mapthematics Geocart 3 user's manual. http://www.mapthematics.com/Downloads/Geocart_Manual.pdf

Tobler WR (1973) The hyperelliptical and other new pseudo cylindrical equal area map projections. J Geophys Res 78(11):1753–1759

Tobler WR (1986) Polycylindric map projections. Am Cartographer 13(2):117–120

Wagner K (1949) Kartographische Netzentwürfe. Bibliographisches Institut, Leipzig

Winkel O (1921) Neue Gradnetzkombinationen. Petermanns Mitteilungen 67:248–252

Chapter 9
A Guide to Selecting Map Projections for World and Hemisphere Maps

Bernhard Jenny, Bojan Šavrič, Nicholas D. Arnold, Brooke E. Marston and Charles A. Preppernau

Abstract Selecting the most suitable projection can be challenging, but it is as essential a part of cartographic design as color and symbol selection and should be given the same degree of consideration. A poorly chosen projection can result in misinterpreted information and impact the effectiveness of a map. This chapter provides guidance in selecting projections for world and hemisphere maps.

9.1 Introduction

Hundreds of different projections for world maps have been developed over the centuries as cartographers sought a projection that best minimized distortion for various applications. Each projection distorts space in a different way whether it is the size of areas, directions, the distances between places, or a combination of these properties. With so many projections to choose from, it can be challenging to select one projection that is most appropriate for what is being mapped and the purpose of the map.

B. Jenny
School of Science, Geospatial Science, RMIT University, Melbourne, Australia

B. Jenny (✉) · B. Šavrič · N.D. Arnold · B.E. Marston · C.A. Preppernau
College of Earth, Ocean and Atmospheric Sciences, Oregon State University, Corvallis, OR, USA
e-mail: bernhard.jenny@rmit.edu.au

B. Šavrič
Esri Inc., Redlands, CA, USA

N.D. Arnold
Arnold Geographics, Portland, OR, USA

B.E. Marston
U.S. Department of State, Washington, DC, USA

C.A. Preppernau
National Geographic, Washington, DC, USA

M. Lapaine and E.L. Usery (eds.), *Choosing a Map Projection*,
Lecture Notes in Geoinformation and Cartography,
DOI 10.1007/978-3-319-51835-0_9

Guidelines exist to help the mapmaker select an appropriate projection. The most commonly recommended and applied guide was published by John P. Snyder (1926–1997), an American cartographer (Snyder 1987). Snyder created a hierarchical tree organized by the extent of the region that was to be mapped and the desired property of the map (e.g., equal-area, conformal, or equidistant). The first selection in Snyder's decision tree is among three different geographic extents: (1) world maps; (2) maps showing a hemisphere; and (3) maps showing a continent or smaller areas. This chapter focuses on projections for world and hemisphere maps. Snyder's selection guideline is not deterministic. Discretion must still be exercised when selecting one of the suggested projections.

9.2 Criteria for the Selection of a Map Projection

Distortion is an inevitable consequence of transforming a sphere or an ellipsoid onto a plane. It is impossible to cut and unfold the surface of the Earth onto a planar surface, like a map, without distortion (Fig. 9.1). Distortion is easy to spot on small-scale world maps where landmasses are considerably altered in geometric appearance.

Because map projections cannot preserve all properties of the original sphere, cartographers must consider which properties are the most important to retain. This consideration is based on the purpose of the map and the cartographic technique used to visualize information. Additional criteria are to be taken into account such

Fig. 9.1 The earth is impossible to unfold onto a planar surface without distortion

as aesthetic appearance of the map, aspect ratio, whether available software applies the projection, and users' preferences.

9.2.1 Deciding What Property to Preserve

Because the primary objective for most maps is to minimize geometric distortion, it is important to determine which geometric properties to preserve. This decision is based on the purpose of the map, that is, what mapping technique is applied and for what applications the reader will use the map.

Equal-area projections maintain the size of map elements relative to one another. This is an essential property to preserve when comparing the size of objects on a map. For example, a choropleth map should be equal-area because the size of enumeration units is being compared.

Conformal projections preserve local angles about any point on a map. Generally, this property is only required for large-scale maps when angles are to be measured from maps, for example when reading an angle from a map using a protractor for navigation or surveying. All conformal maps showing the entire world or large portions of it grossly enlarge or reduce some areas.

Equidistant projections preserve distances between points along some directions. This property is important for comparing distances between locations. For example, a map showing concentric circles that denote the distance from a point requires the distance relative to the center of the circles to be true. However, only some distances can be preserved; it is impossible to correctly display distances between all points on a flat map.

Compromise projections do not preserve area, local angles, or distance. As the name suggests, a compromise projection is an attempt at balancing the distortion. Compromise projections are generally only useful for maps showing the entire world or large portions of it.

9.2.2 Aesthetics and Aspect Ratio

It is possible for the projection selection process to yield several potential projections for maps showing the entire world or a hemisphere. A cartographer should consider making a selection based upon aesthetic preference. This is a subjective decision made by the cartographer or the client. In this case, personal taste is a major selection criterion (Šavrič et al. 2015a).

There are often constraints on the size and format of the available space for a map. Choosing a projection that fits the aspect ratio efficiently can greatly improve the page layout.

9.2.3 Mapping Software

Another factor in the selection of a map projection is its availability in mapping software. Many projections are supported across various packages, but some useful projections are difficult to find. A few of the projections recommended and discussed in this chapter are currently unavailable in some of the more commonly used GIS software.

9.2.4 User Preferences

When selecting a world map projection, cartographers can base their selection on map-reader preferences for world maps. So far, three user studies (Gilmartin 1983; Werner 1993; Šavrič et al. 2015a) have assessed user preferences. Gilmartin (1983) found a preference for elliptical maps over rectangular maps. Gilmartin also found that projections with a distinctively longer width than height are preferred. A user study by Werner (1993) showed that the most preferred projections are uninterrupted pseudocylindrical, followed by interrupted projections, and the least favored were rectangular maps. Šavrič et al. (2015a) confirmed that map-readers prefer uninterrupted projections. They also found that map-readers dislike world map projections with curved parallels, as well as pseudocylindrical projections with bulging meridian curves.

9.3 Projections for World Maps

The main distinguishing characteristics of maps showing the entire world are the distortion properties (equal-area, compromise, or equidistant), the curvature of parallels and meridians, and the representation of poles as either points or lines. In this section, recommendable equal-area and compromise projections are discussed first. Then, projections with other properties are discussed, as well as oblique aspects and interruptions. This section concludes with a summarizing selection-tree table for world map projections.

Conformal projections are not discussed here as they are not useful for world maps. The major disadvantage of conformal projections when applied to world maps is their gross distortion of areas.[1]

[1]Despite their gross area distortion, Snyder also includes conformal projections in his guideline. He suggests the Mercator projection in the normal, transverse, or oblique aspect, for when scale has to be preserved along the equator, a meridian, or an oblique great circle, respectively. Snyder also recommends the conformal Lagrange, August, and Eisenlohr projections.

Most projections for world maps recommended here are pseudocylindrical projections. When centered on the equator, these projections have straight parallels and curved meridians.

9.3.1 Equal-Area World Map Projections

Equal-area world maps projections are required for area comparisons, such as with choropleth and dasymetric maps, or when the number of features per area unit is estimated, as is the case with dot maps.

The cartographer can select between world map projections representing poles as either points or lines. The criteria for this selection are the cartographer's aesthetic preferences and the degree of acceptable deformation. When representing poles as points, landmasses located close to both a pole and an edge of the map are greatly deformed. When representing poles as lines, features located at these extreme locations are also deformed, but to a lesser degree. Features are vertically compressed when using an equal-area projection with a pole line (Fig. 9.2).

Commonly used projections representing the poles as points are the Mollweide, Hammer, Boggs Eumorphic, and sinusoidal projections (Fig. 9.3). The Mollweide projection is very popular for atlases (Canters and Decleir 1989) and is more aesthetically pleasing to many cartographers than other projections due to its elliptical shape (Snyder 1993, p. 112). The sinusoidal projection has significant angular distortion near the edges, especially in higher latitudes (Canters and Decleir 1989, p. 81). The Boggs Eumorphic projection is a blend between the sinusoidal and Mollweide projections that better represents areas near the poles than the sinusoidal projection (Snyder 1993, p. 200). The Hammer projection represents the entire world in an ellipse with curved parallels and unequally distributed meridians.

Recommendable equal-area projections showing the poles as lines are the Eckert IV, Eckert VI, Wagner IV, Wagner VII, and McBryde-Thomas Flat-Polar

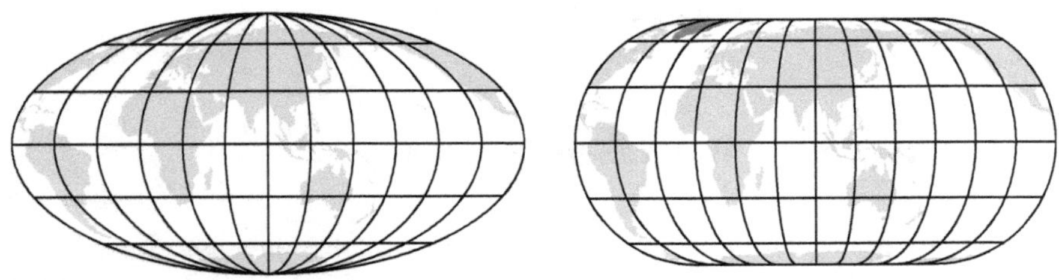

Fig. 9.2 With a projection showing poles as points, the shape of Greenland tends to be more deformed (*left*) than with a projection showing poles as lines (*right*). Mollweide (*left*) and Eckert IV (*right*) projections centered on 90°E

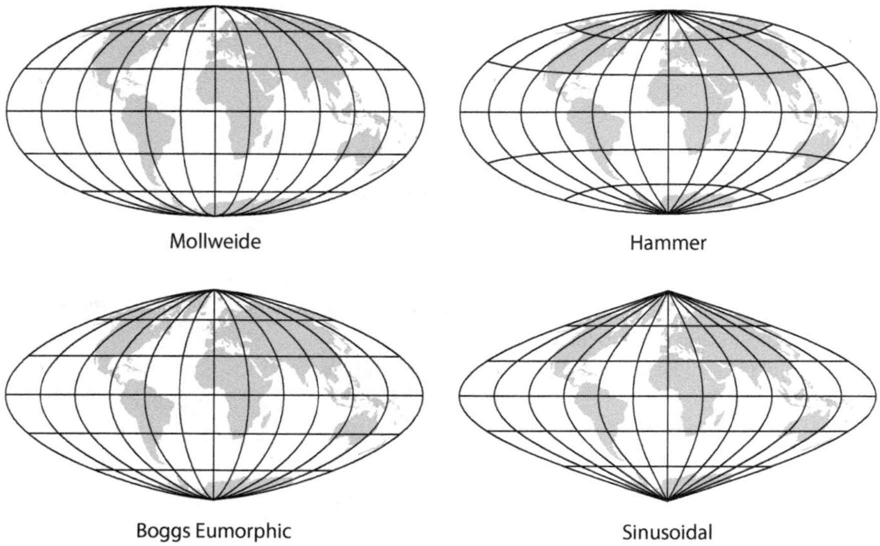

Fig. 9.3 Equal-area world map projections with poles represented as *points*

Quartic projections (Fig. 9.4).[2] Eckert IV is an equal-area projection with very low mean angular distortion (Canters and Decleir 1989). Its rounded corners, where meridians meet the pole line, suggest that the projection represents a spherical Earth. However, Max Eckert preferred his sixth projection with sinusoidal meridians, the Eckert VI projection (Canters and Decleir 1989). The Wagner IV projection is similar in appearance and distortion characteristics to the Eckert IV projection, but the Wagner IV projection does not have the rounded corners of the Eckert IV projection. The Wagner VII projection has curved parallels and pole lines. Unfortunately, only a few software packages include this projection. Snyder includes all projections by McBryde and Thomas in his selection tree, but out of their nine projections, only the McBryde-Thomas flat-polar quartic projection (Fig. 9.4) is commonly used by cartographers (Snyder 1993).

9.3.2 Compromise World Map Projections

Compromise projections with low areal distortion are commonly used for a variety of thematic world maps. These projections include the Robinson, Winkel Tripel,

[2]Snyder states that any pseudocylindrical equal-area projection is suitable. He does not include the Wagner VII projection, which is not a pseudocylindrical projection.

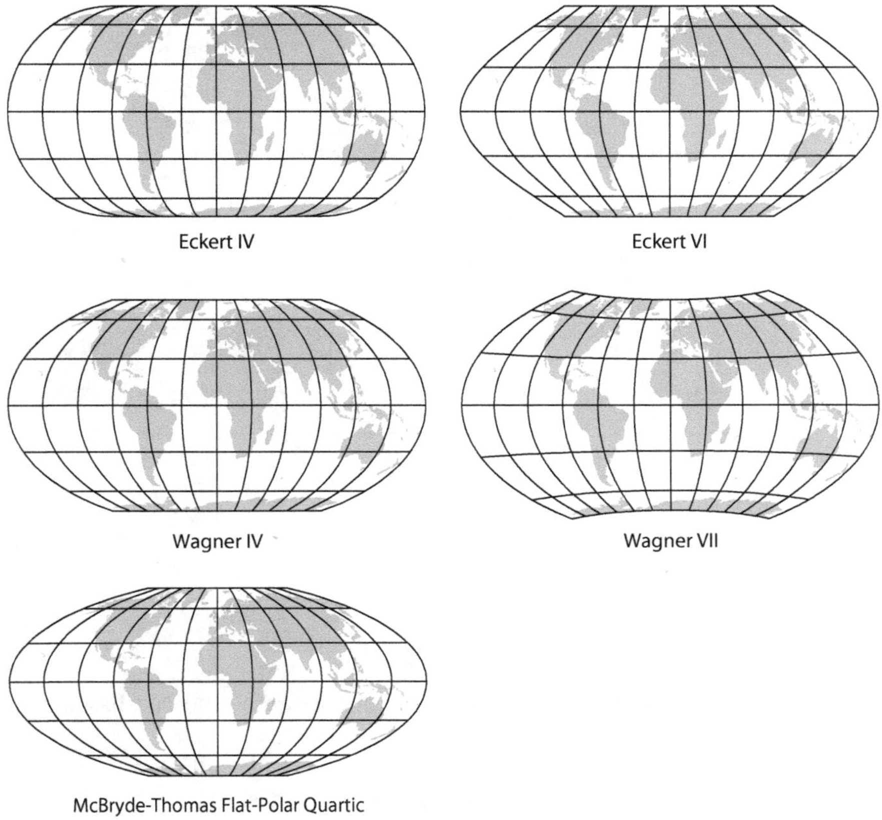

Fig. 9.4 Equal-area world map projections with poles represented as *lines*

Natural Earth, and Wagner V projections (Fig. 9.5).[3] The Robinson projection is perhaps the most used compromise projection for world maps. National Geographic used the Robinson projection for their world maps for about a decade until the Winkel Tripel projection replaced it in 1998 (Slocum et al. 2009). The Winkel Tripel projection has curved parallels and a comparatively large height-to-width ratio. It also has low area and scale distortion compared to other compromise projections. The Natural Earth projection has characteristic rounded corners where parallels meet the pole lines (Jenny et al. 2008; Šavrič et al. 2011). The Natural Earth II projection (not depicted here) is an alternative pseudocylindrical projection with meridians steeply bending toward short pole lines resulting in a map with highly rounded corners (Šavrič et al. 2015b). The Wagner V projection is similar to

[3]Snyder's selection guideline contains only two compromise projections: the Robinson and the Miller cylindrical projections. Snyder's list of compromise projections is extended here with projections with similar distortion characteristic as the two mentioned by Snyder.

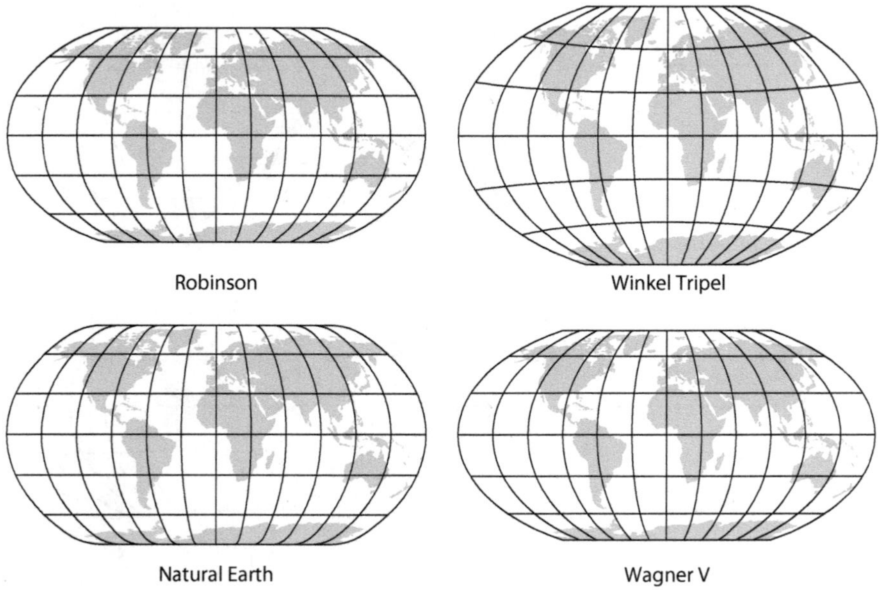

Fig. 9.5 Compromise world map projections

the Robinson projection, but has a smaller height-to-width ratio and has slightly more curved meridian lines.

Many cartographers do not recommend projections that transform the entire Earth into a rectangle because they greatly distort space (Canters 2002, p. 263) and because of the idea that rectangular world maps mislead the map reader's interpretation of the world's shape. This selection guideline, therefore, does not recommend using rectangular projections for most world maps. However, there are some rare phenomena based on longitude that are best represented by a map with straight meridians, such as a map showing world time zones. Also, rectangular projections may be preferred for aesthetic reasons. The Miller cylindrical projection is a compromise projection that is recommended for these applications (Fig. 9.6). When using the Miller cylindrical projection, one has to be aware of the fact that this projection greatly exaggerates the size of Polar areas. The Plate Carrée (also known as the equidistant cylindrical or geographic) is an alternative cylindrical projection that takes up less space vertically than the Miller projection and has less areal distortion in polar areas. The Plate Carrée is a common projection for exchanging geospatial data, because its equations are trivial. The Patterson cylindrical projection is nearly identical in tropical and mid-latitude regions to the Miller cylindrical projection, but polar regions are less exaggerated (Fig. 9.6, Patterson et al. 2014).

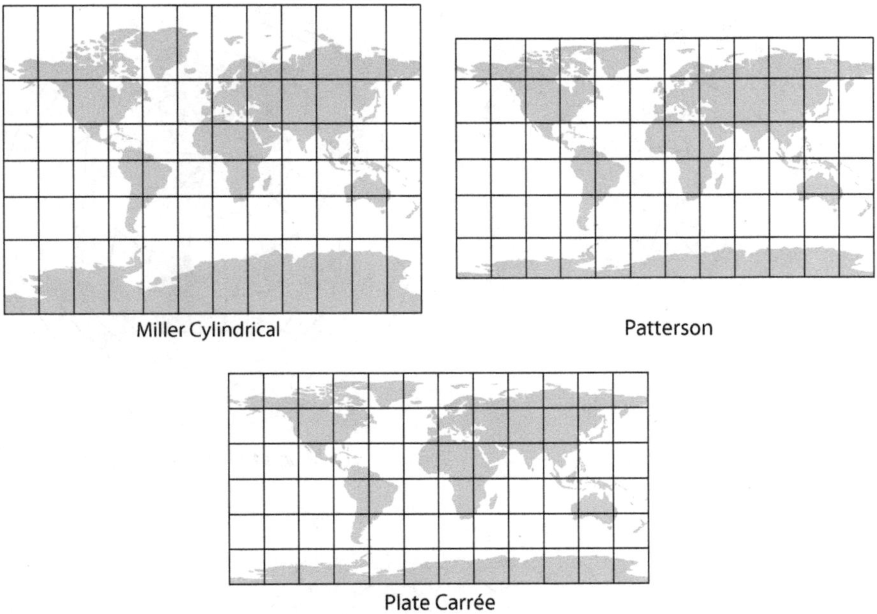

Fig. 9.6 Cylindrical world map projections show polar areas with considerable distortion

9.3.3 Equidistant World Map Projections

On some occasions, such as mapping airport connections where the cartographer would like to preserve distances from the airport, an equidistant projection is essential (Fig. 9.7). The azimuthal equidistant projection is the only projection that preserves all distances relative to its center. Only distances along straight lines passing through the center are portrayed correctly. When the cartographer selects one of the poles as the center, parallels are equally spaced concentric circles (Fig. 9.7).

The only projection that preserves distances relative to two points on a flat map is the two-point equidistant projection (Fig. 9.7 bottom). The cartographer can define the two points. Distances measured along lines passing through either point are mapped without distortion. The two-point equidistant projection is appropriate, for example, for determining the distance from a ship at a known location to the start and end of its voyage (Snyder and Voxland 1989).

The Plate Carrée (Fig. 9.6 bottom) and the more general equirectangular projection preserve distances along all meridians and are useful when differences in latitude are measured. The sinusoidal projection preserves distances along all parallels (Fig. 9.3).

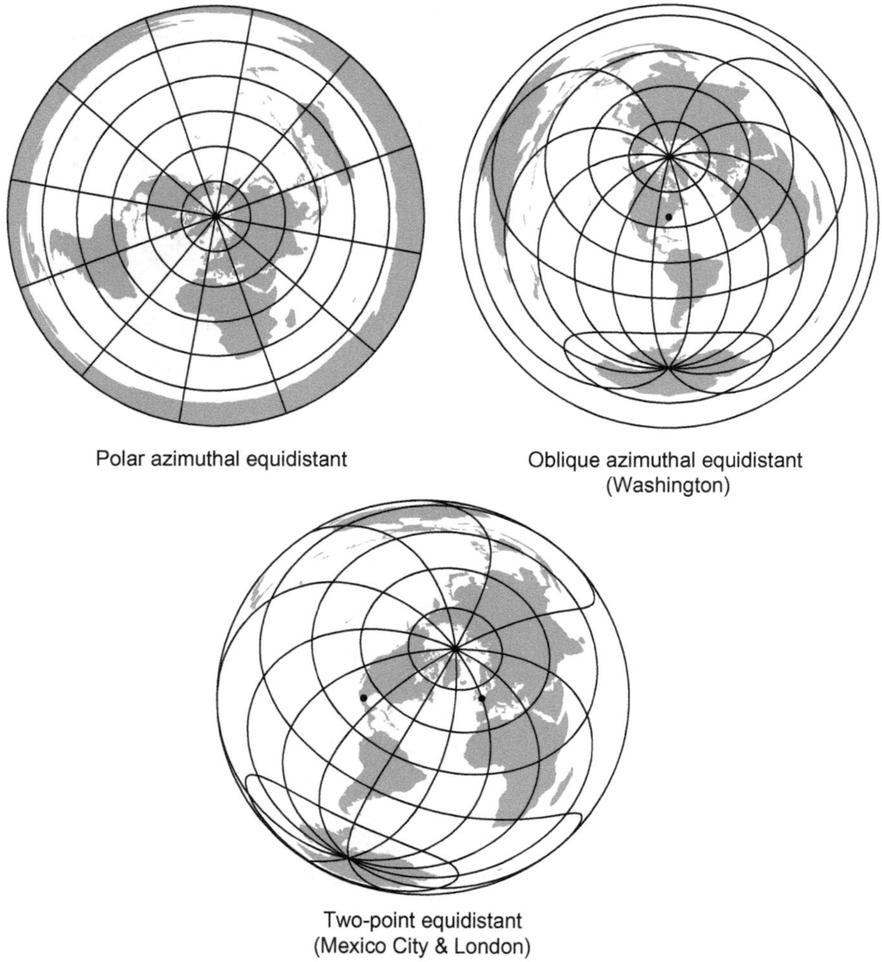

Fig. 9.7 Equidistant map projections

9.3.4 Interrupted World Map Projections

Small-scale world maps usually represent the world on a continuous space without interruptions. An interrupted projection is an alternative (Fig. 9.8), however interrupted projections are not frequently used, and they seem to be disliked by many map readers (Šavrič et al. 2015a). Interruption can be applied to most of the equal-area and compromise world map projections suggested in the sections above. Depending on the purpose of the map (i.e., whether showing land or ocean), the locations of intersections and central meridians are adjusted.

The McBryde S3 projection is sometimes recommended as an interrupted equal-area projection with a pole line (Fig. 9.8). The Goode homolosine is an equal-area projection that is often used in interrupted form. A disadvantage of the Goode homolosine is the discontinuities at approximately ±41°N and S latitudes, which can be visually disruptive. The interrupted form of the Mollweide projection is similar in appearance, but does not show the same discontinuities.

9.3.5 *World Maps with Shifted Central Meridian and Oblique Aspect*

Many world maps are centered on the Greenwich meridian, but there are often good reasons to center the map on a different meridian, mainly to better show spatial relations of the centered area to neighboring areas (Fig. 9.9 left). Shifted meridians can be applied to any projection.

World maps with an oblique aspect are centered on any point on Earth's surface (Fig. 9.9 right). While a shifted meridian only relocates the mapped features in an

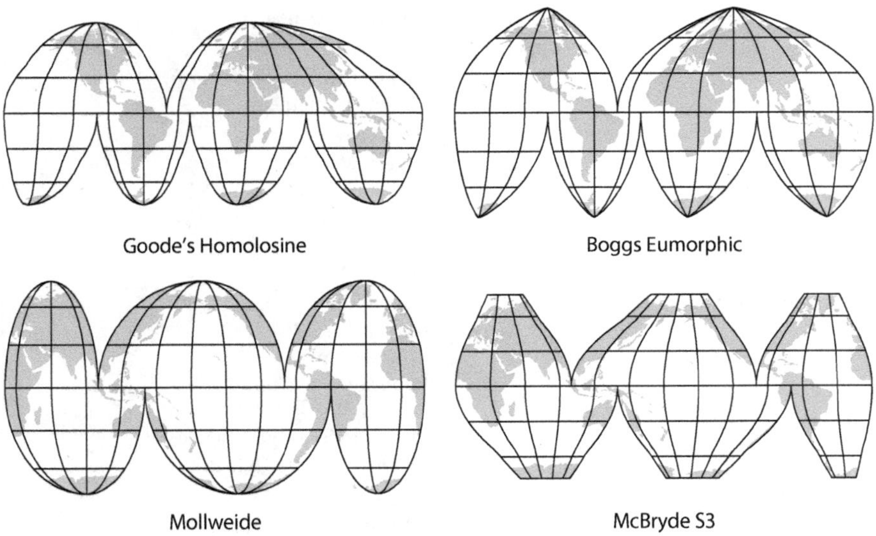

Fig. 9.8 Examples of interrupted projections for mapping continents (*top*) and oceans (*bottom*)

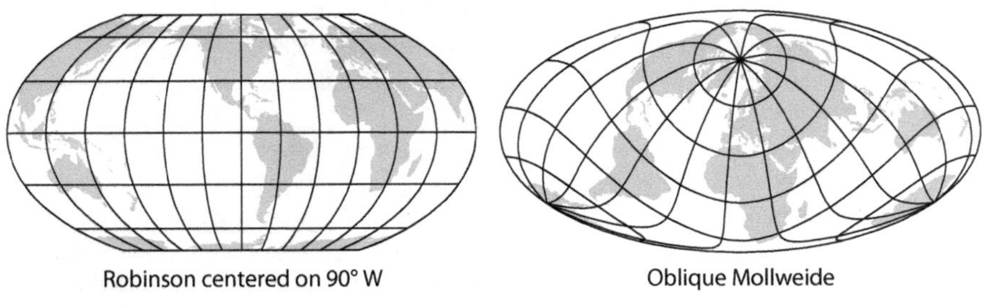

Fig. 9.9 Examples of world maps with shifted central meridian and oblique aspect

east-west direction, an oblique aspect better represents spatial relations of features near one of the poles because the pole is moved toward the center of the map. Oblique aspects can be applied to any world map projection.

9.3.6 Selection Tree for World Maps

Table 9.1 presents a selection tree for world map projections summarizing the discussion above. The cartographer will reach a list of recommended projections after considering a projection's property (first column) and its pole representation (second column).

Table 9.1 Selection tree for projections for mapping the entire sphere (adapted from Snyder 1987)

Property/characteristic		Poles as	Named projection
Equal-area		Points†	Mollweide Hammer (or Hammer-Aitoff) *Boggs Eumorphic Sinusoidal
		Lines†	Eckert IV Wagner IV (or Putnins P2′) Wagner VII (or Hammer-Wagner) McBryde-Thomas flat-polar quartic Eckert VI
Compromise distortion		Lines†	Natural Earth Natural Earth II Winkel Tripel Robinson Wagner V Plate Carrée (or geographic) Miller (or Miller Cylindrical I)
Interrupted (equal-area)		Points	*Mollweide *Boggs Eumorphic Goode homolosine
		Lines	*McBryde S3 *Any of the equal-area projections with a pole line listed above
Equidistant	Centered on a pole	Points	Polar azimuthal equidistant
	Centered on arbitrary point		Oblique azimuthal equidistant
	Relative to two points		Two-point equidistant
	Along meridians	Lines	Plate Carrée (or geographic)

An *asterisk* (*) marks projections that are often not available in GIS and mapping software. Synonyms are given in *parentheses*. An *obelisk* (†) marks a group of projections ordered by mean scale distortion, from least to greatest weighted mean error in the overall scale distortion index (after Canters 2002)

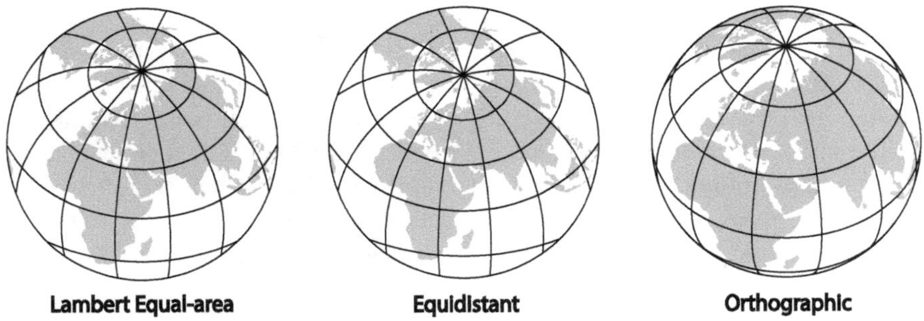

Fig. 9.10 Azimuthal projections for hemisphere maps (centered on 45°N and 45°E)

9.4 Projections for Hemisphere Maps

Three recommendable projections exist for creating maps depicting a hemisphere with a circular boundary. The three projections, which all belong to the azimuthal class of projections, are the Lambert azimuthal equal-area projection, the azimuthal equidistant projection, and the orthographic projection (Fig. 9.10).[4]

Adjusting the central longitude and latitude of the three recommendable projections can shift the area of interest to the center of the map. Polar aspect, that is, maps showing one of the poles in the center of the map, can be created with the three projections (for an example, see Fig. 9.7 top left).

Of the three recommendable projections, the Lambert azimuthal projection is the only projection that preserves areas. It is, therefore, a useful projection for hemisphere maps where the size of countries or other features are compared. The Lambert azimuthal projection is not limited to a single hemisphere, but can show the entire globe in a circle. However, the projection is rarely used to show the entire sphere because of the considerable distortion of angles and distances close to the edge of the circle.

The orthographic projection shows the globe as seen from space at an infinite distance. The orthographic projection is particularly useful for locator maps in combination with large- and medium-scale maps to show the location and extent of a mapped area because it creates pictorial views of the Earth, accentuating its round shape.

The azimuthal equidistant projection preserves distances relative to the central point of the map. Consequently, it is a useful projection for reading distances to a point of interest placed in the center of a map. For example, a map showing the

[4]Snyder also includes the conformal azimuthal stereographic projection for hemisphere maps in his selection schema. The azimuthal stereographic projection is not generally useful for mapping a hemisphere as it grossly distorts shape and area along the border of the projected hemisphere. Preserving angles is rarely needed for hemisphere maps and certainly not for locator inset maps, a typical use case for these projections. The azimuthal stereographic projection is therefore not recommended for maps showing an entire hemisphere in a circle.

relative distance of a country to immediate and distant neighboring countries, or the location of a tsunami wave over time, should use the azimuthal equidistant projection. Because distances are only correct along straight lines when they pass through the projection center, it is crucial to define the central point accordingly.

The visual difference between the Lambert azimuthal equal-area projection and the equidistant projection are small when used for a hemisphere (Fig. 9.10). For locator inset maps, the selection of a projection is a matter of personal taste. The authors recommend using the orthographic projection when the spherical shape of the globe is to be accentuated in inset maps.

9.5 Selecting a Projection with *Projection Wizard*

Projection Wizard is an online map projection selection tool available at projectionwizard.org that assists cartographers in selecting projections (Fig. 9.11) (Šavrič et al. 2016). *Projection Wizard* integrates Snyder's selection guideline with the extensions suggested in this book chapter. It includes the projections for world

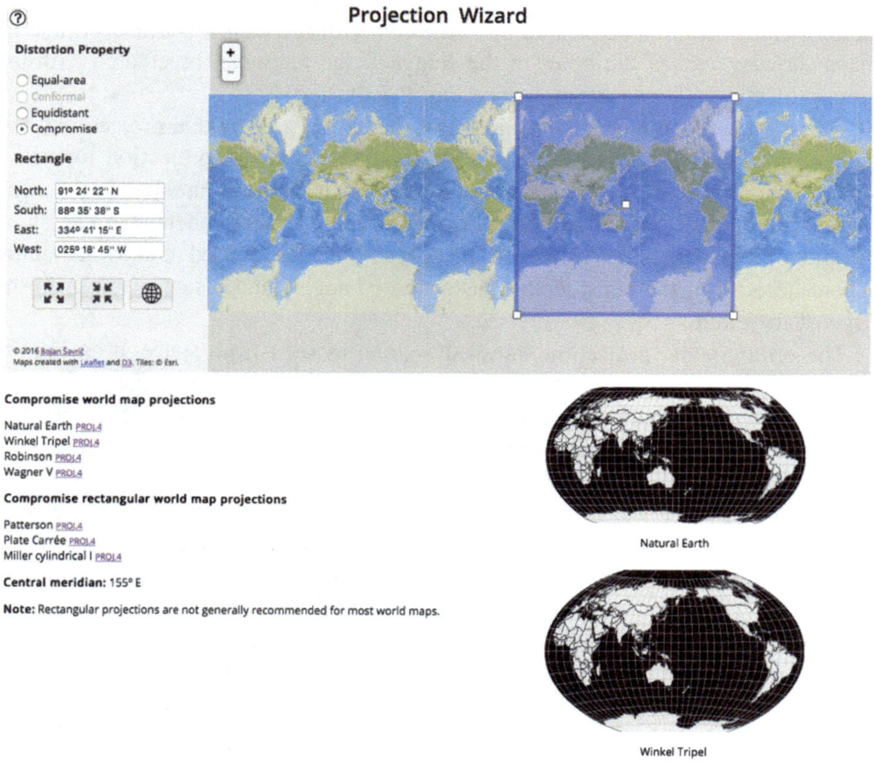

Fig. 9.11 Selecting a projection with *Projection Wizard* at projectionwizard.org

maps listed in Table 9.1 and the projections for hemisphere maps discussed above. The user selects the desired distortion property, and a projection is then suggested and a preview map generated. *Projection Wizard* is simple to use and can also suggest projections for maps at larger scales, showing continents or smaller areas. For these maps, the user defines the area to be mapped on an interactive web map.

9.6 Conclusion

Although the selection of map projections is one of the most important aspects of mapping, it is often given little consideration. The map projection is the basis for the rest of the map and is therefore, an important component of cartographic design. Map projections must be selected carefully according to criteria such as the map's extent, purpose, and cartographic visualization technique. An appropriate projection can minimize distortion when measuring distances, angles, or areas. Most importantly, a poorly chosen projection can result in misinterpreted information and impact the effectiveness of a map.

This chapter only includes a selection of projections for world and hemisphere maps. Many alternative projections exist with similar distortion properties and visual appearance that may also be appropriate for small-scale maps.

Snyder's selection guideline relies on the answers to two main questions: what is the geographic extent mapped, and, which distortion properties need to be preserved? The selection guideline then leads the cartographer to a set of appropriate projections for world and hemisphere maps.

Snyder published his selection guideline in 1987 at a time when cartographers were using computer algorithms for creating maps with a variety of projections, but before Web mapping services, such as Google Maps, existed. Google introduced their Web mapping service in 2005 and the underlying Mercator projection quickly became the de facto standard for Web maps (Battersby et al. 2014). While the Mercator projection is included in Snyder's selection tree for some maps, it is a poor choice for world maps because it shows polar areas with enormous areal distortion. Web mapping services will hopefully include more alternative projections for world maps in the future. A viable option could be adaptive composite map projections (Jenny 2012) that automatically adjust the map's geometry in accordance with Snyder's selection guideline.

References

Battersby SE, Finn MP, Usery EL, Yamamoto KH (2014) Implications of Web Mercator and its use in online mapping. Cartogr Int J Geogr Inf Geovisualization 49(2):85–101. doi:10.3138/carto.49.2.2313

Canters F (2002) Small-scale map projection design. CRC

Canters F, Decleir H (1989) The world in perspective—a directory of world map projections. Wiley, Chichester

Gilmartin PP (1983) Aesthetic preferences for the proportions and forms of graticules. Cartogr J 20(2):95–100. doi:10.1179/caj.1983.20.2.95

Jenny B (2012) Adaptive composite map projections. In: IEEE Trans Visualization Comput Gr (Proc Sci Visualization/Inf Visualization) 18(12):2575–258. doi:10.1109/tvcg.2012.192

Jenny B, Patterson T, Hurni L (2008) Flex Projector—interactive software for designing world map projections. Cartogr Perspect 59:12–27. doi:10.14714/cp59.245

Patterson T, Šavrič B, Jenny B (2014) Introducing the Patterson cylindrical projection. Cartogr Perspect 78:77–81. doi:10.14714/cp78.1270

Šavrič B, Jenny B, Patterson T, Petrovič D, Hurni L (2011) A polynomial equation for the Natural Earth projection. Cartogr Geogr Inf Sci 38(4):363–372. doi:10.1559/15230406384363

Šavrič B, Jenny B, White D, Strebe DR (2015a) User preferences for world map projections. Cartogr Geogr Inf Sci 42(5):398–409. doi:10.1080/15230406.2015.1014425

Šavrič B, Patterson T, Jenny B (2015b) The Natural Earth II world map projection. Int J Cartogr 1(2):123–133. doi:10.1080/23729333.2015.1093312

Šavrič B, Jenny B, Jenny H (2016) Projection Wizard—an online map projection selection tool. Cartogr J 53(2):177–185. doi:10.1080/00087041.2015.1131938

Slocum TA, McMaster RB, Kessler FC, Howard HH (2009) Thematic cartography and geovisualization, 3rd ed. Pearson education, Inc

Snyder JP (1987) Map projections—a working manual. US government printing office

Snyder JP (1993) Flattening the earth: two thousand years of map projections. University of Chicago Press

Snyder JP, Voxland PM (1989) An album of map projections. In: U.S. Geological Survey professional paper 1453. U.S. Government Printing Office

Werner RJ (1993) A survey of preference among nine equator-centered map projections. Cartogr Geogr Inf Sci 20(1):31–39. doi:10.1559/152304093782616733

Chapter 10
The Logic of Selecting an Appropriate Map Projection in a Decision Support System (DSS)

Michael P. Finn, E. Lynn Usery, Laura N. Woodard and Kristina H. Yamamoto

Abstract There are undeniable practical consequences to consider when choosing an appropriate map projection for a specific region. The surface of a globe covered by global, continental, and regional maps are so singular that each type distinctively affects the amount of distortion incurred during a projection transformation because of the an assortment of effects caused by distance, direction, scale, and area. A Decision Support System (DSS) for Map Projections of Small Scale Data was previously developed to help select an appropriate projection. This paper reports on a tutorial to accompany that DSS. The DSS poses questions interactively, allowing the user to decide on the parameters, which in turn determines the logic path to a solution. The objective of including a tutorial to accompany the DSS is achieved by visually representing the path of logic that is taken to a recommended map projection derived from the parameters the user selects. The tutorial informs the DSS user about the pedigree of the projection and provides a basic explanation of the specific projection design. This information is provided by informational pop-ups and other aids.

10.1 Introduction and Objectives

A map projection is a transformation from geographical spherical or ellipsoidal coordinates to a plane coordinate system. It is an essential requirement for geospatial work to place all applicable datasets into a common coordinate system framework. Further, for many datasets already in a plane coordinate system, inverse

M.P. Finn (✉) · K.H. Yamamoto
Center of Excellence for Geospatial Information Science,
U.S. Geological Survey (CEGIS), Denver Federal Centre,
Box 25046, Mail Stop 510, Denver, CO 80225-0046, USA
e-mail: mfinn@usgs.gov

E. Lynn Usery · L.N. Woodard
Center of Excellence for Geospatial Information Science,
U.S. Geological Survey CEGIS, Rolla, MO, USA

M. Lapaine and E.L. Usery (eds.), *Choosing a Map Projection*,
Lecture Notes in Geoinformation and Cartography,
DOI 10.1007/978-3-319-51835-0_10

projection to spherical or ellipsoidal coordinates and forward projection to a different plane system are required. The execution of geospatial analysis commonly involves map projection, a subject that is effectively unknown to most users of geographic information systems (GIS) and Web-based mapping and geoprocessing systems.

To aid users of map projections, the USGS developed a Decision Support System (DSS) for Map Projections of Small Scale Data (http://cegis.usgs.gov/projection/index.html) to guide the user through the process of selecting an appropriate projection based on user needs, data type, volume, region of the world, and other parameters. This system is available online and can be used in combination with other software packages. To aid the user in understanding the operation of the DSS and to support the education of users concerning map projections, the USGS has developed a tutorial to accompany the DSS.

This tutorial is an enhancement to the system previously described in Finn et al. (2004). We include in this current work a summary of map projection theory. The objective of including a tutorial to accompany the DSS is to enhance the interactive interface and aid in technical and societal decision-making and to help explain map projection terminology, concepts, and selection to the user. It informs the user of the characteristics of the projection that is "recommended" based on user input parameters. This paper, thus, includes details about how the tutorial aspect of the software works. The tutorial visually represents the path of logic leading to a suggested map projection based on the parameters the user selects.

The second section of this paper provides relevant background on map projection concepts and factors that influence the choice or decision of an appropriate projection. The third section documents the creation and structure of the decision tree that is used in the selection process. The fourth section provides the underlying logic and corresponding data structure. The fifth section includes information on the methods of implementation of the tutorial with a final section drawing some conclusions from this work.

10.2 Background

There is an extensive literature on methods to guide cartographers in choosing an appropriate projection when creating a map. The selection of an appropriate map projection depends on a number of factors such as the objective or purpose of the map, the resolution of the data and the size of the area of coverage, and the type of data, among other criteria. These criteria are documented in projections manuals and textbooks and research concerning distortion from projection transformations (Robinson et al. 1978; Snyder 1987; Usery et al. 2003; Slocum et al. 2009).

There have been attempts to provide guidance on map projections selection of this type in textbooks and anthologies of map projections (e.g., Robinson et al. 1978; Snyder 1987, 1993; De Genst and Canters 1996; Yang et al. 2000; Canters 2002). There are no detailed formulas that can be specified that will direct the accurate selection because each map is a complex composite of objectives and limitations, but projections may be grouped together that have specific arrangements of the deformation values to assist this selection process (Robinson et al. 1978).

Some specific criteria that affect the appropriate selection of a projection are the size and location of the region to be mapped; specific properties of the projection to be maintained, such as conformality, equivalence of areas, and equivalence of distances; specific characteristics, such as constant scale along the Equator or along a meridian, or centered on a specific location (Snyder 1987; Slocum et al. 2009). There has been work to create a DSS for map projection selection based on these guides (Snyder 1987; Nyerges and Jankowski 1989). Despite that fact that this body of work is available, it is not available online in a quick, easy-to-use manner.

Zhao et al. (2007) wrote about using an expert system as a part of a GIS but no method was shown, other than to say it used Component Object Model technology. Yet, this type of usage is probably beyond most naive users to integrate in their GIS software. Therefore we undertook an effort to provide this service online through a DSS. Our approach is different than Zhao et al., and provides the user not only a recommended choice, but the logic behind the choice. In addition, our approach supports educating GIS users about map projections, especially since we consider the type of data (raster or vector) and the statistical scaling of the data (nominal, ordinal, interval or ratio) which we handle as categorical or continuous. Further it does not appear that the data type or data scaling enters Zhao et al. system to affect the selection process.

A DSS delivers information that can assist in policies, priorities, or processes (Lee and Birk 2003). The DSS that this tutorial accompanies poses questions interactively, allowing the user to select the parameters, which in turn determines the logic path to a solution. The solution is an appropriate projection, or alternate projection, derived from the selected parameters. We focus on the parameters that are used to determine the most appropriate projection. They are divided into two categories, each with special sub-categories, as follows:

- Areal Extent to be Mapped
  - Global
  - Continental
  - Regional
- Property to be Preserved
  - Area
  - Shape (Conformality)
  - Compromise

In addition, Finn and Woodard (2005) provide a user's guide to aid in the use of the DSS.

10.3 Map Projection Theory and Sources

Some projections are more appropriate than others when considering geospatial representation of an area of interest (AOI) or trying to preserve specific geometric properties related to that AOI. Thus, there are specific details to consider when deciding on the best projection. The details and information that are used in the DSS are primarily derived from the projection information provided by Snyder (1987, 1993).

The essential problem in organizing map projections is the splitting of the infinite number of potential distinct projections into a logical and useful finite number of complete classes (Brown 1935; Tobler 1962). For this project, the choice was to group map projections in two basic ways (Usery et al. 2003; Savard 2012):

1. The aspect of a map defines the layout of the projection. The four main projection classes are cylindrical, conical, azimuthal, and pseudo-cylindrical. A developable surface is a surface that can be flattened onto a plane without distortion (Stoker 1961; Slocum et al. 2009). Cylindrical, conical, and azimuthal projections are based on developable surfaces, whereas the pseudo-cylindrical projections on non-developable surfaces (Tobler 1962, 1963, 1979; Snyder 1987).
2. The two most important map projection properties are conformality (shape stays the same) and areal equivalence, which are mutually exclusive in the same projection.

Guided by the knowledge provided by Snyder (1987, 1993), and Usery et al. (2003), we created an appropriate decision tree of the features that need to be considered to determine the map projection most suitable for a specific objective. The fidelity of this decision tree is limited to accomplishing the purpose of informing and educating a general practitioner of cartography or GIScience and not particularly an expert in map projections. The recommendations provided do not differentiate whether the data type is vector or raster with the current release of the software [although the design is in place to handle this additional information (parameter)]. The type of data is inherent based on the user's choice of area or shape preservation (i.e. the assumption is that for raster data (cells—units of area) the user will choose area preservation.) However, a user can choose to preserve shape and actually be using a raster dataset, which becomes an issue of the fidelity of the DSS and, further, brings up a separate series of problems related to projection/re-projection that we address with a different program and in the literature (see Finn et al. 2012).

The outline and reasons the projection type is selected include the following:

1. Global: Global data sets of small-scale data are projected onto every continent and region. The preservation of area or shape with respect to the shape of the Earth are two common criteria (but not the only ones for global datasets) to be considered (Snyder 1987).

(a) Area—To preserve area, an equal area projection is typically used. In this implementation of a decision process, we chose to use pseudocylindrical (with a pole line or with a point) or cylindrical equal area projections because they are designed to preserve the area of regions with greater latitudinal range than longitudinal range. Some examples of pseudocylindrical with a pole line are the Eckert IV and Eckert VI; examples of a pseudocylindrical with a point are Mollweide and sinusoidal; and examples of normal cylindrical equal area projections are the Lambert, Behrmann, Gall, and Trystan-Edwards. The cylindrical equal area projections are effective due to their ability to accurately show the extent of land area, even though these projections change the shapes of features, due to keeping the ratios between the areas or portions of the sphere equal to the ratios between the areas of the corresponding portions of the projected image.

(b) Shape—To preserve conformality, the Mercator projection (a cylindrical projection) is chosen. The meridians of longitude of the Mercator projection are vertical, parallel, and equally spaced lines that cut at right angles by horizontal straight parallels, which are increasingly spaced toward each pole so that conformality exists (Snyder 1987).

(c) Compromise: A compromise projection is one that is neither equal area nor conformal. It is a projection that has value for data presentation and its general look instead of its efficacy for analyses (Finn et al. 2004). These projections have significant applications for world maps. Among those recommended are the Robinson, Van Der Gritten, Eckert V, and Winkel tripel [which has been demonstrated to be one of the most suitable projections for world maps (Capek 2001; Goldberg and Gott 2007)].

2. Continental: Continental projections are based on symmetry with respect to various latitudinal and longitudinal axes and shape of the continent with respect to its coastline (Snyder 1987).

2.1 North America/South America/Africa: These three continents are grouped together due to their greater latitudinal range.

(a) Area: The preservation of area of a continent with greater latitudinal range is treated similar to the preservation of area of a global projection. Thus, a cylindrical equal area projection is designated.

(b) Shape: To preserve the shape of a continent one needs a conformal map; thus we chose the Transverse Mercator. The Mercator has little distortion near the Equator. It is useful in the transverse form, and it can be rotated 90° to coincide with the desired central meridian. The central meridian can be made true to scale, despite how far north and south the map extends, and regions near the central meridian are mapped with low distortion.

2.2 North or South Poles: Due to their location on the geographic grid, the poles are especially difficult to represent in a reprojection unless the projection is centered at either pole. To display one of the poles, it is best to use an azimuthal projection since it is true scale at a point which can correspond to the pole and handles many special attributes of polar areas. Azimuthal projections are radially symmetric in all directions from the center point of the map (Savard 2012).

(a) Area: To preserve the area of the poles, the Lambert Azimuthal equal area projection, centered on the pole, is chosen. Its equal area and azimuthal properties allow for the showing of true dimensions from the center of this projection.

(b) Shape: The Stereographic projection, another azimuthal projection, is also centered on the pole, was chosen to preserve the shape around the poles due to its ability to preserve circular shapes and angles (conformality).

2.3 Asia/Europe/Australia: The greater longitudinal range and mid-latitude $[(>25.0^\circ \text{ and } <60.5^\circ) \text{ or } (<-25.0^\circ \text{ and } >-60.5^\circ)]$ position of these three continents prompts for the use of a conic projection since normal conic projections have constant scale along any given parallel.

(a) Area: To preserve the area of these continents we choose from a variety of equal area conic projections, particularly, Albers Equal Area Conic, or Lambert Equal Area Conic. For the Albers Equal Area Conic projection the "scale factor along the meridians is the reciprocal of the scale factor along the parallels in order to maintain equal area."

(b) Shape: The Lambert Conformal Conic projection is the projection suggested for preserving the shape of the continents with greater longitudinal range. For the Lambert Conformal Conic projection, the scale is constant along any given parallel and is the same in all directions at a given point. Its distortion is constant along any given parallel and it is conformal everywhere except at the poles. In addition, it is normally only used for a single hemisphere.

3. Regional: Geographic location and directional extent must be considered for projecting large regional areas (Finn et al. 2004).The geographical extent is a bit arbitrary, obviously, but we consulted some physical geography sources [see for example Strahler and Strahler (1979) and Trewartha and Horn (1980)] as well as iterative visual inspection (including employing our cartographic license) to define the extents as shown below.

3.1 North/South: These regions have a greater latitudinal range than longitudinal range and are treated similar to the continents with greater latitudinal range if examining areas toward the center of the map. The extreme northern and extreme southern areas on the other hand are treated similar to the continental Polar Regions.

3.1.1 Extreme North (>60.5°) and Extreme South (<−60.5°): Generally, an azimuthal projection is used for the extreme latitudes due to their proximity to the poles.

(a) Area: To preserve the area in the extreme north or south the Lambert Azimuthal Equal Area projection is chosen.
(b) Shape: The stereographic projection, also centered on the pole, conformal azimuthal projection, is chosen.

3.1.2 Mid-Latitude

(a) Area: the cylindrical equal area projection is designated.
(b) Shape: To preserve the shape of an area in a mid-latitude region we decided to use the Transverse Mercator projection.

3.2 East/West

3.2.1 Extreme North and Extreme South:

(a) Area: To preserve an area elongated in the east/west direction in the extreme north and extreme south the Lambert Azimuthal Equal Area projection was chosen. This is because this region is at, or very near, the Polar Regions. Its equal area and azimuthal properties allow for the showing of true dimensions from the center of this projection (Snyder 1987).
(b) Shape: The stereographic projection, conformal azimuthal projection, is chosen.

3.2.2 Mid-North and Mid-South: Because of the east/west orientation, the region has a greater longitudinal range so the most suitable type of projection is a conical projection.

(a) Area: The projection chosen for preserving area in the mid regions is Albers Equal Area Conic.
(b) Shape: The Lambert Conformal Conic projection is the projection that is suggested for preserving the shape of the regions with greater longitudinal range.

3.2.3 Low-Latitude:

(a) Area: Because of the vast extent of area around the low-latitude region of the world, an Equal Area or Cylindrical Equal Area projection, with standard parallel(s) near the equator, would be appropriate when wanting to preserve its area. A few examples of these are Lambert Azimuthal Equal area, and Mollweide.
(b) Shape: To preserve the shape of an east/west region in low latitudes (>−25.0° and <25.0°), a Mercator projection is suggested.

Many of these explanations are given in the "Usage" section at the bottom of each projection page in the DSS; see the "Usage" section of this paper for more details.

10.4 The Logic and Corresponding Data Structure

The DSS relies on conventional theory for the selection of projections (Maling 1973; Voxland 1978; Snyder 1987; Snyder and Voxland 1989), and the design utilizes a specific decision tree that handles regional and global raster projection. We supplemented these selection criteria with results from research performed by Usery et al. (2003). Thus, the structure that was chosen for the tutorial is the tree structure.

One special node called the "root" node exists in the tree structure, and each node may be associated with one or more different nodes, called its children (Main and Savitch 2001). If a node *c* is the child of another node *p*, then we say, "*p* is *c*'s parent." Each child can only have one parent whereas each parent can have multiple children. If one continually moves to a node's parent, the root eventually will be reached. In addition, the line that connects the parent to each of its children is considered a branch of the tree. The tree structure efficiently represents hierarchical dependencies and allows for a clear overview of a large amount of information. It is easy to navigate since once a parameter ("branch") is chosen; the next parameter to be chosen is always the current node's direct descendant along one of its "branches." There is no need to search other "branches" that are not directly linked to the current node.

10.5 Tree Layers

The "root" of the tree is the DSS. This is because it ties all the components together. The first layer (of children) is "Global," "Continental," and "Regional" (Fig. 10.1) since the "Type of Area" is the primary dividing feature when deciding the most appropriate projection. The area covered by global, continental, and regional maps are so vastly different that each type uniquely affects the amount of distortion (often to a large degree) during a projection because of the various effects caused by distance, direction, scale, and area.

Each of the nodes branching from the "root" node continues to divide the elements that influence a projection of that type. "Global" is the first branch of the "root" node (Fig. 10.2). Since a global area is the largest possible area to re-project, it cannot be further divided into additional categories, as can smaller areas. For

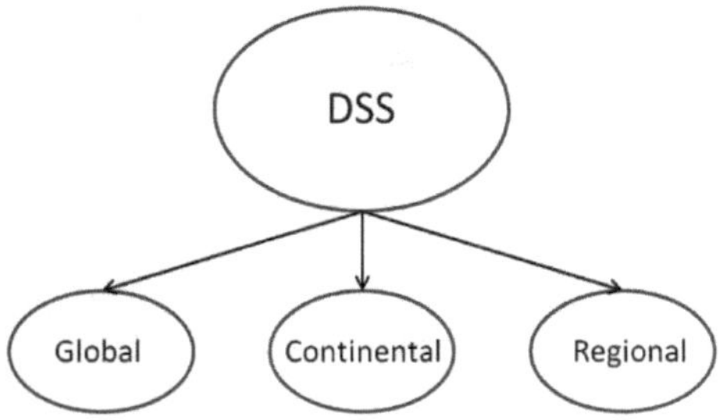

Fig. 10.1 Root node and the first layer of the decision tree structure

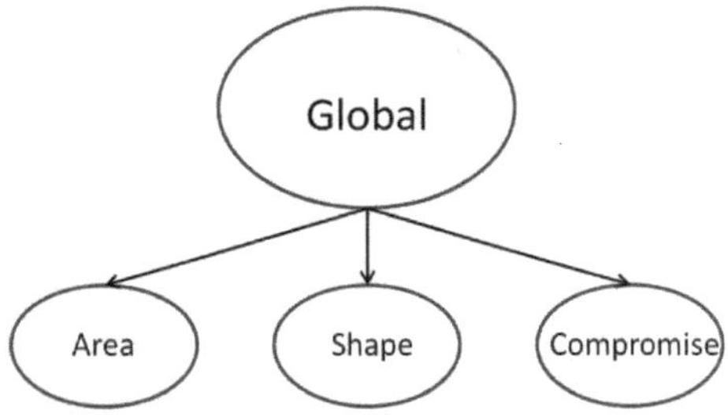

Fig. 10.2 Global branches

these smaller areas, possible divisions can be based on shape, greater height than width or vice versa, or location (for example, north, mid-to-low latitudes or south) in relation to the surrounding elements. Thus, it is wisest to allow the user to decide which type of preservation is more important, area or shape. Alternatively, the user may decide to use a compromise projection, to generate a map that is neither equal-area nor conformal, but some combination of both.

In general, cylindrical projections (and sometimes pseudo-cylindrical) are used for maps on a global scale because they are most likely to preclude the extreme deformation that occurs on azimuthal and conic projections in locales that are distant from the standard point or line. However, cylindrical projections should not be used for wall maps of global data. As a result of exorbitant claims by Arno Peters for a cylindrical projection, Gall's cylindrical which Peter's rediscovered and claimed as his, the American Cartographic Association (now the Cartography and Geographic Information Society) issued a resolution against using cylindrical projections for wall maps of global data (Usery 2001).

"Continental" is another branch from the "root" (Fig. 10.3) and has sub nodes for "Area" and "Shape" since these are the properties that can be preserved. Then, the continents are handled based on their geographic location and directional extent. For example, the DSS handles continents with greater longitudinal range differently from those with greater latitudinal range or those in a polar region, for example, Antarctica (Finn et al. 2004).

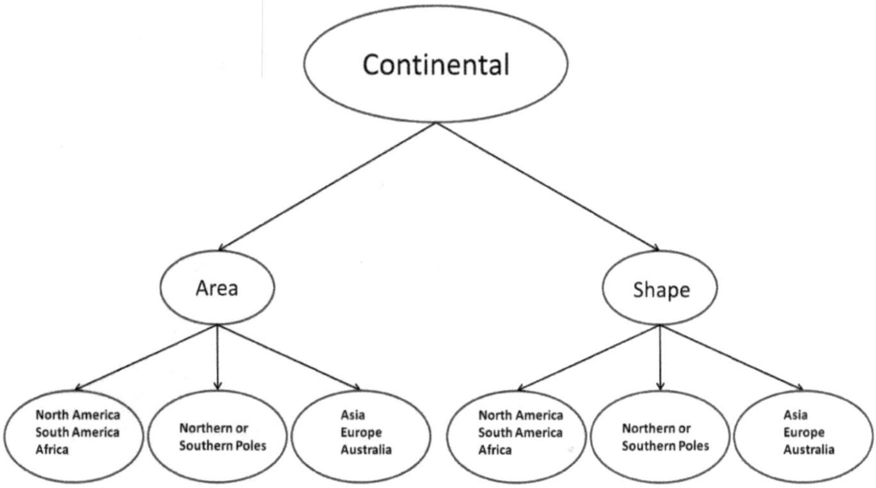

Fig. 10.3 Continental branches

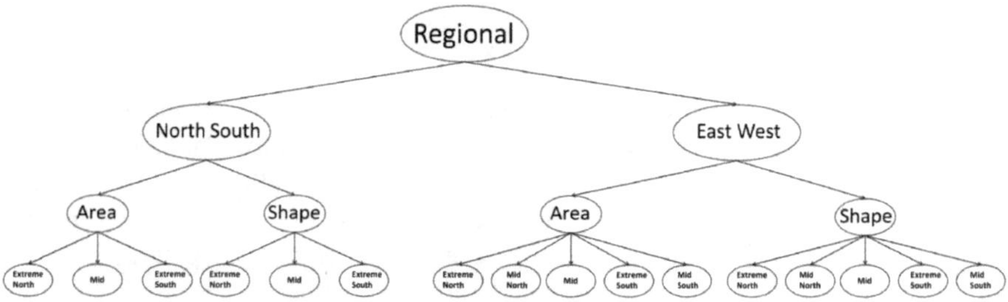

Fig. 10.4 Regional branch. Note "mid" is equivalent to low latitude

"Regional" is the third branch of the "root" node (Fig. 10.4). Since the shape of the region is determined by the user by specifying rectangular areas of different size and orientation, the regions can be split into those that are "North South" or "East West." Then each of those nodes can be further divided into "Area" and "Shape" for an equal-area or conformal projection.

Furthermore, the "Area" and "Shape" nodes can be subdivided since regional projections are also based on the dimensions of the region being mapped, and the location of the region with respect to its the proximity to the Earth's poles. Due to the shape of the region and its direction, the "Area" and "Shape" nodes under the "North South" direction can be subdivided into "Extreme North", "Mid," and "Extreme South". Since an "East West" region can exclusively cover a more defined region, its "Area" and "Shape" can be further divided into "Extreme North," "Mid North," "Mid," "Mid South", and "Extreme South."

10.6 Enhancing Qualities

The tutorial has unique features that enrich the user's experience and allow a user to understand the reasons the system suggested a specific map projection. For example, to emphasize which element is the most recent parameter in the decision tree, its node is colored green fading to yellow in the center. This green hue makes it more prominent than the other nodes, since the rest are light blue. In addition, the path can be easily followed because it is outlined with a thick red line. Thus, these special features make the current appropriate node easily discernable from the surrounding background and nodes.

Another enhancing feature of the tutorial is the availability of more information. When the mouse is hovered over the current node, colored green and yellow, a message box appears with more information about that layer and a link to a Website, obtained by clicking on the "More ⋙" label at the end of the message (Fig. 10.5). The Websites provide some of the same information as the "Usage" information (see next section and Fig. 10.7), but also provide a greater overview of projections in general.

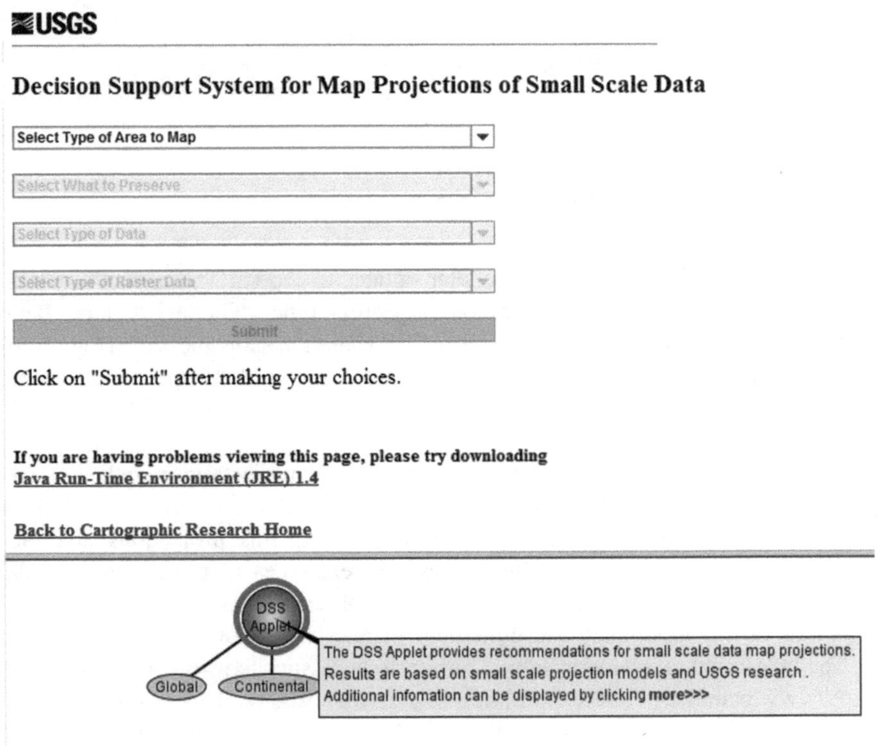

Fig. 10.5 Message box with information for a current node, with an option (more ⋙) for even more information

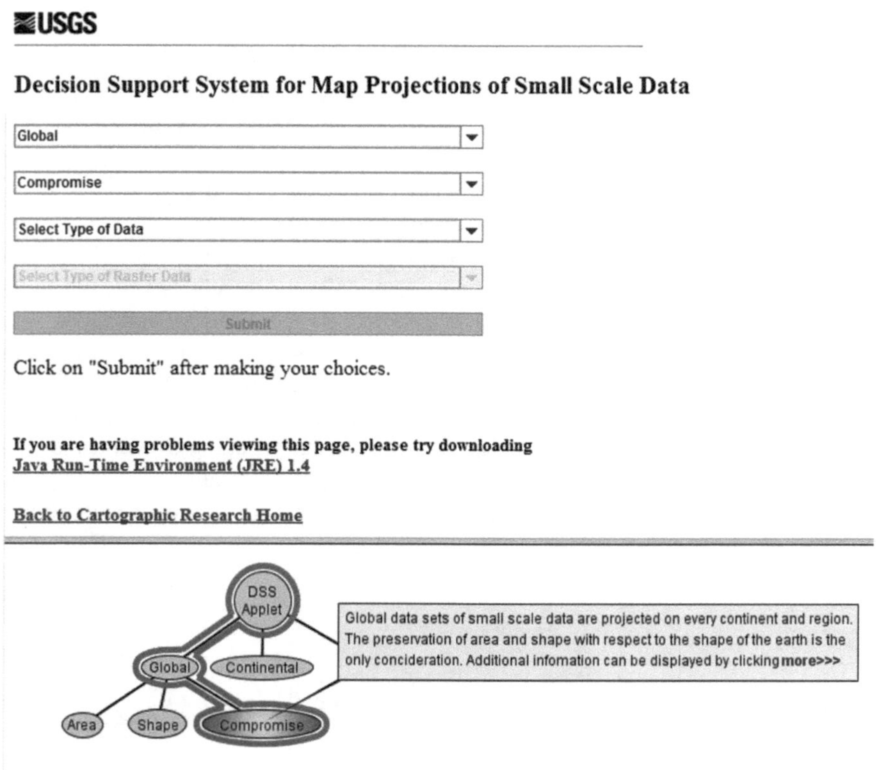

Fig. 10.6 More detailed message for sub-nodes

The cursor will change from a regular pointer to a pointing finger when more information is available. The common projection type, "Global" and its "leaf" nodes "Area," "Shape," and "Compromise" all share the same message (Fig. 10.6). The same applies to "Continental" and its nodes "Area" and "Shape," and their sub nodes, all share the same message, as do "Regional" and its nodes "North South" and "East West," and their sub-nodes.

The link found in the message box for the "Global" projection group leads to the Website http://www.mapthematics.com/Essentials.php (Strebe 2010). This Website covers mapping and projection vocabulary, and explains projections and their importance to mapping. It also introduces different types of projections, the differences between them, and the encountered problems. This link also covers the essentials of the importance and difficulties of map projections.

The link found in the message box for the nodes associated with "Continental" is http://www.pathaway.com/mapproj.htm (PathAway 2004). This Website mainly provides the differences between cylindrical, conical, and azimuthal projections, and other miscellaneous projections. It also provides examples of each type of projection.

The link found in the message box of the "Regional" group leads to the Website http://erg.usgs.gov/isb/pubs/MapProjections/projections.html which provides further details about different projections. The Website provides summary tables of a projection's properties, such as its type, whether it is conformal, equal area, equidistant, true direction, perspective, or compromise, and if it has straight rhumb lines. In addition, there is a summary of areas suitable for mapping that type of area, such as world, hemisphere, continent/ocean, region/sea, medium scale, or large scale. Furthermore the Website contains a summary of the projection's general use, for example in topographic maps, geological maps, thematic maps, presentations, navigation, or USGS maps.

10.7 Usage Information

Another addition to the DSS that provides direct information about a projection is the usage section at the bottom of the projection pages. In general, the usage section will name the projection's creator and the date created, and give a basic description of the projection's nature (for example, that the shape is more stretched as you reach the poles). Some pages will also provide examples of the users of that particular projection and the purpose for which it is used, such as the Albers projection is exclusively used by the USGS for sectional maps of all 50 states. The usage information annotated in the tutorial is extracted primarily from Snyder (1987), and Snyder and Voxland (1989). Though these two books served as the basis for the projection information, some projections are not included and their usage information came from various map projection Websites (Savard 2012; Strebe 2010). This "usage" information comprises the heart of the tutorial (Fig. 10.7).

10.8 Implementation

The tutorial is implemented using a Java applet and requires the Java Runtime Environment (JRE) on the user's computer. A copy of the JRE is available, free, at http://www.java.com/en/download/manual.jsp.

Each node in the complete decision tree is assigned a number starting at the top, moving left to right, down each layer of the tree (Fig. 10.8). The numbers are collected into sets based on the part of the tree that is required for display as follows:

- The "DSS Applet," "Global," "Continental," and "Regional" nodes, 0, 1, 2, and 3 respectively, have their own set of numbers and are also in every other set since they are always present (See Fig. 10.8).
- The second set adds nodes "Area," "Shape," and "Compromise," 4, 5, and 6, to nodes 0 and 1.

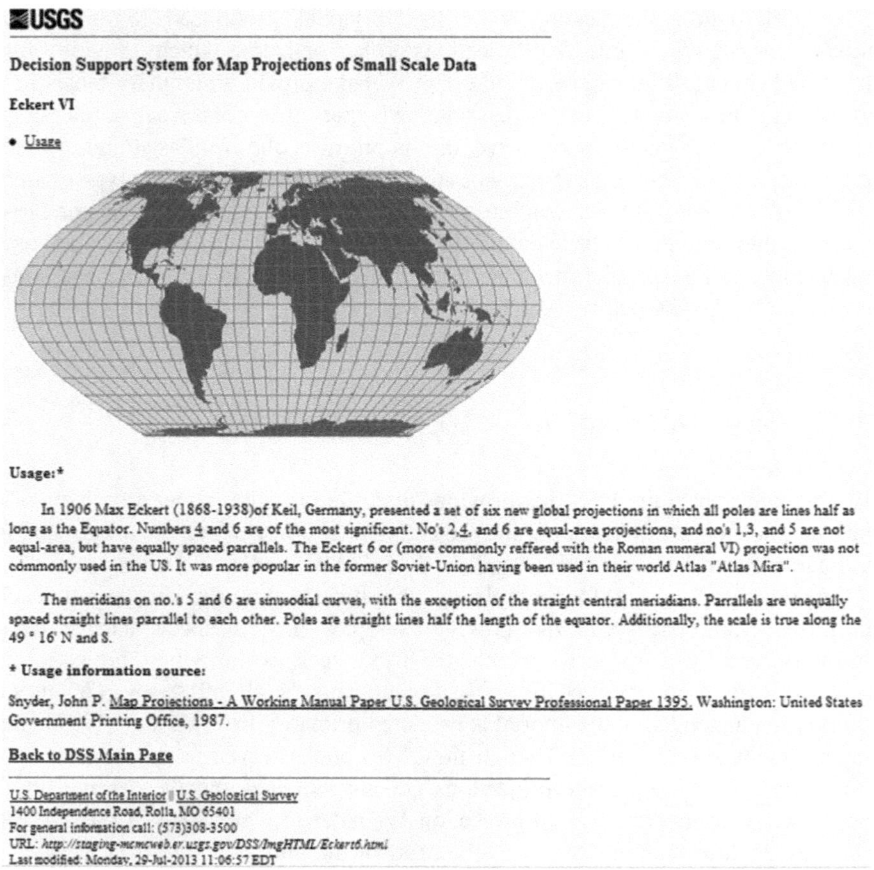

USGS

Decision Support System for Map Projections of Small Scale Data

Eckert VI

- Usage

Usage:*

In 1906 Max Eckert (1868-1938)of Keil, Germany, presented a set of six new global projections in which all poles are lines half as long as the Equator. Numbers 4 and 6 are of the most significant. No's 2,4, and 6 are equal-area projections, and no's 1,3, and 5 are not equal-area, but have equally spaced parrallels. The Eckert 6 or (more commonly reffered with the Roman numeral VI) projection was not commonly used in the US. It was more popular in the former Soviet-Union having been used in their world Atlas "Atlas Mira".

The meridians on no.'s 5 and 6 are sinusodial curves, with the exception of the straight central meriadians. Parrallels are unequally spaced straight lines parrallel to each other. Poles are straight lines half the length of the equator. Additionally, the scale is true along the 49 ° 16' N and S.

*** Usage information source:**

Snyder, John P. Map Projections - A Working Manual Paper U.S. Geological Survey Professional Paper 1395. Washington: United States Government Printing Office, 1987.

Back to DSS Main Page

U.S. Department of the Interior | U.S. Geological Survey
1400 Independence Road, Rolla, MO 65401
For general information call: (573)308-3500
URL: http://staging-mcmcweb.er.usgs.gov/DSS/ImgHTML/Eckert6.html
Last modified: Monday, 29-Jul-2013 11:06:57 EDT

Fig. 10.7 A screen shot of suggested projection with an example of the "usage" information provided

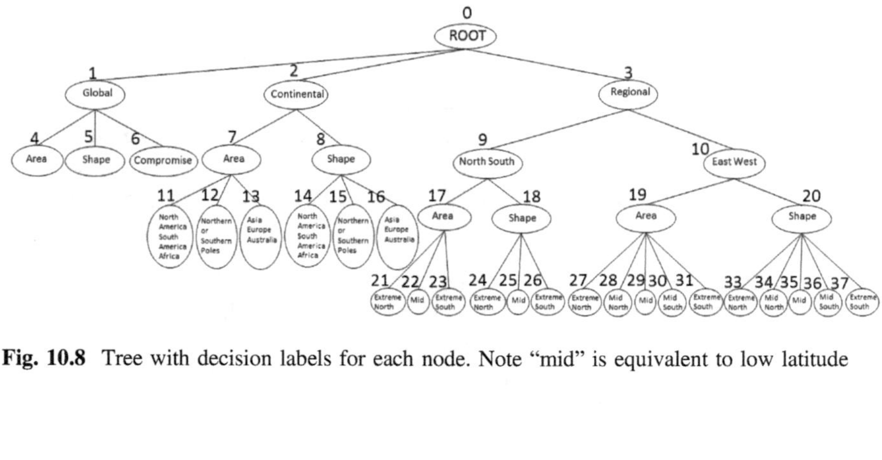

Fig. 10.8 Tree with decision labels for each node. Note "mid" is equivalent to low latitude

- A third set includes "Area" and "Shape," 7 and 8, nodes of "Continental" and adds them, along with the sub nodes under "Area," 11, 12, and 13, to nodes 0 and 2.
- The process continues in this pattern until eight sets are collected following the structure in Fig. 10.8 and are ready for display. Splitting the tree into sets simplifies the organization of the nodes.

An important feature of the DSS is the informational pop-up. One difficulty during the construction of the tutorial was to determine the most appropriate timing of the pop-up's appearance. In the execution of the DSS, it was decided that the pop-up would activate when the cursor is inside the area highlighted in red around a node that is colored green.

Another difficulty while implementing the tutorial was the synchronization with the DSS global, continental, and regional applets. The synchronization requires that there is communication between the different applets. Threading was used in order to achieve the required communication between the DSS applets and the tutorial. A thread is a "single sequential flow of control within a program," but more than one thread can run at once and perform different tasks in a single program (Sun Microsystems 2005). The thread connects the applets so that they are running concurrently and using the same information. Thus, when information changes based on the mouse movement on one of the DSS global, continental, or regional applets, it also changes the information for the tutorial applet as they share the same information. This signal allows the tutorial applet to determine that a change in the displayed or highlighted nodes may be necessary.

10.9 Conclusions

The tutorial is used to enhance understanding of the logic used to derive the most appropriate projection for a specific area based on the desired wants of the user and support basic education about map projections. In addition, the added features such as "Usage" supplement the understanding of the properties of a type of projection and its uses, and allow easy access to projection information on its usage or its special features. These capabilities address the problem that the myriad of projection choices available can be overwhelming to most users.

The ability for a user to identify the most appropriate projection or projections for the user's data using a DSS can help decrease unintentional distortions by narrowing the list of projections from which to choose. However, without this further documentation of the DSS logic, the user will not have the benefit of learning why specific projections are more appropriate than others. This tutorial is used to enhance understanding of the logic used to derive the most appropriate projection for a specific area based on the desired wants of the user. The added features such as "Usage" supplement the understanding of the properties of a type of projection and its uses, and allow easy access to projection information on its usage or its special features. Thus, the user is able to more fully understand the projection choices, and therefore become more fully informed.

Acknowledgements The authors acknowledge Mr. Jason Trent and Mr. James Nelson, two former USGS Computer Scientist Trainees, for their work on this version of the DSS. In particular, we are most grateful to Mr. James Nelson for his outstanding efforts on the tutorial and the corresponding Java development. In addition, we thank Misters David M. Mattli and Brian Passuello, USGS Computer Scientists for assistance in keeping the DSS current. Further, we thank Professor Emeritus Nedjeljko Frančula, University of Zagreb, for his valuable comments during the review of the manuscript.

References

Brown BH (1935) Conformal and equiareal world maps. Am Math Mon 42:212–223

Canters F (2002) Small-scale map projection design. Taylor & Francis, Loondon

Capek R (2001) Which is the best projection for the word map? In: Proceedings of the 20th international cartographic conference (Beijing, China), vol 5, pp 3084–93. http://icaci.org/documents/ICC_proceedings/ICC2001/icc2001/file/f24014.doc

De Genst W, Canters F (1996) Development and implementation of a procedure for automated map projection selection. Cartography Geogr Inf Syst 23(3):145–171

Finn MP, Woodard LN (2005) User's guide for the decision support system for map projections of small scale data, ver. 2.0. U.S. Geological Survey Open File Report 2005-1335

Finn MP, Usery EL, Posch ST, Seong JC (2004) A decision support system for map projections of small scale data. U.S. Geological Survey. Department of the Interior. Scientific Investigations Report 2004-5297

Finn MP, Steinwand DR, Trent JR, Buehler RA, Mattli D, Yamamoto KH (2012) A program for handling map projections of small scale geospatial raster data. Cartographic Perspect 71:53–67

Goldberg DM, Gott JR III (2007) Flexion and skewness in map projections of the earth. Cartographica 42(4):297–318

Hu Z, Zhu H, Li L, Xing Y (2007) COM-based expert system for map projection selection. In: Li M, Wang J (eds) Geoinformatics 2007: cartographic theory and models, proceedings of the SPIE, vol 6751. doi:10.1117/12.759722

Lee T, Birk RJ (2003) Connecting earth observation to decision making. EOM: Earth Obs Mag 12 (2):2–4 (Professional Surveyors Publishing Company, Inc., Fredrick, MD)

Main M, Savitch W (2001) Data Structures and other objects using C++, 2nd edn. Addison Wesley, Boston, p 459

Maling DH (1973) Coordinate systems and map projections. George Philip and Son Ltd, London

Nyerges TL, Jankowski P (1989) A knowledge-base for map projection selection. Am Cartographer 16(1):29–38

PathAway (2004) Map projections description. http://www.pathaway.com/mapproj.htm. Accessed 24 Oct 2012

Robinson A, Sale R, Morrison J (1978) Elements of cartography, 4th edn. Wiley, New York

Savard JJG (2012) Map projections. http://www.quadibloc.com/maps/mapint.htm. Accessed 15 Jan 2013

Slocum TA, McMaster RB, Kessler FC, Howard HH (2009) Thematic cartography and geovisualization, 3rd edn. Upper Saddle River, Prentice Hall, NJ

Snyder JP (1987) Map projections—a working manual. U.S. Geological Survey Professional Paper 1395. Washington, DC

Snyder JP (1993) Flattening the earth: two thousand years of map projections. The University of Chicago Press, Chicago

Snyder JP, Voxland PM (1989) An album of map projections. U.S. Geological Survey Professional Paper 1453. Washington, DC

Stoker JJ (1961) Developable surfaces in the large. Commun Pure Appl Math 14(3):627–635. doi:10.1002/cpa.3160140333

Strahler AN, Strahler AH (1979) Elements of physical geography, 2nd edn. Wiley, New York

Strebe D (2010) Map projection essentials. http://www.mapthematics.com/Essentials.php. Accessed: 4 Oct 2012

Sun Microsystems, Inc. (2005) The Java$^{(TM)}$ Tutorials. http://docs.oracle.com/javase/tutorial/. Accessed 24 Aug 2010

Tobler WR (1962) A classification of map projections. Ann Assoc Am Geogr 52(2):167–175

Tobler WR (1963) Geographical area and map projections. Geogr Rev 53(1):59–78

Tobler WR (1979) A transformational view on cartography. Am Cartographer 6(2):101–106

Trewartha GT, Horn LH (1980) An Introduction to climate, Fitfth edn. McGraw-Hill Book Company, New York

Usery EL (2001) Peters' projection. American Congress on Surveying and Mapping Bulletin, No. 193, Nov 2001

Usery EL, Seong JC, Steinwand DR, Finn MP (2003) Accurate projection of small-scale raster datasets. In: Proceedings 21st international cartographic conference, Durban, South Africa. International Cartographic Association

Voxland PM (1978) World mapping and projection program: reference manual. Regents of the University of Minnesota, Minneapolis, MN

Yang Q, Snyder JP, Tobler W (2000) Map projection transformation: principles and applications. Taylor & Francis, London

Chapter 11
Short History on Map Projections

Miljenko Lapaine

Map projections have been developed in parallel with the development of map production and cartography in general. The development of many sciences, technical achievements and the needs of everyday life have gradually initiated wider and wider demands for the production of various topographic and thematic maps in various scales and for various purposes, which required continuous growth of map projections and improvement of mathematical basis of maps.

The beginnings of map projections date as far as two thousand years ago, originating from the time when the old Greek scientists introduced mathematical principles into the basis of projecting the Earth and starry sky and started to apply the graticule. The works of Anaximander, Eratosthenes, Apollonius and Hipparchus played an important role in the development of cartography (Snyder 1993).

It is believed that Thales of Milet made the first map in some projection 600 years B.C. It was a map of the heavenly sphere in gnomonic projection. Stereographic and orthographic projections belong to the oldest projections and were used by the Greek astronomer and mathematician Hipparchus for the purpose of making maps of heavenly sphere about 150 B.C. Ever since there have been hundreds of map projections invented up to the present day.

In the 2nd century Ptolemy wrote a capital work *Geography* and included into it the description of map compilation and determination of the Earth's dimensions, as well as the construction of map projections. The period of Middle Ages in Europe was characterised by the so-called monastery maps reflecting a religious image of the world.

Special development of cartography started during the renaissance—the period of great geographic discoveries. Accurate, reliable maps were created to be used for state government and military purposes, for the development of trade and maritime

M. Lapaine (✉)
Faculty of Geodesy, University of Zagreb, Zagreb, Croatia
e-mail: mlapaine@geof.hr

M. Lapaine and E.L. Usery (eds.), *Choosing a Map Projection*,
Lecture Notes in Geoinformation and Cartography,
DOI 10.1007/978-3-319-51835-0_11

affairs. Such maps could be produced only through the application of mathematical basis and land survey results. The first to have appeared were topographic maps.

At the end of the 16th and 17th centuries, the compilation and publication of geographic atlases were a very significant event in further development and popularisation of cartography, which was carried out by the famous Dutch cartographers Ortelius and Mercator. Mercator (Fig. 11.1) was the first who ever applied conformal cylindrical projection used successfully for maritime navigation charts up to now.

In the production of world maps and the maps of larger territories, trapezoidal projection and Apianus projection used to have large application in that time, and they were used as archetype for later more detailed pseudocylindrical projections. In the 17th century a new sinusoidal pseudocylindrical projection for the world map was suggested by the French cartographer N. Sanson.

Detailed work on the scientific basis of cartography and the beginnings of topographic studying of the Earth, and as a result, further growth of accuracy and

Fig. 11.1 Title page of Mercator's *atlas sive* ..., Duisburg 1595

reliability of maps are regarded as the characteristics of the 18th century. A series of new projections suggested by R. Bonne, J.H. Lambert. J.L. de Lagrange, L. Euler and others, was introduced into cartographic practice.

The renaissance came to an end with the introduction of elementary mathematical analysis into the development of map projections. Such analysis was applied especially to the Mercator projection in which the rhumb lines were shown as straight lines, and to projections where all parallels are divided by meridians in true interspaces, regardless of the fact whether the parallels are circular arcs (Werner's projections) or straight lines (sinusoidal). Map projections become more complex: instead of those having graticule simply drawn because they consist of circular arcs and straight lines, there are such that are delineated by means of tables of trigonometric functions.

Murdoch (1758) required consciously that the total area of the projected territory should be correct, but he did not insist on the constant local scale of the area. The first intentional preservation of area in each point was made by Lambert (1772) by inventing cylindrical, azimuthal and conical equal-area projections. Halley proved geometrically the conformity of stereographic projection, and Lambert did it with differential calculus (Fig. 11.2), and invented also three new conformal projections. Many map projections were created also in the 19th and 20th centuries, but the basic principals were developed by 1772, and especially in that year.

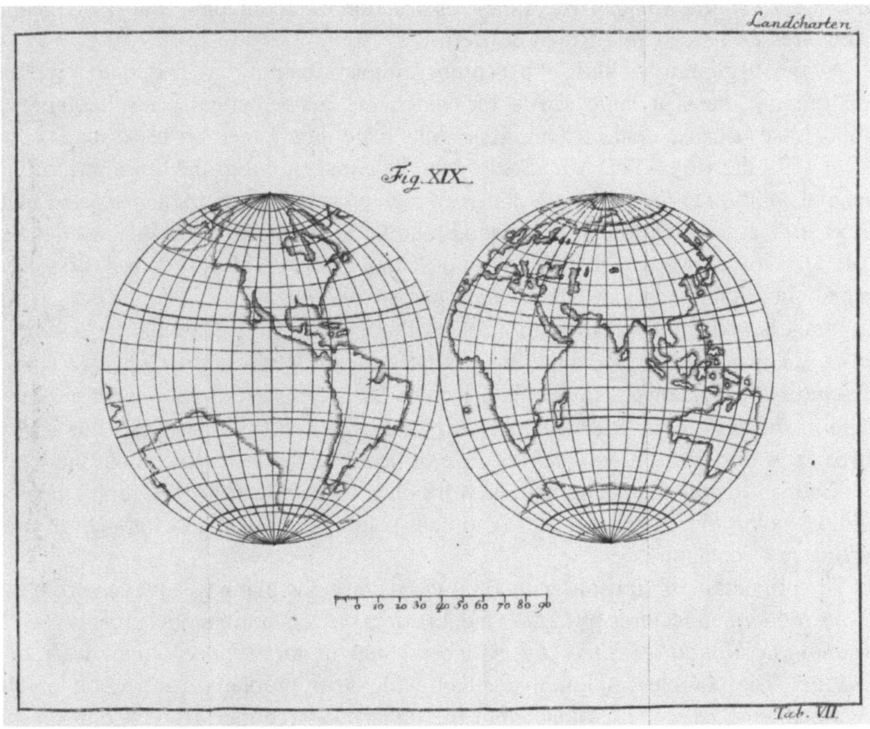

Fig. 11.2 Fig. XIX of *Lambert's Beiträge zum Gebrauche der Mathematik und deren Anwendungen, Dritter Theil*, Berlin 1772

The most significant contribution of cartographers, geodesists and mathematicians of the 19th century to the theory of map projections was the establishment of firm mathematical principals. Lambert and Lagrange made an important beginning in 1770, but especially Gauss and Tissot gave large contributions prior to 1880. Other researchers, such as e.g. Airy, Clarke, Schwartz and Pierce, concentrated on more specific, but more complex tasks, and Germain, Gretschel and Craig gathered various works by other scientists into special monographs. New projections by Mollveide, Albers, Gall and others contain simpler application of mathematics, but still deserve the titles according to their inventors. Rapid development of the theory of map projections in this period is demonstrated by a number of new projections and a number of published books and articles (Snyder 1993).

According to Frischauf (1905), the beginning of the theory of projecting one surface onto another belongs to J.H. Lambert, who dealt with generally given problem of projecting a sphere and spheroid into the plane in his *Anmerkungen und Zusätze zur Entwerfung der Land- und Himmelscharten* (Remarks and Additions to the Establishment of Land and Sky Maps) in the third part of his *Beiträge zum Gebrauch der Mathematik und deren Anwendung* (Contributions to the Usage of Mathematics and its Applications, 1772). Lambert's colleague, J.L. Lagrange (1781) was inspired by his separation of variables in the expression for the arc length differential in conformal projection and he solved the problem of conformal projection of rotational surfaces publishing it in two treatises *Sur la construction des cartes géographiques* (About the Construction of Geographic Maps, Nouveaux Mémoires de l'Académie Royal de Berlin, 1779).

At the beginning of the 19th century military institutions started to produce topographic maps at large scales for which the mathematical basis has special importance because distances and directions were determined on these maps.

In 1822 the Royal Scientific Society in Copenhagen raised the question: to find general solution to the problem—a part of a given surface should be projected onto another given surface so that the image is similar to the origin in its smallest details. C.F. Gauß was awarded for the solution of this problem. His work was first published in Schumacher's *Astronomische Abhandlungen* in 1825 (Fig. 11.3). Lagrange's and Gauß' treatise are in the volume No. 155 of Ostwald's classics of exact sciences. C.G.J. Jacobi (1886) noticed in his *Vorlesungen über Dynamik* (Lectures on Dynamics), in the 28th lecture *Die kürzeste Linie auf dem dreiachsigen Ellipsoid. Das Problem der Kartenprojektion* (The shortest line on a three-axes ellipsoid. The problem of map projection), the following about the work by Gauß: "It contains Langrange's work that is only a little bit supplemented without being mentioned." Apart from that, Gauß gave only the examples of projecting rotational surfaces.

The projection of ellipsoid with various axes into the plane was processed in the above mentioned lectures by C.G.J. Jacobi. He gave the first announcement of such solution in *Monatsberichten der Akademie* and in the 19th volume of *Crelles Journal*. The complete solution together with other problems (projection of the rotational surface, cone, cylinder into the plane) was reported by L. Cohn on the basis of Jacobi's heritage in the 59th Volume of Crelles Journal under the title *Über*

[5]

Allgemeine Auflösung der Aufgabe: Die Theile einer gegebnen Fläche auf einer andern gegebnen Fläche so abzubilden, dass die Abbildung dem Abgebildeten in den kleinsten Theilen ähnlich wird.

Von

C. F. Gauss.

(Astronom. Abhandlungen, herausgegeben von H. C. Schumacher. 3. Heft. Altona 1825. S. 5—30.)

1.

Die Natur einer krummen Fläche wird durch eine Gleichung zwischen den sich auf jeden Punkt derselben beziehenden Coordinaten x, y, z bestimmt. Vermöge dieser Gleichung kann jede dieser drei veränderlichen Grössen wie eine Function der beiden andern betrachtet werden. Noch allgemeiner ist es, noch zwei neue veränderliche Grössen t, u einzuführen, und jede der x, y, z als eine Function von t und u darzustellen, wodurch, wenigstens allgemein zu reden, bestimmte Werthe von t und u allemal einem bestimmten Punkte der Oberfläche angehören, und umgekehrt.

2.

In Beziehung auf eine zweite krumme Fläche sollen X, Y, Z, T, U ähnliche Bedeutungen haben, wie resp. x, y, z, t, u in Beziehung auf die erstere.

Fig. 11.3 Title page of Gauss's *Allgemeine Auflösung der Aufgabe* ..., after Ostwald's Klassiker der exakten Wissenschaften, Nr. 155, 57–101, Verlag von Wilhelm Engelmann, Leipzig

die Abbildung eines ungleichachsigen Ellipsoides auf einer Ebene, bei welcher die kleinsten Teile ähnlich bleiben (About the projection of the ellipsoid with unequal axes into the plane, whereby the smallest parts remain similar). This work contains also a very simple theory of conformal projection of one surface onto another. In the lectures about dynamics Jacobi explains: "The successful solving of the problem of conformal ellipsoid projection is achieved by means of adequate substitution and method leading to one partial differential equation. The noticed ellipsoid point will be determined by intersection of two curves of curvature". Disintegrating the ellipsoid surface into the elements limited by curves of curvature, Legendre already determined this surface (*Exercises du calcul integral*, Exercises in integral calculus, 1811) and found the corresponding variables by means of which the known projection problem could be solved. Ernst Schering gave a complete solution of the problem on the basis of Jacobi's announcement in his work *Über die konforme Abbildung des Ellipsoides auf der Ebene* (About the conformal projection of ellipsoid into the plane) for which he was awarded in 1858 with the reward of the Faculty of Philosophy.

Surrounded by numerous works on cartography the majority of which are dealing with the methods of constructing graticules, I should by all means look back on the classical work, the corner stone of the theory of map projections by N.A. Tissot *Mémoire sur la représantation des surfaces et les projections des cartes géographiques* (Treatise on surface presentation and geographic map projections, Paris 1881) the basic part of which, chapters from I to IV, appeared already between 1878 and 1880 in Nouvelles Annales des Mathématique, 2e série. According to K. Zöppritz, the work by Tissot was noticed in Germany and Austria as well, and E. Hammer translated it into German under the title *Die Netzentwürfe geographischer Karten nebst Aufgaben über Abbildung beliebiger Flächen aufeinander* (Presentations of graticulesof geographic maps including the problems on projecting one arbitrary surface onto another, Stuttgart 1887), in which many presentation of graticules and tables were added.

After 1900 about fifty monographs on map projections in about ten languages were published (Snyder 1993). There are a few in English and are very often quoted. In chronological order related to the first publication these are Hinks (1912) *Map projections*, Melluish (1931), *An Introduction to the Mathematics of Map Projections*, Deetz and Adams (1934) *Elements of Map Projections*, Steers (1970) *An Introduction to the Study of Map Projections*, Richards and Adler (1972) *Map Projections for Geodesists, Cartographers and Geographers*, Maling (1973) *Coordinate Systems and Map Projections*, Snyder (1982) *Map Projections Used by the U.S. Geological Survey* and Snyder (1987) *Map Projections: A Working Manual*. There is also the latest manual by Bugayevskiy and Snyder (1995) *Map Projections—A Reference Manual*. One should mention also longer treatises with special topics about map projections by Adams (1919, 1949), Young (1920), Thomas (1952), Lee (1976) and Snyder (1985).

The following monographs in German should be mentioned: Maurer (1935) *Ebene Kugelbilder*, Wagner (1949) *Kartographische Netzentwürfe*, Merkel (1956) *Grundzüge der Kartenprojektionslehre*, Hoschek (1969, 1984) *Mathematische Grundlagen der Kartographie*, Kuntz (1983) *Kartennetzentwurfslehre*, then a series of works by Bulgarian geodesist Hristow, published in Zeitschrift für Vermessungswesen, i.e. in books (1943, 1955) and the German translation from Czech Fiala (1957) *Mathematische Kartographie*.

In French there are: Driencourt and Laborde (1932) *Traité des projections des cartes géographiques* and Reignier (1957) *Les systèms de projection et leurs applications.*

In the former Soviet Union many monographs were written, and the most prominent authors are: Kavrayskiy (1958, 1959, 1960), Solov'ev (1946, 1969), Graur (1956), Urmayev (1941), Ginzburg and Salmanova (1957), Meshcheryakov (1968), Pavlov (1972, 1974), and Vakhrameyeva et al. (1986). The book by Bulgarian geodesist Hristov (1957) was also published in Russian.

To meet the requirements of teaching and production, many educational materials, such as Mathematical Cartography (Wu 1961), Map Projections (Li et al. 1993), Mathematical Basis for Charts (Hua 1985), and Principles of Mathematical Cartography (Wu 1989) were edited and published by Chinese colleges and universities of surveying and mapping, and by secondary technical schools of surveying and mapping. These monographs systematically expounded the general theory of map projections and gave a representation of new research results on map projections.

The application of map projections has a long history. Especially with the development of radio navigation and the technology of radio positioning, a series of new research topics has become the subject of map projection research. Map projections, as a means of establishing the relation between space and a plane, have been widely used to solve some geometric problems of spherical geometry, astronomy, crystallography, and geology in graphical form. The application of Landsat to mapping introduced completely new concepts for map projections. Time has now become a parameter in mapping. This is quite different from conventional static mapping in which the relations among the Earth's shape, perspective center and projection plane are fixed. The entirely new research topic of studying a projection suitable for satellite mapping confronts map projection science. In recent years, electronic computers, especially personal computers, have been widely applied to all aspects of map projections and have thoroughly changed the look of map projection science. Examples are the applications of computers to the calculation of coordinates, to the automatic creation of the mathematical foundation of maps, and to the automatic plotting of thematic mathematical elements on a map. Computer-aided map projection transformation is even more of a leap for cartography. To meet the need of computer cartography, it is a pressing task to study the theory and methods of map projection transformation, to study topographic data processing, spatial information positioning, and transformation in information systems (Yang et al. 2000).

Geoinformation systems (GIS) enable today that some problems that have been so far solved only on geographic maps (various cartometric problems) are now solved directly from the databases. On the basis of that fact we could come to a conclusion that GIS lessen the significance of geographic maps. If we look upon it from that point of view, then it really is true, but on the other hand geographic maps are very important for every GIS. They have a very important role in creation of databases, but also as one of the forms in presenting the output data (Sijmons 1992).

Since each geographic map is made in a certain map projection, one can come to a conclusion that map projections are especially important in creating GISs. In the creation of national digital bases of geodetic, topographic and cartographic data that must make the foundations of each GIS being prepared for the territory of the entire state, a method of digitising existing maps is very important (Sowton 1991; Teng 1991). Therefore, the majority of GIS software contain also a module for digitising. In the application of this module it is necessary to be familiar with the map projection of the origin and projection constants (e.g. geographic longitude of the central meridian or the latitude of the standard parallel, and linear scales along them).

These data are indispensable in order to transform the co-ordinates from local digitizer system into the system of source map projection and then by means of inverse equations of map projections into the system of geographic coordinates. It would thus be possible for GIS software as one of the presentation forms for output data to offer a geographic map in one of a great number of the most important map projections. In order to draw such a map, it is necessary to calculate rectangular co-ordinates x, y in the selected map projection from the geographic co-ordinates.

The production of software for any GIS requires thus the direct and inverse equations for a large number of map projections. Hence, the computer aided method in the map production and first of all GISs have not reduced, but increased the importance of map projections. More than 1000 works on map projections published after 1960 and registered in the bibliography of Snyder and Steward (1988) prove that this statement is correct.

According to Yang et al. (2000) the major tasks in front of map projection science can be summed up as follows:

- To explore the theory and method of creating geographic networks for different purpose maps, their calculating and plotting, as well as to solve some problems about map disposition and sheet separation
- To study the projection of selected curves from an ellipsoid or sphere to a plane and their expression thereon, as well as problems involving the measurement of elements on a map
- To delve into the theory and method of map projection transformation meeting the requirements of automatic cartography and the transformation between different projections
- To develop a projection method fit for solving the problems of planar and linear geometry in order to meet the needs of other sciences such as astronomy, crystallography and geology.
- To research celestial space and satellite projections to meet the need of developing of space technology
- To probe digital data processing in map databases, spatial information positioning systems, and map projection transformation systems to meet the requirement of building different specific information systems and developing spatial information science.

References

Adams OS (1919) A study of map projections in general. U.S. Coast and Geodetic Survey Special Publication 57. Reprinted 1934

Adams OS (1949) Latitude developments connected with geodesy and cartography. Special Publication No. 67, Department of Commerce, U.S. Coast and Geodetic Survey, Washington

Bugayevskiy LM, Snyder JP (1995) Map projections—a reference manual. Taylor & Francis, London

Deetz CH, Adams OS (1934) Elements of map projection with application to map and chart construction. U.S. Coast and Geodetic Survey Special Publication 68, Washington

Driencourt L, Laborde J (1932) Traité des projections des cartes géographiques à l'usage des cartographes et des géodésiens. Hermann et Cie, Paris

Fiala F (1957) Mathematische kartographie. VEB Verlag Technik, Berlin

Frischauf J (1905) Die Abbildungslehre und deren Anwendung auf Kartographie und Geodäsie. Druck und Verlag von B. G. Teubner, Leipzig

Gauß CF (1825) Allgemeine Auflösung der Aufgabe: Die Theile einer gegebnen Fläche auf einer andern gegebnen Fläche so abzubilden, dass die Abbildung dem Abgebildeten in den kleinsten Theilen änlich wird. Astronomischen Abhandlungen, herausgegeben von H. C. Schumacher, Heft 3, 1-30, Altona. Also in: Gauß' Werken Bd. IV, 189–216, Göttingen, 1873; Ueber Kartenprojection. Abhandlungen von Lagrange (1779) und Gauß (1822). Herausgegeben von A. Wangerin. Ostwald's Klassiker der exakten Wissenschaften, Nr. 155, 57–101, Verlag von Wilhelm Engelmann, Leipzig

Ginzburg GA, Salmanova TD (1957) Atlas dlya vybora kartograficheskikh proyektsiy. Trudy Central'nogo nauchno-issledovatel'skogo instituta geodezii, aeros'emki i kartografii, Vypusk 110. Geodezizdat, Moscow

Graur AV (1956) Matematicheskaya kartografiya. Izdatel'stvo Leningradskogo Universiteta, Leningrad

Hinks AR (1912) Map projections. Cambridge University Press, Cambridge

Hoschek J (1969, 1984) Mathematische Grundlagen der Kartographie. Bibliographisches Institut, Mannheim, Zurich

Hristow WK (1943) Die Gauss-Krüger'schen Koordinaten auf dem Ellipsoid. B. G. Teubner, Leipzig, Berlin

Hristow WK (1955) Dei Gaußschen und geographischen Koordinaten auf dem Ellipsoid von Krassowsky. VEB Verlag Technik, Berlin

Hristov VK (1957) Koordinaty Gaussa-Krjugera na ellipsoide vrashcheniya. Geodezizdat, Moscow

Hua Tang (1985) Mathematical basis for charts. The Navigation Guarantee Department of the Chinese Navy Headquarters (in Chinese)

Jacobi CGJ (1886) Vorlesungen über Dynamik, nebst fünf hinterlassenen Abhandlungen desselben, herausgegeben von A. Clebsch, Berlin, Druck und Verlag von Georg Reimer, 28. Vorlesung: Die kürzeste Linie auf dem dreiaxigen Ellipsoid. Das Problem der Kartenprojection, 212–220

Kavrayskiy VV (1958) Izbrannye trudy, Tom II, Matematicheskaya kartografiya, Vypusk 1, Obshchaja teorija kartograficheskikh proyekcij. Izdanie Upravleniya nachal'nika Gidrograficheskoy sluzhby VMF

Kavrayskiy VV (1959) Izbrannye trudy, Tom II, Matematicheskaya kartografiya, Vypusk 2, Konicheskie i cilindricheskie proekcii, ikh primyenenie. Izdanie Upravleniya nachal'nika Gidrograficheskoy sluzhby VMF

Kavrayskiy VV (1960) Izbrannye trudy, Tom II, Matematicheskaya kartografiya, Vypusk 3, Perspektivnye, krugovye i drugie vazhnejshie proekcii. Navigacionnye zadachi. Izdanie Upravlenija nachal'nika Gidrograficheskoj sluzhby VMF

Kuntz E (1983) Kartennetzentwurfslehre. Herbert Wichmann Verlag, Karlsruhe

Lagrange JL de (1781) Sur la construction des cartes géographiques, Nouveaux Mémoires de l'Académie Royale des Sciences et Belles Lettres de Berlin, Année 1779, 161–210 (Berlin 1781). Also in: Oeuvres de Lagrange, T. IV, 635–692, and in: Ueber Kartenprojection. Abhandlungen von Lagrange (1779) und Gauß (1822). Herausgegeben von A. Wangerin. Ostwald's Klassiker der exakten Wissenschaften, Nr. 155, Verlag von Wilhelm Engelmann, Leipzig, 3–56

Lambert JH (1772) Beiträge zum Gebrauche der Mathematik und deren Anwendungen, Dritter Theil, im Verlag der Buchhandlung der Realschule, Berlin. Translated in English with an introduction by Tobler WR: Notes and Comments on the Composition of Terrestrial and Celestial Maps, Michigan Geographical Publication No. 8, Department of Geography, University of Michigan, Ann Arbor, 1972; republished by Esri Press, 2011

Lee LP (1976) Conformal projections based on elliptic functions. Canadian Cartographer 13(1) (Cartographica Monograph 16)

Li G, Yang Q, Hu D (1993) Ditu touying (Map projections). Chinese People's Liberation Army Press

Maling DH (1973) Coordinate systems and map projections. George Philip and Son Ltd., 2nd edn. Pergamon Press, Oxford

Maurer H (1935) Ebene Kugelbilder. Ein Linnésches System der Kartenentwürfe. Petermanns Mitteilungen, Ergänzungsheft 221

Melluish RK (1931) An introduction to the mathematics of map projections. Cambridge University Press, Cambridge

Merkel H (1956, 1958) Grundzüge der Kartenprojektionslehre, Teil I, Die theoretischen Grundlagen, Teil II, Abbildungsverfahren. Deutsche Geodätische Kommission, Reihe A, Heft 17 und 17/II

Meshcherjakov GA (1968) Teoreticheskie osnovy matematicheskoy kartografii. Izdatel'stvo Nedra, Moscow

Murdoch P (1758) On the best form of geographical maps. Philos Trans Roy Soc 50(2):553–562

Pavlov AA (1972) Nekatorye voprosy matematicheskoy kartografii. Geodeziya i kartografiya 2:65–68

Pavlov AA (1974) Prakticheskoe posobie po matematicheskoy kartografii. Izdatel'stvo Leningradskogo universiteta, Leningrad

Reignier F (1957) Les systèmes de projection et leurs applications, Tome I, Texte, Tome II, Graphiques. Publications techniques de l'Institute Géographique National, Paris

Richardus P, Adler RK (1972) Map projections for geodesists, cartographers and geographers. North Holland Publishing, Amsterdam

Sijmons K (1992) Cartographic representation and production of GIS data. In: Proceedings of the first international conference on surveying and mapping, Tehran, vol 2, pp 117–125

Snyder JP (1982) Map projections used by U.S. geological survey. U.S. Geological Survey Bulletin 1532, 2nd edn, Washington

Snyder JP (1985) Computer-assisted map projection research. U.S. Geological Survey Bulletin 1629, Washington

Snyder JP (1987) Map projections: a working manual. U.S. Geological Survey Professional Paper 1395, Washington

Snyder JP (1993) Flattening the earth, two thousand years of map projections. The University of Chicago Press, Chicago, London

Snyder JP, Steward H (eds) (1988) Bibliography of map projections. U.S. Geological Survey Bulletin 1856, Washington

Solov'ev MD (1946) Kartograficheskie proyektsii. Izdatel'stvo geodezicheskoy i kartograficheskoy literatury. GUGK pri SNK SSSR, Moscow

Solov'ev MD (1969) Matematicheskaya kartografiya. Izdatel'stvo Nedra, Moscow

Sowton M (1991) Development of GIS-related activities at the ordnance survey. In: Maguire J, Goodchild MF, Rhind DW (eds) Geographical information systems, vol. 2—application. Longman, London, pp 23–38

Steers JA (1970) An introduction to the study of map projections, 15th edn. University of London Press, London
Teng AT (1991) Building geographic data bases with input from federal digital data files. Technical papers of 1991 ACSM-ASPRS annual convention, Baltimore 1991, vol 4, GIS, pp 208–216
Thomas PD (1952) Conformal projections in geodesy and cartography. U.S. Coast and Geodetic Survey Special Publication 251, Washington
Tissot A (1881) Mémoire sur la représentation des surfaces et les projections des cartes géographiques. Gauthier-Villars, Paris
Tissot A (1887) Die Netzentwürfe geographischer Karten (Autorisierte Deutsche bearbeitung mit einigen Zusätze besorgt von E. Hammer). J. B. Metzlersche Buchhandlung, Stuttgart
Urmayev NA (1941) Matematicheskaya kartografiya. Izdatel'stvo Vojennoinzhenernoy Akademii, Moscow
Vakhrameeva LA, Bugayevskiy LM, Kazakova ZL (1986) Matematicheskaya kartografiya. Nedra, Moscow
Yang Q, Snyder JP, Tobler WR (2000) Map projection transformation, principles and applications. Taylor & Francis, London
Young AE (1920) Some investigations in the theory of map projections. Royal geographical society, Technical Series 1, London
Wagner K-H (1949) Kartographische Netzentwürfe. Bibliographisches Institut, Leipzig
Wu Z (1961) Mathematical cartography. Zhengzhou Institute of Surveying and Mapping (in Chinese)
Wu Z, Yang Q (1989) Shuhxue zhitu xue yuanli (Principles of mathematical cartography). China Surveying and Mapping Press (in Chinese)

Chapter 12
Famous People and Map Projections

Miljenko Lapaine and Ana Kuveždić Divjak

Abstract Map projections have been developed in parallel with the development of map production and cartography in general. Development of sciences, technical achievements and needs of everyday life have gradually increased demands for production of various topographic and thematic maps in various scales and for various purposes, which required continuous improvement of map projections and mathematical basis of maps. Beginnings of map projections date as far as two thousand years ago, when ancient Greek scientists applied mathematical principles to projecting Earth and the starry sky and started applying the graticule. Hundreds of map projections have been invented since the antique. Many people have been interested in the theory of map projections and have written about them. Since there are so many of them, we decided to make a narrower selection of abut 40 people. Those people are presented in chronological rather than alphabetical order in this chapter. Their most important contributions to map projections are described and illustrated.

Beginnings of map projections date as far as two thousand years ago, when ancient Greek scientists applied mathematical principles to projecting Earth and the starry sky and started applying the graticule. Hundreds of map projections have been invented since the antique. Development of sciences, technical achievements and needs of everyday life have gradually increased demands for production of various topographic and thematic maps in various scales and for various purposes, which required continuous improvement of map projections and mathematical basis of maps.

Many people dealt with theory of map projections and wrote about them (Snyder and Steward 1988). Since there are too many of them to be dealt in details in one chapter, we decided to make a narrower selection. This selection is shown in Table 12.1.

Small maps of the world in different projections have been prepared with the publicly-available resampled *Shuttle Radar Topographic Mission* (SRTM) digital elevation data to 250 m resolutions for the entire globe (http://srtm.csi.cgiar.org/)

M. Lapaine (✉) · A. Kuveždić Divjak
Faculty of Geodesy, University of Zagreb, Zagreb, Croatia
e-mail: mlapaine@geof.hr

M. Lapaine and E.L. Usery (eds.), *Choosing a Map Projection*,
Lecture Notes in Geoinformation and Cartography,
DOI 10.1007/978-3-319-51835-0_12

for use in the sections below. We used either geoinformation software *Grass GIS* or map projection software *G.Projector* to transform an equirectangular map image into any of target map projections. Resulting images were further processed with vector graphics editor *Inkscape* and raster graphics editor *GIMP*, where the graticules have been drawn and the oceans lightened to enhance image legibility and contrast.

Hereafter follow basic information about persons from Table 12.1 in cronological order.

Table 12.1 Alphabetical list of persons presented in more detail in this chapter, which are extremely important for development of map projections

Airy, George Biddell—British astronomer, mathematician and cartographer
Aitoff (Aitov, Aitow), David Aleksandrovich—Russian cartographer
Albers, Heinrich Christian—German cartographer
August, Friedrich Wilhelm Oscar—German professor of mathematics
Boggs, Samuel Whittemore—American geographer
Bonne, Rigobert—French engineer, mathematician and cartographer
Briesemeister, William A.—American cartographer
Bugayevskiy, Lev Moiseyevich—Russian cartographer professor, colonel
Cassini, César François Cassini de Thury—French astronomer and cartographer
Clarke, Alexander Ross—British geodesist, mathematician and officer
Eckert-Greifendorf, Max—German geographer and cartographer
Eisenlohr, Friedrich Eisenlohr—German mathematician
Euler, Leonhard—Swiss mathematician, physicist, astronomer, logician and engineer
Gall, James—Scottish clergyman, cartographer and astronomer
Gauss, Carl Friedrich—German mathematician, astronomer and geodesist
Gilbert, Edward Nelson—American mathematician and coding theorist
Goode, John Paul—American geographer and cartographer
Hammer, Ernst Hermann Heinrich von—German geodesist and cartographer
Hipparchus—Old Greek astronomer and mathematician
Jordan, Wilhelm—German geodesist and mathematician
Kavrayskiy, Vladimir Vladimirovich—Russian cartographer, professor, engineer-rear-admiral
Khristov (Hristov, Hristow), Vladimir Kirilov—Bulgarian geodesist, astronomer and cartographer
Krüger, Johann Heinrich Louis—German mathematician and geodesist
Lagrange, Joseph Louis—French mathematician and astronomer
Lambert, Johann Heinrich—German physicist, mathematician and astronomer and cartographer originating from France
Lee, Laurence Patrick—New Zealand surveyor and cartographer
McBryde, Felix Webster—American geographer, cartographic consultant and educator
Mercator, Gerhard Kremer—Flemish geographer, cartographer
Miller, Osborn Maitland—American cartographer and geographer
Mollweide, Karl Brandon—German mathematician and astronomer
Nicolosi, Giambattista—Italian priest and cartographer
Postel, Guillaume—French linguist, geographer, astronomer, diplomat, professor of mathematics

(continued)

Table 12.1 (continued)

Ptolemy, Claudius—Egyptian astronomer, mathematician, and geographer
Putniņš, Reinholds V.—Latvian mathematician
Robinson, Arthur H.—American geographer and cartographer
Sanson, Nicolas—French cartographer and geographer
Snyder, John Parr—American cartographer
Thales—Old Greek philosopher and mathematician
Tissot, Nicolas Augustes—French cartographer
Van der Grinten, Alphons J.—American cartographer of German origin
Vitkovskiy, Vasiliy Vasil'evich—Russian topographer and geodesist, cartographer, professor, lieutenant general
Wagner, Karlheinz (Karl Heinrich)—German cartographer
Winkel, Oswald—German cartographer

Thales (Greek Θαλῆς)
(Miletus, ca. 625. B.C.–ca. 547. B.C.)
Old Greek philosopher and mathematician

Thales of Miletus was the first Greek philosopher, scientist and mathematician, one of seven Wise Greek Men. Unfortunately, no writings by Thales were preserved, so it is hard to determine his ideas or to be completely certain about his mathematical conclusions. In many books on the history of mathematics, Thales is credited with these theorems: a circle is bisected by any diameter, the base angles of an isosceles triangle are equal, the angles between two intersecting straight lines are equal (ceiling angles are considered), two triangles are congruent if they have two angles and one side equal, an angle in a semicircle is a right angle. Last mentioned theorem is today called Thales theorem. He used the properties of similarity and in that way he measured the height of pyramids and the distance of a boat on the open sea. It is reported that Thales predicted an eclipse of the Sun in 585 B.C. Prediction of Moon eclipse was well known at this time, but it was hard to say when the eclipse of Sun would occur, since this phenomenon could not have been seen from all the parts on Earth. Still, the most important thing mathematicians attribute to him is the fact that Thales was the first to give logical foundations for proving the theorems. In other words, he was first to emphasize that it is not sufficient just to observe the phenomena, but they must be proven.

He believed the Earth was a flat disc floating on water, i.e. on an infinite ocean and that all things come to be from water. Despite those, today unacceptable theses,

Thales' greatness is that he was the first recorded person who tried to explain his attitudes by rational rather than by supernatural means, as many did before him.

It is believed that Thales made the first map in a projection 600 B.C. It was a map of the heavenly sphere in gnomonic projection.

Hipparchus (Greek Ἵππαρχος, Hipparkos)
(Nicaea, now Iznik, ca. 190. B.C.–Rhodes, ca. 120. B.C.)
Old Greek astronomer and mathematician

Fig. 12.1 Celestial map of the northern sky by Albrecht Dürer, 1515. This, together with its southern sky companion, were the first printed star charts. National Gallery of Victoria, Melbourne, Felton Bequest, 1956

He is considered the greatest astronomical observer, and by some, the greatest overall astronomer of antiquity. He had been conducting very precise measurement of stars' positions and their apparent sizes, which he defined in a scale from 1 to 6. He measured the length of tropical year and synodic month; he discovered precession of equinoces and non-uniformities in lunar motions. He created the first big catalogue with 850 stars, was the first to determine positions on Earth with usage of geographical latitudes and longitudes, and he founded trigonometry. Ptolemy included Hipparchus' results in his works. Hipparchus' synthesis of astronomy excelled his work. Although he had written at least 14 books, only his comments on Arat's popular astronomic epic remained preserved from later scribes (Kovačec 2002). Around 150 B.C., he used the stereographic and the orthographic projection, which belong to the oldest projections, to create a map of celestial sphere (Fig. 12.1).

Claudius Ptolemy (Greek: Κλαύδιος Πτολεμαῖος, *Klaúdios Ptolemaîos*)
(Alexandria, Egypt ca. 100 CE–Alexandria, Egypt, ca. 170 CE)
Egyptian astronomer, mathematician, and geographer

He was a Greek-Roman citizen and lived in Alexandria, Egypt. In several fields his writings represent the culminating achievement of Greco-Roman science, particularly his geocentric (Earth-centred) model of the universe now known as the Ptolemaic system. Virtually nothing is known about Ptolemy's life except what can be inferred from his writings.

His first major astronomical work, the *Almagest*, was completed about 150 CE and contains reports of astronomical observations that Ptolemy had made over the preceding quarter of a century. The size and content of his subsequent literary production suggests that he lived until about 170 CE.

Ptolemy has a prominent place in the history of mathematics primarily because of the mathematical methods he applied to astronomical problems.

Ptolemy's fame as a geographer is hardly less than his fame as an astronomer. *Geōgraphikē hyphēgēsis* (*Guide to Geography*) provided all the information and techniques required to draw maps of the portion of the world known by Ptolemy's contemporaries. By his own admission, Ptolemy did not attempt to collect and sift all the geographical data on which his maps were based. Instead, he based them on the maps and writings of Marinus of Tyre (c. 100 CE), only selectively introducing more current information, chiefly concerning the Asian and African coasts of the Indian Ocean.

Ptolemy's most important geographical innovation was to record longitudes and latitudes in degrees for roughly 8000 locations on his world map, making it possible

Fig. 12.2 A printed map from the 15th century depicting Ptolemy's description of the *Ecumene*, (1482, Johannes Schnitzer, engraver)

to make an exact duplicate of his map. Hence, we possess a clear and detailed image of the inhabited world as it was known to a resident of the Roman Empire at its height—a world that extended from the Shetland Islands in the north to the sources of the Nile in the south, from the Canary Islands in the west to China and Southeast Asia in the east. Ptolemy's map is seriously distorted in size and orientation compared to modern maps, a reflection of the incomplete and inaccurate descriptions of road systems and trade routes at his disposal.

Ptolemy also devised two ways of drawing a grid of lines on a flat map to represent the circles of latitude and longitude on the globe (Snyder 1993). His grid gives a visual impression of Earth's spherical surface and also, to a limited extent, preserves the proportionality of distances. The more sophisticated of these map projections, using circular arcs to represent both parallels and meridians, anticipated later area-preserving projections. Ptolemy's geographical work was almost unknown in Europe until about 1300, when Byzantine scholars began producing many manuscript copies, several of them illustrated with expert reconstructions of Ptolemy's maps. The Italian Jacopo d'Angelo translated the work into Latin in 1406. The numerous Latin manuscripts and early print editions of Ptolemy's Guide to Geography, most of them accompanied by maps, attest to the profound impression this work made upon its rediscovery by Renaissance humanists (Fig. 12.2).

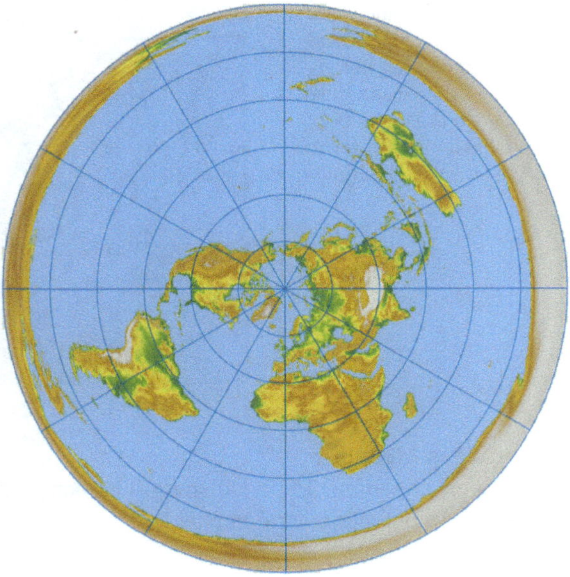

Fig. 12.3 Map of the world in Postel projection

Gillaume Postel
(Barenton, France, 1510–Paris, 1581)
French linguist, geographer, astronomer, diplomat, professor of mathematics

French linguist, adept at Semitic languages (Arabic, Hebrew and Syriac), as well as Classical languages (Ancient Greek and Latin), and in 1538 in his work *Linguarum Duodecim Characteribus Differentium Alphabetum Introductio* he gave an introduction to alphabetic characters of twelve different languages. In 1544, he published *De orbis terrae concordia* in which he advocated a universalist world religion. He was extremely tolerant to other religions at the time when such a tolerance was not common. It is believed that he spent the years 1548–1551 travelling to Israel and Syria, to collect manuscripts. After this trip, he earned the title of Professor of Mathematics and Oriental Languages at the Collège de France in Paris. After several years, Postel resigned his professorship and travelled all over central Europe, including Austria and Italy.

He was considered an originator of the equidistant azimuthal projection. Although this projection was possibly developed by the Egyptians for star charts, Postel was the first one to use it in 1581 and it was named after him the *Postel projection* (Figs. 12.3 and 12.4).

Fig. 12.4 Map of the world in normal aspect equal-area azimuthal projection (the projection of the map extends from the North Pole to 60° south latitude) is used for the emblem of the United Nations

Gerardus Mercator
(Flanders, Belgium, 1512–Duisburg, Germany, 1594)
Flemish geographer, cartographer

Out of need for more accurate mapping of larger Earth territories on marine charts and related to distortions appearing during this process, map projections and practical cartography developed in the 16th century. Mercator was educated in 's-Hertogenbosch in the Netherlands. He studied mathematics and astronomy at the Belgian University of Leuven and in 1532 he obtained the title of master. He was educated for engraver and globe maker. He travelled a lot and began to be interested in geography. He returned to Leuven and started to learn and work with Gemma Frisius (astronomer and mathematician) and Gaspar Myrica (engraver and goldsmith). They worked together to construct globes, maps and astronomical instruments.

The conformal cylindrical projection is named *Mercator projection* after him. The normal aspect of the Mercator projection has special importance in navigation, because the rhumb lines are represented as straight lines in this projection. The Transverse Mercator projection is used in many countries for official cartography. The Universal Transverse Mercator (UTM) coordinate system based on the transverse Mercator projection is used in military (NATO). The Web Mercator projection is a slight variant of the Mercator projection that is widely used on internet. Despite its obvious scale variation at small scales, this projection is well-suited as an interactive world map that can be zoomed seamlessly to large-scale (local) maps,

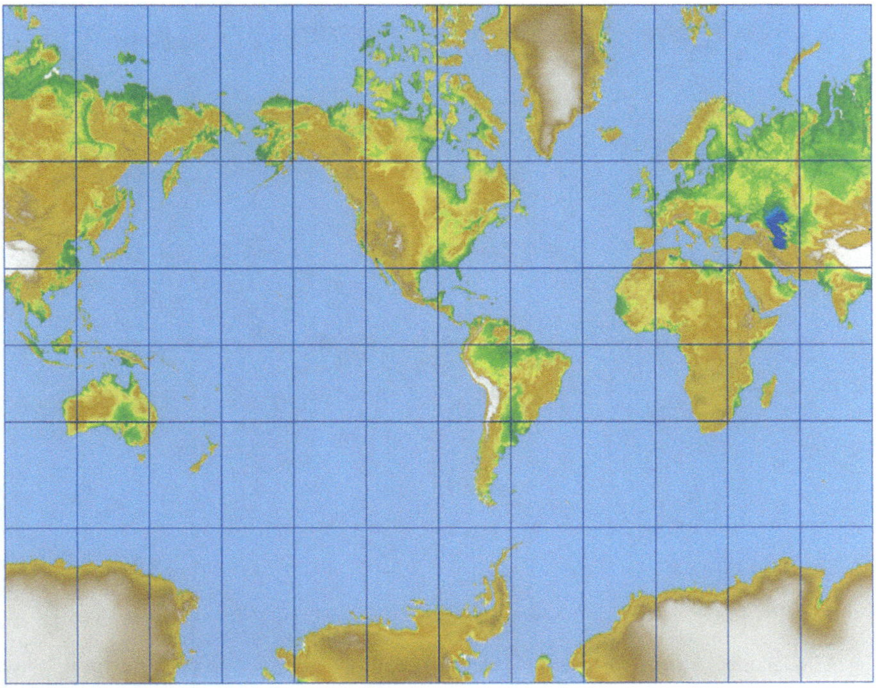

Fig. 12.5 Map of the world in the Mercator projection. It should not be used (although it frequently is) for depicting general information or any area-related subjects. Because of the excessive distortion of area it presents a misleading view of the world (Monmonier 2004)

where there is relatively little distortion due to the variant projection's near-conformality (Figs. 12.5 and 12.6).

Nicolas Sanson
(Abbeville, 1600–Paris, 1667)
French cartographer and geographer

He was born in Abbeville, where as a young man he studied history. He moved to Paris and founded craft in Rue d'el Arbe, St. Germain. He was a Royal Geographer from 1630 to 1665. He lectured geography both to Luis XIII and Luis XIV. In period from 1618 to 1667, he made atlases and illustrated texts.

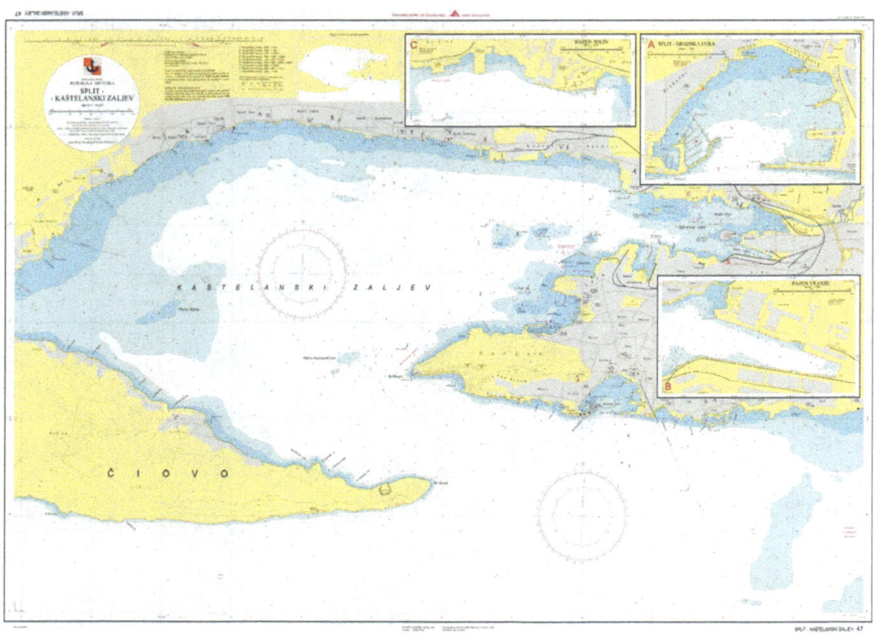

Fig. 12.6 Marine charts are usually made in normal aspect Mercator projection. Sheet Split-Kaštelanski zaljev 47, publisher: Hydrographic Institute of the Republic of Croatia, 2002, original scale 1:15,000

He produced about 300 maps in total, out of which two maps of North America are particularly important: *Septentrionale* (1650) and *Le Canada ou Nouvelle France* (1656).

He proposed a new sinusoidal pseudocylindrical projection for world maps. This projection is named after him the *Sanson projection*, and it is an equal-area sinusoidal pseudocylindrical projection in which all the parallels and the central meridian are mapped in real size (Baily 1886; Curie 1900).

In fact, the projection was developed in the 16th century. It was used by J. Cossin in 1570 and by J. Hondius in Mercator atlases of the early 17th century. It is often called the sinusoidal or Sanson-Flamsteed projection after later users. This is the oldest current pseudocylindrical projection (Fig. 12.7).

Giambattista Nicolosi

(Paternò, 1610–Roma, 1670)

Italian priest and cartographer

He was a priest and cartographer for Pope Gregory XV's Sacred Congregation for the *Propagation of the Faith*, established by the pope to promote missionary work.

In 1652 the *Propagation of the Faith* commissioned him to produce an atlas. The result, eight years later, was the very rare *Dell'Hercole* which contained maps of the world and the continents, the latter in four sheets each. It is based on Sanson's 1650

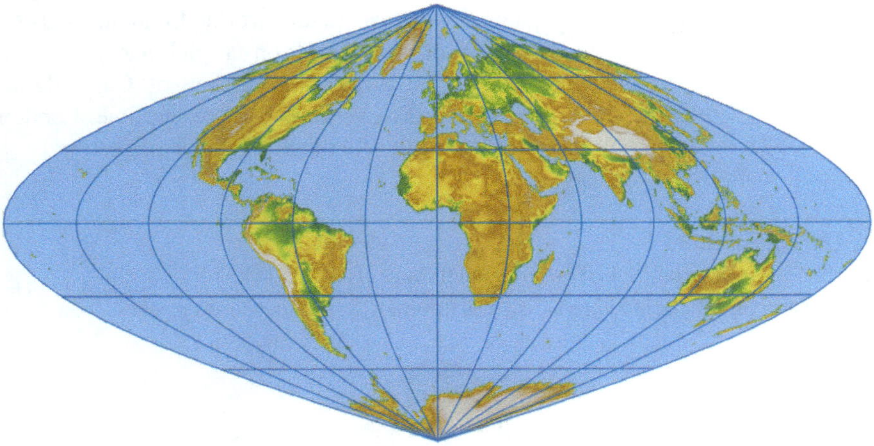

Fig. 12.7 Map of the world in the Sanson or Flamsteed-Sanson or sinusoidal projection

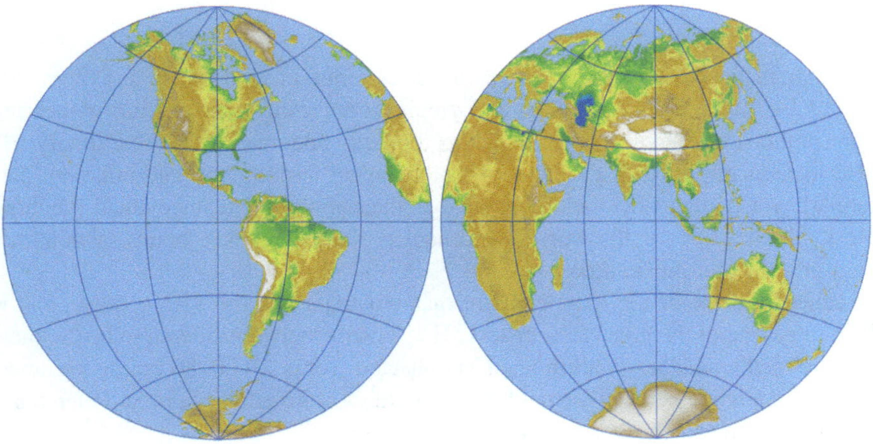

Fig. 12.8 Map of the world in Nicolosi globular projection

map, but follows Ramusio and earlier Italian cartographers in that south is on top. The geographical information is Ptolemaic and somewhat conservative for the time in that few place names are provided. The second state of the map (1670), with additional geographical information was published posthumously by Joanne Baptista Nicolosi.

The *Nicolosi globular projection* was used on hemispheric maps, common in atlases between 1850 and 1925 (Snyder 1993). This projection was first presented by al-Biruni about A.D. 1000, but reinvented by Giambattista Nicolosi in 1660 (Fig. 12.8).

Arrowsmith A (1794) A companion to a map of the world: London, G. Bigg. [Relates to his Map of the world on a globular projection, exhibiting particularly the nautical researches of Capn. James Cook, with all the recent discoveries to the present time: London, 1794. Discusses al-Bīrūnī (or Nicolosi) Globular projection.]

Leonhard Euler
(Basel, Switzerland, 1707–Saint Petersburg, Russia, 1783)
Swiss mathematician, physicist, astronomer, logician and engineer

Euler made important and influential discoveries in many branches of mathematics like infinitesimal calculus and graph theory while also making pioneering contributions to several branches such as topology and analytic number theory. He also introduced much of the modern mathematical terminology and notation, particularly for mathematical analysis, such as the notion of a mathematical function. He is also known for his work in mechanics, fluid dynamics, optics, astronomy, cartography, and music theory.

Euler was one of the most eminent mathematicians of the 18th century, and is held to be one of the greatest in history. He is also widely considered to be the most prolific mathematician of all times. His collected works fill 60–80 quarto volumes, more than anybody else in the field. He spent most of his adult life in St. Petersburg, Russia, and in Berlin, then the capital of Prussia.

The oldest proof that one can not map a sphere into a plane without distortion originates from Euler. He published three papers on map projections in 1777. Among them is the first formal proof about impossibility to map a sphere into a plane without any distortions. In the original form his papers are written in Latin, then published again in 1955. In Switzerland within Euler's *Opera omnia*, translated in German and published in 1898. And in Russia in 1959.

In order to decrease the angular distortions in map projections, Euler suggested the use of an equal-area projection with orthogonal graticule, known today as Euler projection.

Cartography was another area in which Euler became involved during his stay in St. Petersburg. In 1753, he was appointed director of the St. Petersburg Academy's geography section, and his task was to help French astronomer and cartographer Joseph Nicolas De l'Isle in preparing a map of the whole Russian Empire titled

Fig. 12.9 General map of the Russian Empire (Mappa Generalis Totius Imperii Russici). David Rumsey Map Collection www.davidrumsey.com

Mappa Generalis Totius Imperii Russici (1:8,9 Mill., Fig. 12.9). In his autobiographical writings, Euler says his eyesight problems began in 1738 with overstrain due to his cartographic work. However, it is believed that blindness was consequence of poisoning because of a boil. Despite everything, he continued with his work with great passion and devoted himself to the creation of an atlas. Atlas of the Russian Empire (*De L'Isle – Atlas Rvssicvs ... Vastissimvm Imperivm Rvssicvm cum adiacentibvs Regionibvs*), consisting of 20 maps, was published in 1745 and in spite of several shortcomings (small number of astronomically determined points, low degree of accuracy of the maps, etc.); it represents an important contribution to Russian cartography.

Euler was involved in cartography during his stay at the Royal Academy of Sciences in Berlin. He is the author of a school atlas first published in 1753 under the title *Atlas geographicus omnes orbis terrarum regiones in XLI tabulis*. The next edition of the atlas with a title and foreword in German, French and Latin, was published in 1760 and it contained 44 maps (Fig. 12.10).

Euler continued his cartographic work during his second stay in St. Petersburg. In 1777, the St. Petersburg Academy of Sciences published three important Euler's works on cartography: *De repraesentatione superficiei sphaericae super plano, De projectione geographica superficiei sphaericae* and *De projectione geographica de Lisliana in mappa generali Imperii Russici usitata*. These works were later translated into German and were published in vol. 93 of Ostwald's Classics of Exact Sciences (*Ostwald's Klassiker der Exakten Wissenschaften*) in Leipzig in 1898 under the title *Drei Abhandlungen über Kartenprojection* (Three discussions on cartographic projections) and later on also into Russian (Eyler 1959).

Fig. 12.10 Map of the world and map of the Europe from Euler's Atlas, edition from 1760 www.vintage-maps.com

As it is well known, P. Chebyshev (1821–1894) was the first mathematician who investigated the problem of best uniform approximation in depth, and which can be expressed in classical notation in this way:

Let f be continuous function, $a, b \in R, n \in N$. Find a polynomial of degree at most n so that

$$\max_{x \in [a,b]} |f(x) - p(x)|$$

will be minimal for all polynomials of degree at most n.

The most important property of the solution of this problem is the fact that it gives an estimation of the error of approximation for every point of the interval $[a, b]$. But, due to the difficulty of this problem, the first results were presented only in 19th century. Chebyshev himself gave a necessary condition for the solution, stating that there must be at least $n + 2$ points where error function $f - p$ reaches its maximum, but he did not explicitly mention that these deviation points reach the maximum values with an alternating sign. So, the alternation theorem that characterizes the solution was not proven by Chebyshev. Using Chebyshev's results, Kirchberger made the first attempt that was completed by Borel and Young. An algorithm to calculate the best approximation was firstly presented by Remez only in 1934. Here, the alternation property is crucial for defining the iterating procedure that generates the solution.

Still, some special cases had been discussed before Chebyshev, and first of them was Euler's cartographic problem he dealt with during his second stay in St. Petersburg. Namely, Euler analysed local and global accuracy of the De l'Isle conic projection in 1777. This paper was the last of three papers dealing with cartography. It was published in 1777 and it seems that it was made based on his former work at the cartographic department of the Academy of Sciences in St. Petersburg. Euler first proved the fact that a part of the sphere cannot be projected onto a plane while preserving scale in both dimensions. Then he considered the issue of the most convenient projection for a map of all of Russia. Considering several kinds of projections, for example stereographic and polar projections, he determined that De l'Isle projection should be used because of the following important properties:

1. Parallels and meridians intersect in the projection perpendicularly.
2. It gives a good approximation locally.

A map can be used for estimating the distance between any two points.

The mathematical problem that Euler solved was approximating the function $\cos x$ by a linear function. He noted that the best approximation can be characterized by the fact that there must be three points in which error $a - bx - c\cos x$ reaches the maximum value alternating in sign—the alternation theorem in its simplest form. With these settings, he was able to determine parameters for, in that sense, the best map. This work defined the Euler projection that was used for a map of whole Russia until the beginning of the 20th century (Grattan-Guinness and Pulte 2007; Steffens 2007).

Euler's projections are equivalent mappings in which the graticule is being mapped to an orthogonal graticule in the projection plane. Of all equivalent projections, they are closest by character of deformation to conformal projections, so many scientists worked with them, like L. Euler, D.A. Grave, N.A. Urmayev, G.A. Meshcheryakov, K. Frankich, J. Györffy and others.

If we use R to denote the radius of a sphere, then Euler's projections can be described by following differential equations:

$$\left|\frac{dx}{d\phi}\cdot\frac{dy}{d\lambda}-\frac{dy}{d\phi}\cdot\frac{dx}{d\lambda}\right| = R^2\cos\phi \quad \text{(equivalency)}$$

$$\frac{dx}{d\phi}\cdot\frac{dx}{d\lambda}+\frac{dy}{d\phi}\cdot\frac{dy}{d\lambda}=0 \quad \text{(orthogonality of the graticule)}$$

where φ and λ are geographical coordinates of a point on the sphere, and x and y are corresponding coordinates in the projection plane. Different types of Euler's projections can be obtained by solving this system of partial differential equations. It is not hard to show that equivalent normal aspect cylindrical, conical and azimuthal projections are also Euler's projections.

Although interesting from a theoretical viewpoint, Euler's projections also attracted scientists to study these kinds of projections for specific areas. So, for example, Meshcheryakov (1958, 1959, 1963, 1964, 1968), Urmayev (1947) deals with Euler's projection for a world map, Frankich (1982) for Canada, and Györffy (2006) for Europe.

Euler's publications on map projections:

Euler L (1777) De repraesentatione superficiei sphaericae super plano: St. Petersburg, Academia Scientiarum Imperialis Petropolitanae, Acta, part 1, pp 107–132 [Latin. Translated into German as one of Drei Abhandlungen über Kartenprojection, in Ostwald's Klassiker der Exakten Wissenschaften, no. 93: Leipzig, Wilhelm Engelmann, 1898, with editing by Albert Wangerin, pp 3–37]

Euler L (1777) De projectione geographica superficiei sphaericae: St. Petersburg, Academia Scientiarum Imperialis Petropolitanae, Acta, part 1, pp 133–142 [Latin. Translated into German as one of Drei Abhandlungen über Kartenprojection, in Ostwald's Klassiker der Exakten Wissenschaften, no. 93: Leipzig, Wilhelm Engelmann, 1898, with editing by Albert Wangerin, pp 38–52]

Euler L (1777) De projectione geographica de Lisliana in mappa generali Imperii Russici usitata: St. Petersburg, Academia Scientiarum Imperialis Petropolitanae, Acta, part 1, pp 143–153 [Latin. Equidistant conic projection used for map of Russia. Translated into German as one of Drei Abhandlungen über Kartenprojection, in Ostwald's Klassiker der Exakten Wissenschaften, no. 93: Leipzig, Wilhelm Engelmann, 1898, with editing by Albert Wangerin, pp 53–64]

Eyler L (1959) Izbrannye kartograficheskiye stat'i. Tri stat'i po matematicheskoy kartografii. Perevod s nemetskim Bagratuni GV, Geodezizdat, Moscow 80 p. [Russian. Selected cartographic contributions: Three contributions on mathematical cartography. Translation from German.]

César-François Cassini de Thury (also called Cassini III)
(Thury-sous-Clermont, Oise, 1714–Paris, 1784)
French astronomer and cartographer

In 1744, César-François Cassini de Thury began the construction of a great topographical map of France, one of the landmarks in the history of cartography. Completed by his son Jean-Dominique and published by the *Académie des Sciences* from 1744 to 1793, its 180 plates are known as the *Cassini map*.

His chief works are: *La méridienne de l'Observatoire Royal de Paris* (1744), a correction of the Paris meridian; *Description géométrique de la terre* (1775); and *Description géométrique de la France* (1784), which was completed by his son.

The *Cassini projection* is a map projection which he described in 1745. It is the transverse aspect of the equirectangular projection. In this projection, the globe is first rotated so the central meridian becomes the "equator", and then the normal equirectangular projection is applied. In practice, the projection has always been applied to models of the earth as an ellipsoid, which greatly complicates the mathematical development but is suitable for surveying. Nevertheless the use of the Cassini projection has largely been superseded by the Transverse Mercator projection, at least with central mapping agencies (Figs. 12.11 and 12.12).

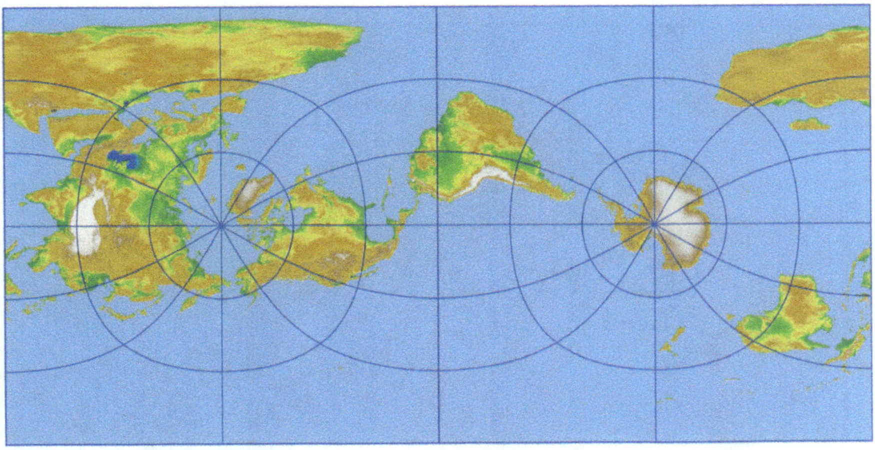

Fig. 12.11 Map of the world in Cassini projection

Rigobert Bonne

(Raucourt, 1727–Paris, 1795)

French engineer, mathematician and cartographer

Rigobert Bonne was a hydrographer at the Royal Court in Paris (*Hydrographe du Roi a Paris*). His main interest was the production of marine charts. He also published several atlases. Of significance is *Atlas Encyclopédique*, produced in collaboration with his son and with Nicolas Demarest and Bory de St. Vincent. In 1752, Bonne proposed a pseudoconical equal area map projection in which all the parallels and the central meridian are projected free of all distortion, for the map of France (Snyder 1993). Although the projection was developed in rudimentary form by Claudius Ptolemy (about year 100) and further developed by Bernardus Sylvanus (1511), it was named after Rigobert Bonne (Fig. 12.13).

Johann Heinrich Lambert

(Mülhausen, 1728–Berlin, 1777)

German physicist, mathematician, astronomer and cartographer originating from France

Johann Heinrich Lambert encompassed algebra, spherical geometry and perspective with his mathematical studies. He was the first to prove (1768) that π is an irrational number and made the first systematic use of hyperbolic functions. His

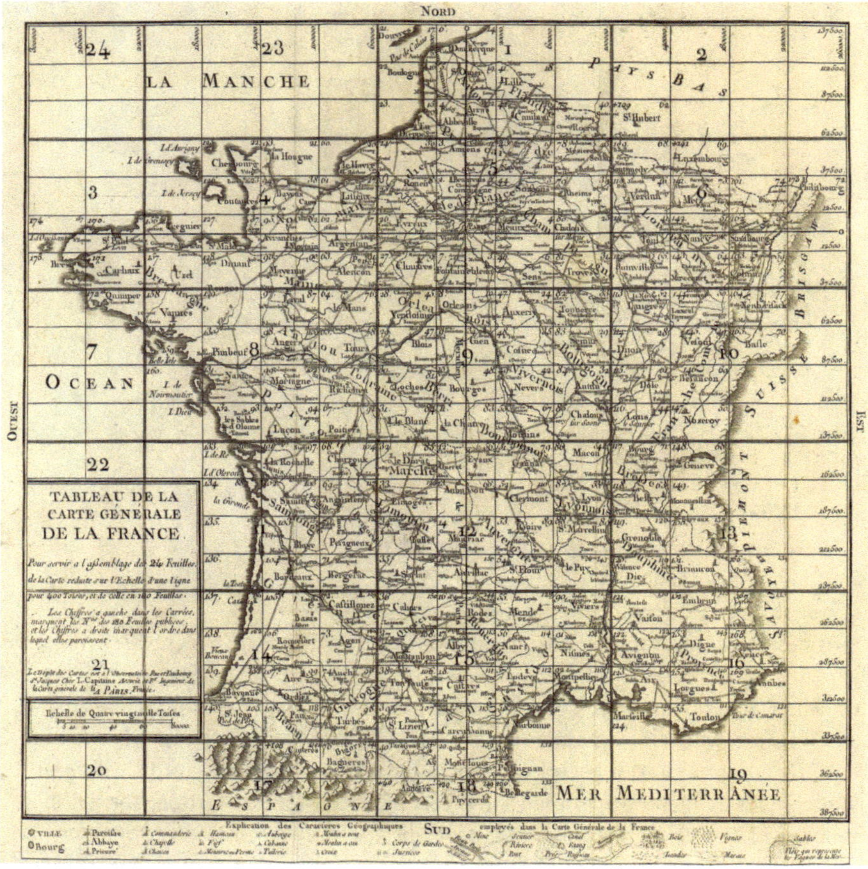

Fig. 12.12 The Carte de France was published by four generations of the Cassini family from 1750 to 1815. It consists of 182 sheets at the same scale 1:86,400, allowing the sheets to be joined together. The map was the first national survey completed systematically, relying on the latest science of its time. David Rumsey Map Collection www.davidrumsey.com

work on theory of parallel lines (1766) is of particular importance. In his work *Photometry* (Photometrie, 1760), he clearly distinguished concepts of brightness and luminance and in that way he set the foundations of photometry. Besides, he researched refraction of light in the atmosphere, comet paths and related to this, he discovered new properties of conics. In his astronomical works, there is first mention of double stars (Bollmann and Koch 2001).

Lambert presented in 1772 a (conformal conical) projection which was named after him *Lambert's conformal conic projection* (Fig. 12.14). This projection is today still in use for the requirements of airplane navigation and in some countries as official state projection. In Croatia, this projection is the official map projection for general topographic maps (Fig. 12.15). The equivalent azimuthal projection was also named after Lambert. The Transverse Mercator projection of sphere is also called the *Lambert-Gauss projection*.

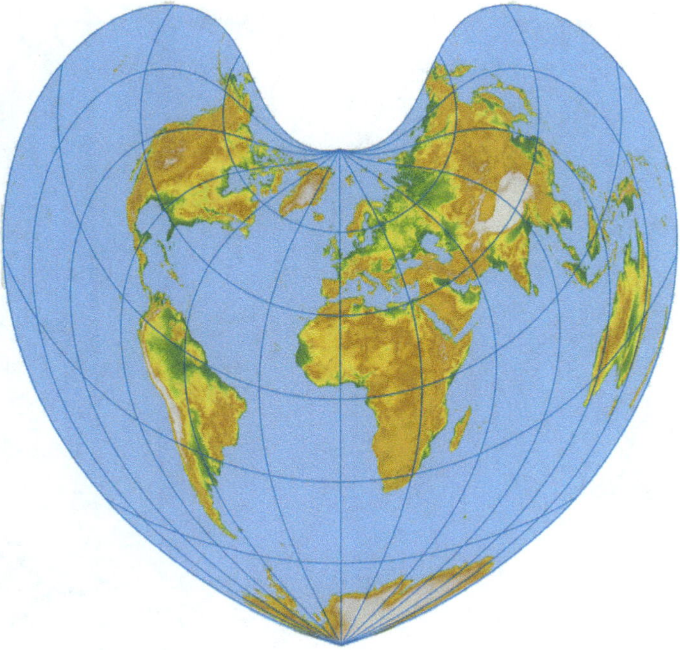

Fig. 12.13 Map of the world in the Bonne projection

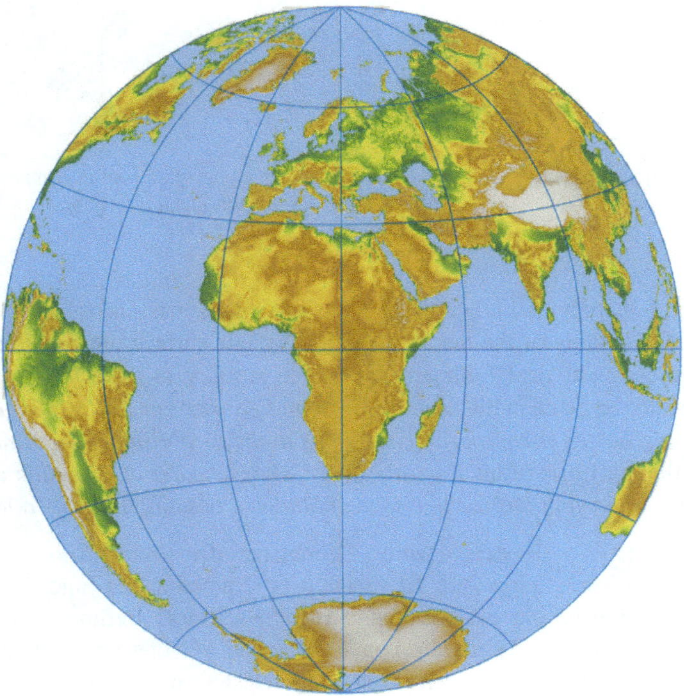

Fig. 12.14 Transverse Lambert azimuthal equal-area projection

Fig. 12.15 Rijeka and surroundings on the section of the topographic map in Lambert conformal conic projection, Republic of the Croatia, sheet 44 Zagreb-Zadar 54, publisher: Ministry of Defence of the Republic of Croatia, 1995, original scale 1:500,000

Lambert's theorem is known in the adjustment calculus. According to Frischauf (1905), the beginning of the theory of projecting one surface onto another belongs to J.H. Lambert, who dealt with generally given problem of projecting a sphere and spheroids into the plane in his *Anmerkungen und Zusätze zur Entwerfung der Land- und Himmelscharten* (*Remarks and Additions to the Establishment of Land and Sky Maps*) in the third part of his *Beyträge zum Gebrauche der Mathematik und deren Anwendung* (*Contributions to the Usage of Mathematics and its Application, 1772*).

Lambert JH (1772) Beyträge zum Gebrauche der Mathematik und deren Anwendung, Part III, section 6: Anmerkungen und Zusätze zur Entwerfung der Land- und Himmelscharten. Berlin [Translated into English and introduced by W.R. Tobler as Notes and comments on the composition of terrestrial and celestial maps: Ann Arbor, Univ.

Michigan, 1972, Mich. Geographical Publication no. 8, 125 p. Also reprinted in German, 1894, Ostwald's Klassiker der Exakten Wissenschaften, no. 54: Leipzig, Wilhelm Engelmann, with editing by Albert Wangerin. Presents Conformal Conic, Equal-Area Conic, including "Isospheric Stenoteric," Transverse Mercator, Cylindrical Equal-Area, Transverse Cylindrical Equal-Area, "Lagrange," Azimuthal Equal-Area projections. See also review of Tobler translation by Wray Th, 1975: Canadian Cartographer, v. 12, no. 2, pp 231–233]

Joseph Louis Lagrange
(Torino, 1736–Paris, 1813)
French mathematician and astronomer

He made significant contributions to the fields of analysis, number theory, and both classical and celestial mechanics.

In 1766, on the recommendation of Euler and d'Alembert, Lagrange succeeded Euler as the director of mathematics at the *Prussian Academy of Sciences* in Berlin, Prussia, where he stayed for over twenty years, producing volumes of work and winning several prizes of the *French Academy of Sciences*. Lagrange's treatise on analytical mechanics (*Mécanique Analytique*, 4. ed., 2 vols. Paris: Gauthier-Villars et fils, 1888–89), written in Berlin and first published in 1788, offered the most comprehensive treatment of classical mechanics since Newton and formed a basis for the development of mathematical physics in the nineteenth century.

In 1787, at age 51, he moved from Berlin to Paris and became a member of the *French Academy*. He remained in France until the end of his life. He was significantly involved in the decimalisation in Revolutionary France, became the first professor of analysis at the *École Polytechnique* upon its opening in 1794, founding member of the *Bureau des Longitudes* and Senator in 1799 (Bollmann and Koch 2001).

The *Lagrange projection* is a conformal projection of the Earth on a circle, except that angles are halved at the poles instead of being faithful there. This projection is named after Lagrange, who generalized Lambert's concept of presenting the world conformally in a circle (Fig. 12.16).

Lagrange JL de (1779) Sur la construction des cartes géographiques: Nouveaux mémoires de l'Académie Royale des Sciences et Belles-lettres de Berlin, pp 161–210 [Also in Oeuvres de Lagrange, 1869:

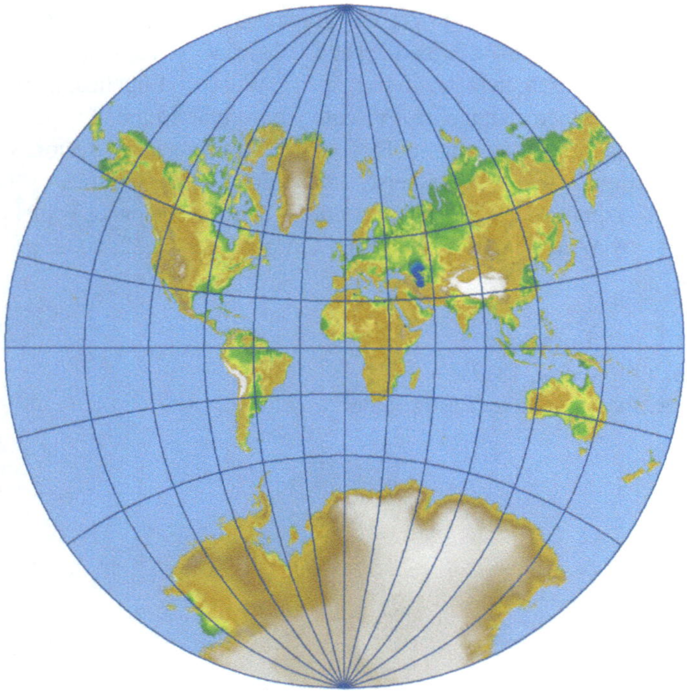

Fig. 12.16 Lagrange projection of a sphere in a circle

Gauthier-Villars, Paris v. 4, pp 635–692. Also in German, 1894, as Ueber die Construction geographischer Karten: Ostwald's Klassiker der Exakten Wissenschaften, no. 55: Leipzig, Wilhelm Engelmann, pp 1–56, with editing by Albert Wangerin, pp 82–97. Analyzes conformal projections with circular arcs for meridians and parallels, including "Lagrange projection" developed by Lambert (1772)]

Heinrich Christian Albers
(1773–1833)
German cartographer

Son of a merchant, native and lifelong resident of Lüneburg, Germany. He derived the formulas for the projection of the sphere using two standard parallels. He published this form of conical figure in a paper in the journal *Monthly Zach's correspondence* (*Zachs Zeitschrift Monatliche Correspondenz*) and has since been considered the inventor of the projection called *Albers Equal-Area Conic Projection.*

The *Albers projection* was used for a German map of Europe in 1817, but it was promoted for maps of the United States in the early part of 20th century by

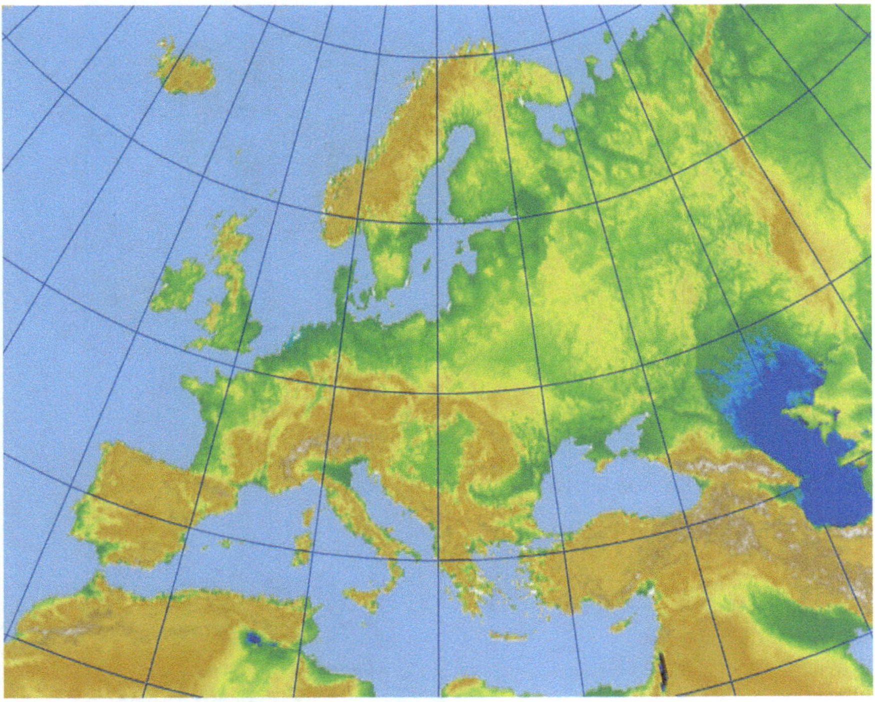

Fig. 12.17 Map of the Europe in Albers projection (standard parallels: 40° and 66°15′)

Oscar S. Adams of the *Coast and Geodetic Survey* as an equal-area representation that is "as good as any other and in many respects superior to all others". It is also used and recommended for equal-area maps of regions that are predominantly eastwest in extent (Fig. 12.17).

Karl Brandan Mollweide
(Wolfenbüttel, 1774–Leipzig, 1825)
German mathematician and astronomer

Mollweide was an observer at the Observatory of the Leipzig University until 1816. In 1812, he obtained the title of a full professor of astronomy, and from 1814, a full professor of mathematics. From 1820 to 1823, he was the Dean of the Faculty of Philosophy. He discovered spherical trigonometric formulae, which were named

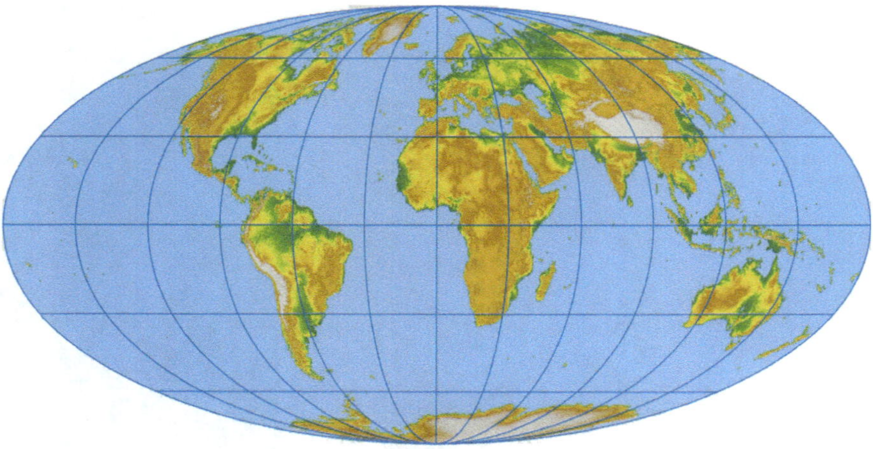

Fig. 12.18 Map of the world in the Mollweide projection

Fig. 12.19 The logo of the International Cartographic Associtaion (ICA) has the world in Mollweide projection in its central part

Mollweide's formulae after him (Lapaine 2011). He discovered and published the pseudocylindrical equal-area projection (1805) that was, in his honor, named the *Mollweide projection* (Figs. 12.18 and 12.19).

Mollweide KB (1805) Ueber die vom Prof. Schmidt in Giessen in der zweyten Abtheilung seines Handbuchs der Naturlehre S. 595 angegeben Projection der Halbkugelfläche, Zach's Monatliche Correspondenz zur Beförderung der Erd- und Himmels-Kunde, vol. 12, Aug., pp 152–163

Carl Friedrich Gauss
(Braunschweig, 1777–Göttingen, 1855)
German mathematician, astronomer and geodesist

A versatile mathematical genius and one of the greatest mathematicians of all time (princeps mathematicorum). He showed his great mathematical talent already in childhood, and he achieved first scientific results as a mathematics student in Göttingen. Relating to theory of circle division, he solved (1796) the problem of construction of regular polygons with a ruler and a compass. He was promoted in 1799 based on his doctoral thesis in which he gave a proof of extremely important, so called fundamental theorem of algebra. In his publication *Investigations in Arithmetic* (*Disquisitiones arithmeicae*, 1801) he set the foundations for the modern theory of numbers. His General *Investigations of Curved Surfaces* (*Disquisitiones generales circa superficies curves*, 1828) represent a new stage in the development of differential geometry and foundation of its progress until the present day. In this work he introduces systematic usage of parametrical representations of surfaces, two basic square forms, spherical projection, and based on this, the concept of curvature in the point of surface. The basic theorem about invariability of curvature of surface during its isometric projection was proved (*Theorema egregium*).

His contribution to theory of errors during measurement is also very important, and it was represented as the theory of least squares in the work *Theory of the Combination of Observations Least Subject to Errors* (*Theoria combinationis observantium erroribus minimis obnoxiae*, I–III, 1821–26), according to which the most adequate value of measured scale is the one in respect of which sum of errors' squares is minimal. His researches in the field of basic geometry are of particular importance, although he did not publish anything about it. He had been managing the observatory in Göttingen for a long time, and he had been calculating mathematical tables for the needs of astronomy for ten years, which were afterwards in use for decades.

Many things are named after him, like Gaussian curve, Gaussian elimination method during the solving of system of linear equations, Gaussian sum mark, Gaussian condition for tetragon with diagonals, Gauss-Krüger projection, etc. Between years 1821 and 1825, during the calculations of Hannover's triangulation for projection of ellipsoid into the plane, Gauss used a projection procedure which is today called the *Gauss-Krüger projection*. Professor Dr. Louis Krüger published a book about that projection in 1912, and in 1919 a collection of formulae for

Fig. 12.20 Section of the topographic map in Gauss-Krüger projection, sheet Varaždin, 271-2-3, publisher: State Geodetic Administration of the Republic of Croatia, 1997, original scale 1:25,000

practical usage. Since then, this projection had been called Gauss-Krüger. During the 20th century, this projection was the official map projection in many countries. It is also known as transverse Mercator projection (Fig. 12.20).

Gauss CF (1816–1827) Conforme Abbildung des Sphäroids in der Ebene: in C. F. Gauss, Werke, Königlich Gesellschaft der Wissenschaften zu Göttingen, Abhandlungen, 1903, v. 9, pp 142–194

Gauss CF (1825) Allgemeine Auflösung der Aufgabe: Die Theile einer gegebnen Fläche auf einer andern gegebnen Fläche so abzubilden, daß die Abbildung dem Abgebildeten in den kleinsten Theilen ähnlich wird: Preisarbeit der Kopenhagener Akademie 1822, Schumachers Astronomische Abhandlungen, Altona, no. 3, pp 5–30 [Reprinted, 1894, Ostwald's Klassiker der Exakten Wissenschaften, no. 55: Leipzig, Wilhelm Engelmann, pp 57–81, with editing by Albert

Wangerin, pp 97–101. Also in Herausgegeben von der Gesellschaft der Wissenschaften zu Göttingen in Kommission bei Julius Springer in Berlin, 1929, v. 12, pp 1–9]

George Biddell Airy
(Alnwick, 1801–Greenwich, 1892)
British astronomer, mathematician and cartographer

He graduated at Trinity College, Cambridge, in 1823. Beginning in 1826, he became a professor of astronomy and mathematics at Cambridge, and in 1828 he became the director of the Cambridge observatory. His main interest was astrometry and optics, especially wave optics. In 1861, Airy suggested the arithmetic mean from the deformation on main directions, for the comparison of two projections to the middle square deformations on the whole area being projected. This criterion is named Airy's criterion after him. In the same year, he presented an azimuthal projection which is named *Airy's projection* after him (Fig. 12.21).

Airy GB (1861) Explanation of a projection by balance of errors for maps applying to a very large extent of the earth's surface and comparison of this projection with other projections, Philosophical Magazine and Journal of Science, 22, pp 409–421

James Gall
(1808–1895)
Scottish clergyman, cartographer and astronomer

James Gall was a Scottish clergyman, but his contribution to astronomy and cartography is also very important. He was a son of a famous publisher, raised in Edinburgh. He joined his father's business in 1838, but he quickly left it to pursue a religious career. He studied at the University in Edinburgh and then New College. He published several religious works, and he excelled with his works in astronomy: *Easy Guide to the Constellations* (1870) and *People's Atlas of the Stars*. Gall is the author of three map projections (*Gall isographic*, *Gall stereographic* and *Gall orthographic projection*), whose purpose was to reduce distortion on constellation maps. He presented his work in Glasgow, at the meeting of British Association for the Advancement of Science and explained it more fully in an article published in the *Scottish Geographical Magazine* entitled *Use of Cylindrical Projections for Geographical, Astronomical and Scientific Purposes* in 1885.

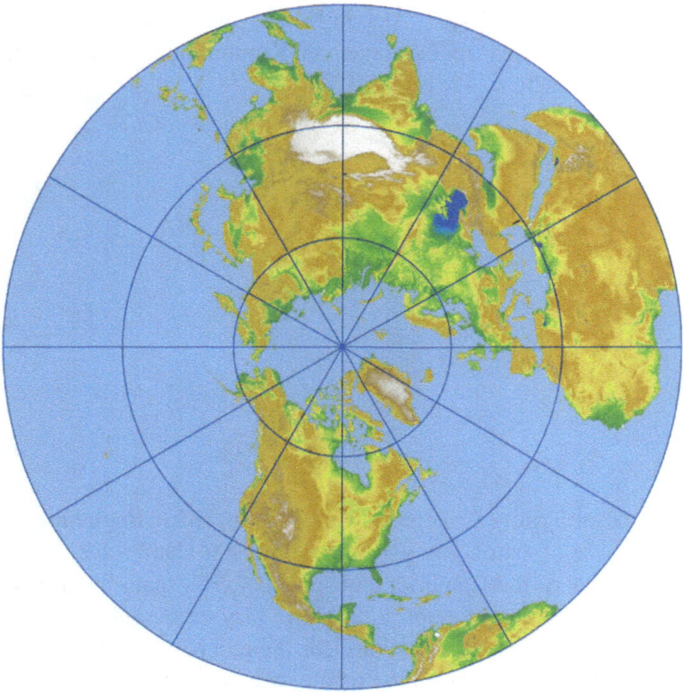

Fig. 12.21 Airy projection

Gall projection excels among perspective cylindrical projections that found greater use in practice (Snyder and Steward 1988). In this projection, the point of view is on the sphere's surface, and this is why it bears the name *Gall stereographic projection*, and the cylinder cuts the sphere alongside the parallels with the latitude $\varphi = \pm 45°$ (Fig. 12.22).

Gall J (1855) On improved monographic projections of the world. British Assn. for the Advancement of Science, Sept., 25th mtg., Report, p. 148. [Two perspective secant cylindrical projections, orthographic and stereographic (the "Gall proj."), and an equirectangular.]

Gall J (1871) On a new projection for a map of the world. Royal Geographical Society, Proceedings, v. 15, July 12, p. 159. [His stereographic projection.]

Gall J (1885) Use of cylindrical projections for geographical, astronomical, and scientific purposes. Scottish Geographical Magazine, v. 1, no. 4, p. 119–123. [Reports his 1855 projections.]

Nicolas Auguste Tissot

(Nancy, Meurthe-et-Moselle, France, 1824–Paris, 1897)

French cartographer

Tissot was trained as an engineer in the French Army, from which he graduated as *capitaine du génie*. In the early 1860s he became an instructor in geodesy at the

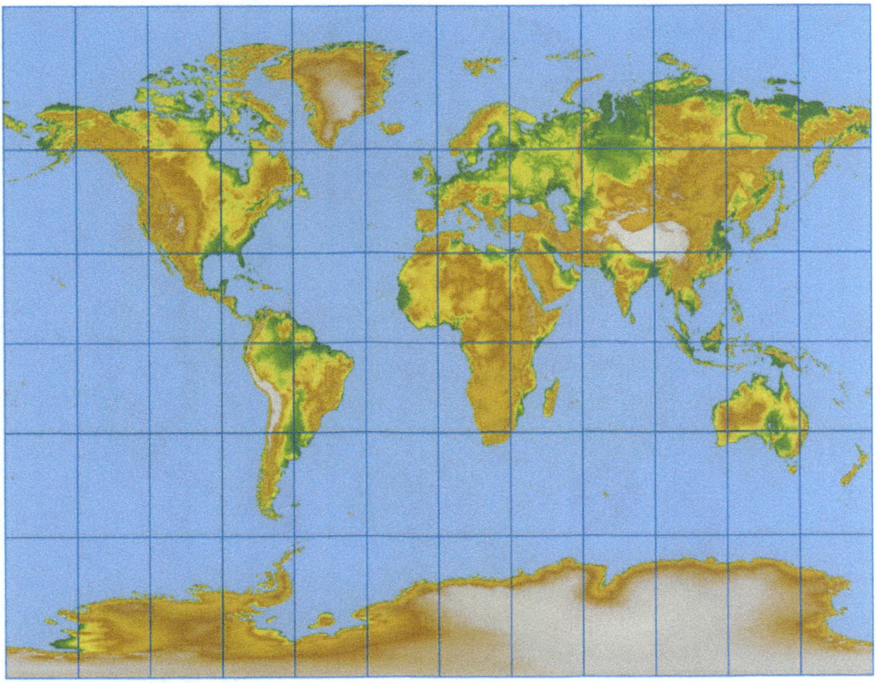

Fig. 12.22 Map of the world in Gall Stereographic projection

well-reputed Ecole Polytechnique in Paris. Around the same time, he indulged a research program meant to determine the best way of cartographic projection for a particular region and presented his findings to the French Académie des Sciences. He published analyses of the distortion that occurs on map projections in 1859 and 1881. He devised the ellipse of distortion which indicates how the scale changes in every direction at a given point. This ellipse was named the *Tissot indicatrix* after him. Additionally, he pursued research studies of the projections which are best suited for representation of a certain part of the Earth's surface (Snyder and Steward 1988). For the representation of relatively small parts of the Earth's surface in a plane, with minimal deformations of angles and lengths, Tissot proposed a specific projection, which was named after him the *Tissot compensational projection* (Fig. 12.23).

Tissot NA (1858) Sur le développement modifié de Flamsteed. Académie des Sciences, Comptes Rendus, v. 46, no. 13, pp 646–648 [Modification of Sinusoidal projection using hyperbolic curves.]

Tissot NA (1859–60) Sur les cartes géographiques. Académie des Sciences, Comptes Rendus, 1859, v. 49 no. 19 pp 673–676; 1860, v. 50 no. 10 pp 474–476; 1860, v. 51 pp 964–969 [See also 1865, v. 60 pp 933–934]

Fig. 12.23 Perspective projection of the Earth and Tissot's indicatrices

Tissot NA (1878) Sur la représentation des surfaces et les projections, des cartes géographiques. Nouvelles Annales deMathématiques, 2nd series, v. 17, pp 49–55, 145–163, 351–366

Tissot NA (1881) Mémoire sur la représentation des surfaces et les projections des cartes géographiques. Gauthier Villars, Paris, 337 p. +60 p. tables. [Details the use of his indicatrix, introduced earlier. Also presents his version of equal-area conic projection. Note: The preamble and first four chapters of this work, pp 1–126, first appeared in Nouvelles Annales de Mathématiques, series 2, 1878, v. 17; 1879, v. 18; 1880, v. 19.]

Tissot NA (1887) Die Netzentwürfe geographischer Karten nebst Aufgaben über Abbildung beliebiger Flächen auf Einander. Autorisierte Deutsche Bearbeitung mit einigen Zusätzen bezorgt von E. Hammer, J. B. Metzlersche Buchhandlung, Stuttgart

Alexander Ross Clarke

(Reading, Berkshire, England, 1828–Strathmore, Reigate, Surrey, 1914)

British geodesist, mathematician and officer

A geodesist whose work is primarily remembered for defining different reference ellipsoids. He was born in England, and he spent his childhood in Jamaica. He

Table 12.2 Clarke ellipsoids parameters

Name	Semi major axis	Flattening of the ellipsoid
Clarke 1858	6,378,293.65	1/294.26
Clarke 1866	6,378,206.4	1/294.978698
Clarke 1880	6,378,249.15	1/293.465

Fig. 12.24 Map of the world in Clarke's Twilight general vertical perspective projection

returned to England and in 1847 joined the British army, and was assigned to the Royal Engineers. He served in Canada from 1851 to 1854. In the year 1856, he became the director of the measurement department, and in 1858 he published his first article on the history of land surveying in Great Britain. In June 1862, he was elected as a member of the Royal Society. His ellipsoids from 1858, 1866 and 1880, which are named after him (for example Clarke 1866 or Clarke 1880) are famous (Table 12.2). He received a gold medal from the Royal Society (1887) for his contribution of determining shape and size of Earth. A.R. Clarke was interested in map projections as well. In the year 1862, he used the least squares method for his perspective projection with minimal deformations for part of terrestrial sphere, margined with a determined spherical circle. He determined parameters for several continental areas, and he published his projection called *Twilight* as well, which presents most of the Earth's land. All these projections are classified as perspective projections with minimum or low-error distortions (Fig. 12.24).

Clarke AR (1879) Geography: mathematical geography. Encyclopædia Britannica, 9th ed., v. 10, pp 197–210 [Same text, vol., and page nos. in 10th ed., 1902]

Friedrich Eisenlohr
(Mannheim, 1831–Heidelberg, 1904)
German mathematician

Eisenlohr studied in Göttingen, Berlin and Heidelberg, where he obtained his PhD in 1852. He finished his habilitation two years later and was appointed a professor in Heidelberg in 1872.

In 1870 he presented a conformal projection, which provides the minimum overall scale variation for a conformal world map, because the boundary of a world map is at a constant scale. Like the August Epicycloidal projection, the Eisenlohr has no "singular" points at which conformality fail. It is used for whole-world maps and was named *Eisenlohr projection* after him (Fig. 12.25).

Eisenlohr F (1870) Ueber Flächenabbildung: Journal für die reine und angewandte Mathematik [Crelle's], v. 72, no. 2, pp 143–151 [His conformal projection of world with minimum overall distortion]
Eisenlohr F (1875) Ueber Kartenprojection: Gesellschaft für Erdkunde zu Berlin, Zeitschrift, v. 10, no. 59, pp 321–334

Friedrich Wilhelm Oscar August
(Berlin, 1840–Berlin, 1900)
German professor of mathematics

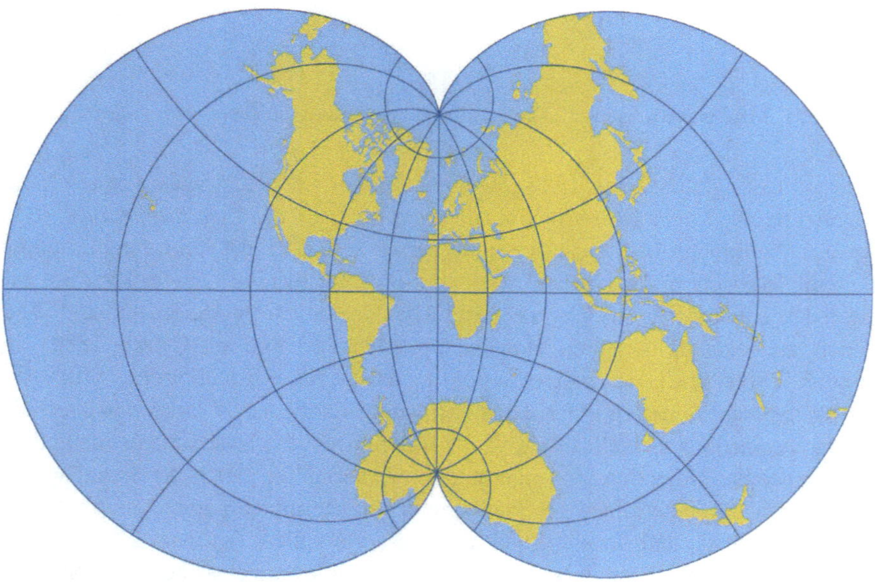

Fig. 12.25 Map of the world in Eisenlohr projection

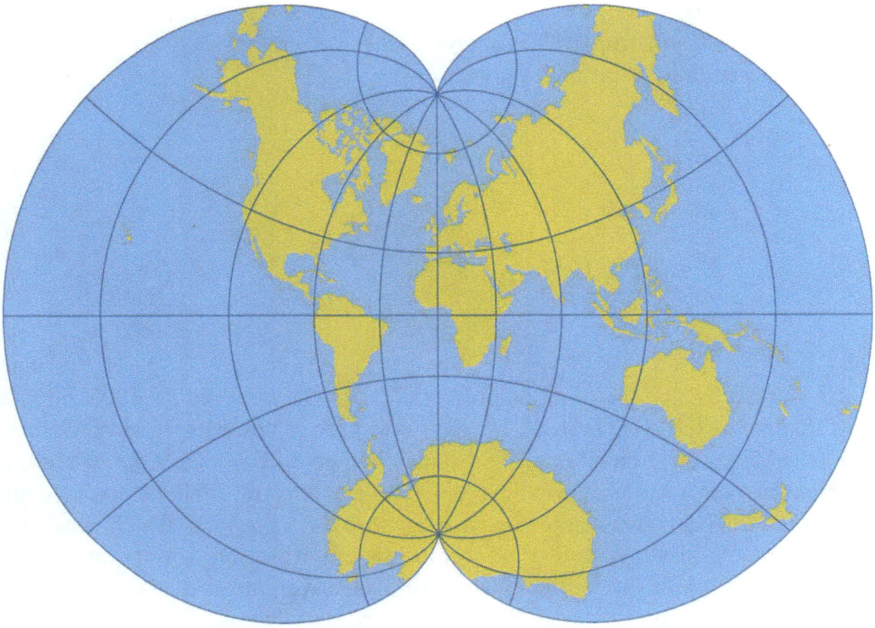

Fig. 12.26 Map of the world in August epicycloidal projection

Dr. Friedrich August was professor of mathematics and son of the mathematician. His publications mainly relate to the field of geometry and mechanics since his position at the Royal Bavarian Artillery and Engineering School (*Königlich Bayerische Artillerie- und Ingenieur-Schule*) encouraged him to handle various questions from these fields.

In book form, he published Research on the imaginaries in Geometry (*Untersuchungen über das Imaginäre in der Geometrie*) in 1872 and the essay "A conformal mapping of the earth in epicycloidal projection" (*Eine konforme Abbildung der Erde nach der epiex cloiden Projektion*) in 1875.

August's epicycloidal map projection is a conformal map projection of the whole sphere into a two-cusped epicycloid. It was designed by Friedrich August and co-developed by Bellermann and published in 1874. In this projection, a world map is bounded by an epicycloid, the shape defined by a point on a circle rolling without sliding around another, fixed, circle (Fig. 12.26).

August F (1874) Ueber eine conforme Abbildung der Erde nach der epicycloidischen Projection: Gesellschaft für Erdkunde zu Berlin, Zeitschrift, v. 9, pp 1–22 [His conformal projection of world bounded by two-cusped epicycloid.]

Wilhelm Jordan
(Ellwangen, Germany, 1842–Hannover, 1899)
German geodesist and mathematician

He worked as a professor at the polytechnic institute in Stuttgart (1865–68) and in Karlsruhe (1868–81). In 1847, Jordan took part in the expedition of Gerhard Rohlfs to Libya. As a member of the German Society of Geometers, he worked on restructuring the German geodetic school system, and in 1887 he founded the *magazine Kalender für Vermessungswesen und Kulturtechnik*, which was published regularly once a year till 1949. From 1881, he was professor of geodesy and practical geometry at the technical university Hannover and he had been working on his most significant work—handbook of geodesy which is today known under the title *Jordan–Eggert–Kneissl: Handbuch der Vermessungskunde* (Fig. 12.27). He is remembered for his algorithm for bringing the matrix to reduced form, which is used for finding the inverse of a matrix. The method was named the Gauss-Jordan algorithm after him and Carl Friedrich Gauss. In 1896, Jordan suggested a formula for determination of mean square distortion in the given point in order to compare two projections according to mean square distortions in the whole area of projection. This criterion is named the *Jordan criterion* after him.

Jordan W (1875) Zur Vergleichung der Soldner'schen rechtwinkligen sphärischen Coordinaten mit der Gauss'schen conformen Abbildungen des Ellipsoids auf die Ebene. Zeitschrift für Vermessungswesen, IV, pp 27–32
Jordan, W. (1896) Der mittlere Verzerrungsfehler, Zeitschrift für Vermessungswesen, XXV, pp 249–252

Alphons J. Van der Grinten
(Kranenburg, Nordrhein-Westfalen, Germany, 1852–Chicago, USA, 1921)
American cartographer

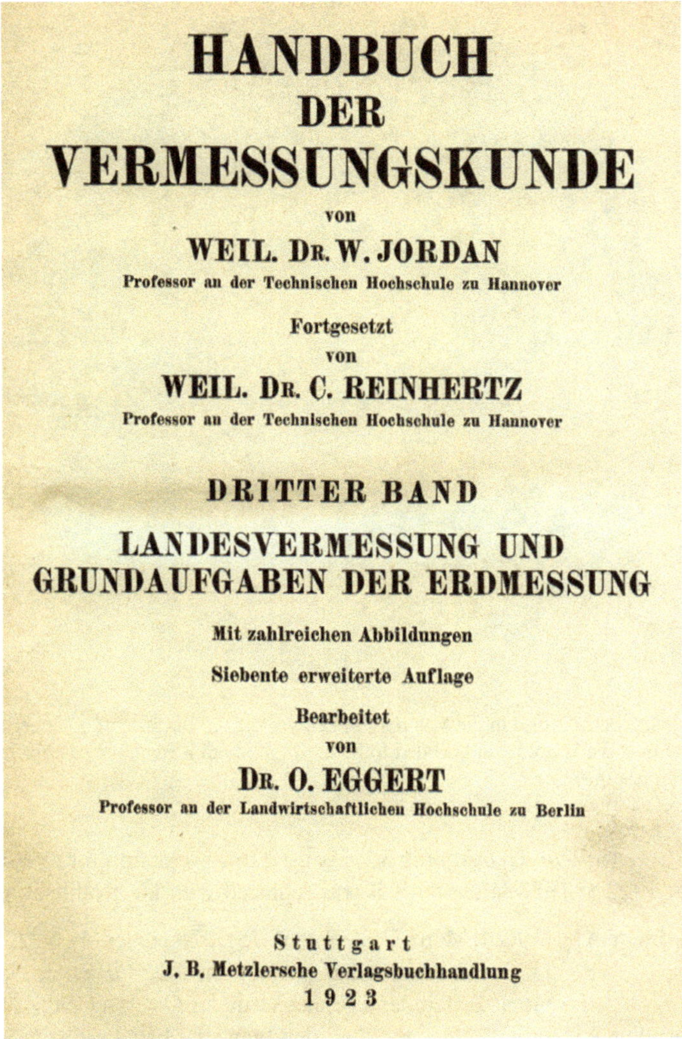
HANDBUCH
DER
VERMESSUNGSKUNDE

von

WEIL. DR. W. JORDAN

Professor an der Technischen Hochschule zu Hannover

Fortgesetzt
von

WEIL. DR. C. REINHERTZ

Professor an der Technischen Hochschule zu Hannover

DRITTER BAND

LANDESVERMESSUNG UND
GRUNDAUFGABEN DER ERDMESSUNG

Mit zahlreichen Abbildungen

Siebente erweiterte Auflage

Bearbeitet
von

DR. O. EGGERT

Professor an der Landwirtschaftlichen Hochschule zu Berlin

Stuttgart
J. B. Metzlersche Verlagsbuchhandlung
1923

Fig. 12.27 Title page of Jordan's work Handbuch der Vermessungskunde, published in 1923

He described, in patent specification from 1904, a graphical way of constructing a projection, which bears his name today (Snyder and Voxland 1989, 1994). The *Van der Grinten projection* is the most famous projection in the group of circular projections (Fig. 12.28). According to distortion characteristics, it belongs to the group of arbitrary projections. Meridians are mapped as circles symmetrical in relation to the central meridian, which is mapped as a straight line. The Van der Grinten projection is often used for political world maps, although it is not suitable for this purpose because of large surface distortions. For example, Greenland is three times smaller than Australia, and in Grinten's projection it is larger than

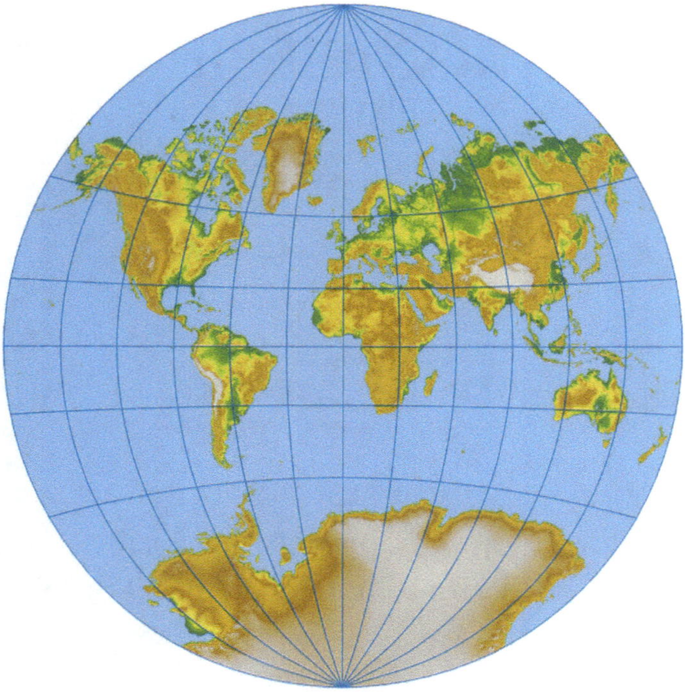

Fig. 12.28 Map of the world in the Van der Grinten projection. Because of the great distortion of area near the poles, it is not recommended to be used (although it frequently is) for depicting the whole world on a map

Australia. The *National Geographic Society* used this projection for reference world maps from 1922 to 1988, afterwards it was replaced with the Robinson projection.

van der Grinten AJ (1904) Map: U.S. Patent 751,226, dated Feb. 2. [His best known circular map projection. Also patented in Canada, Great Britain, and France (van der Grinten (1905, New ….)).]

van der Grinten AJ (1904) Darstellung der ganzen Erdoberfläche auf einer kreisförmigen Projektionsebene. Petermanns Mitteilungen, v. 50 no. 7 pp 155–159. [World map projections bounded by circles and having circles for meridians and parallels. See also corrections, no. 10, p. 250; 1905, v. 51, no. 2, p. 48. See also his supplementary comments, 1905, Zur Verebnung der ganzen Erdoberfläche. Petermanns Mitteilungen, v. 51, no. 10, p. 237; 1906, Zu der zweiten Notiz von van der Grinten über seine Weltprojektion. Petermanns Mitteilungen, v. 52, no. 2, p. 46]

van der Grinten AJ (1905) New circular projection of the whole earth's surface. Amer. Journal of Science, series 4, v. 19, no. 113, pp 357–366

David Aleksandrovich Aitoff (Aitov, Aitow)

(Orenburg, 1854–Paris, 1933)

Russian cartographer

After his emigration to France in 1879 Aitoff worked as a cartographer at Hachette-Verlag, where he dealt mostly with the *Atlas universel* edited by Louis Vivien de Saint-Martin and after by Franz Schrader. Based on the general census of the Russian Empire in 1897 he published the ethnographic general map at the scale of 1:12,500,000 with an explanatory text.

He proposed a suitable for world map in 1889. This projection was named the *Aitoff projection* after him, and it was created by modification of transverse equidistant azimuthal projection. The modification of the projection is, that a perimeter circle of half-sphere map with longitude $\lambda = \pm 90°$ is replaced with an ellipse, within which the whole terrestrial sphere will be mapped. In that projection, the pole is a point, the relation of length of the Equator to the length of central meridian is 2:1, and according to distortion characteristics, the projection is arbitrary.

Three years later, inspired by Aitoff's projection, professor of geodesy Ernst Hermann Heinrich von Hammer devised a projection called the *Hammer-Aitoff projection*, which was created by modification of the transverse equivalent azimuthal projection, in the same manner as the Aitoff projection was created by modification of the equidistant azimuthal projection. The projection is equal-area, the pole is a point, and relation of Equator length to the length of the central meridian is 2:1 (Fig. 12.29).

Aitoff D (1889) Projections des cartes géographiques. In: Atlas de Géographie Moderne. Hachette, Paris. [Introduces his projection adapting the Azimuthal Equidistant projection.]

Aitoff D (1892) Note sur la projection zenithale equidistante et sur les canevas qui en est dérivé. Nouvelles Géographiques, v. 6, June, pp 87–90

Aitoff D (1893) Projection équivalente applicable au continent Américain. Nouvelles Géographiques, series 3, no. 5, pp 72–74

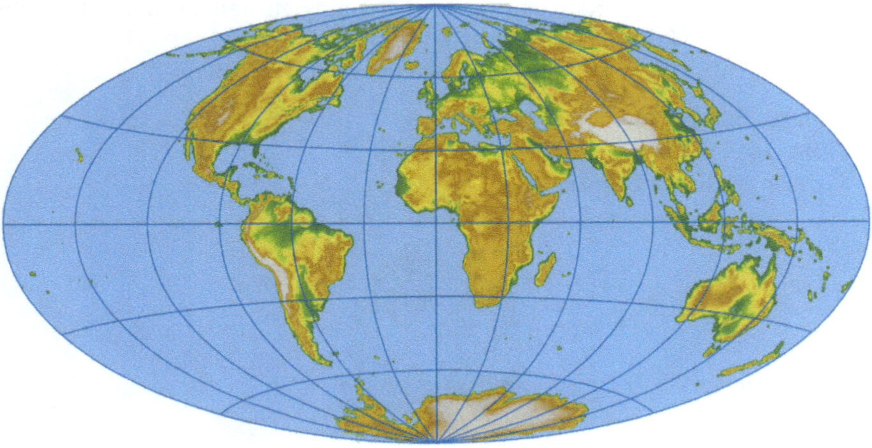

Fig. 12.29 Map of the world in the Aitoff projection

Aitoff D (1913) Projections des cartes géographiques. In: Schrader F et al., Atlas de Géographie Moderne, new ed. Hachette & cie, Paris

Vasiliy Vasil'evich Vitkovskiy
(Novogeorgievsk, 1856–Saint Petersburg, 1924)
Russian topographer and geodesist, cartographer, professor, lieutenant general

He worked in the military and as a lecturer of geodesy and astronomy at several military and civillian higher schools. He was responsible for introducing the Jäderin base-line apparatus to triangulation of Russia. He published scientific papers in Russian journals and translated Clarke's handbook *Geodesy* from English. In 1907 he became a distinguished full professor and had the reward of the Military-Topographic School named after him (Filatov and Potievskiy 2015).

In 1907 he published procedures for minimizing scale variation according to following criteria: the maximum scale error between the standard parallels equals (with opposite sign) the scale error at each of two limiting parallels. Once the standard parallels are selected, Vitkovskiy projection III is constructed by using the same formulas used for the Lambert Conformal Conic with two standard parallels.

Vitkovskiy VV (1900) Vygodneishaya ravnopromezhutochnaya konicheskaya proyektsiya. Izvestija Russkogo geograficheskogo obshchestva, 36, pp 457–462
Vitkovskiy VV (1907) Kartografiya; teoriya kartograficheskikh proyektsiy. Yu. N. Erlikh, St. Petersburg
Vitkovskiy VV (1911) Prakticheskaja geodezija. St. Petersburg

Johann Heinrich Louis Krüger
(Elze, 1857–Elze, 1923)
German mathematician and geodesist

He finished the study of mathematics in Berlin. During his studies, he developed a passion for geodesy and at the age of 27, he became a doctor in that area. At Institute of Geodesy in Berlin he was firstly an assistant, then a professor and after that, the headmaster. The Gauss-Krüger projection (conformal transverse cylindrical projection of ellipsoid into the plane) was named after great German scientist Carl Friedrich Gauss. Professor Dr. Louis Krüger published a book on this projection in 1912, and in 1919 a collection of formulae for practical usage. Since then, this projection has been called Gauss-Krüger. In the *Gauss-Krüger projection*, the central meridian of the given area is mapped as a straight line and serves as the x axis of the rectangular coordinate system in the plane. The central meridian of the given area is mapped in its real length, i.e. without linear distortions or the linear scale along this meridian is constant, and the whole projection is conformal. This projection was used in many countries for official cartography (Fig. 12.30).

Krüger L (1903) Bemerkungen zu C.F. Gauss: Conforme Abbildungen des Sphäroides in der Ebene. In: C.F. Gauss, Werke, Königlich Gesellschaft der Wissenschaften, Göttingen, Abhandlungen, 1903, v. 9, pp 195–204

Krüger L (1912) Konforme Abbildung des Erdellipsoids in der Ebene. Königlich Preußisches Geodätisches Institut, Potsdam, Veröffentlichung, new series, no. 52, 172 p. [He published formulas for the Prussian land survey as Formeln zur konformen Abbildung des Ellipsoids in der Ebene: Berlin, 1919]

Krüger L (1914) Transformation der Koordinaten bei der konformen Doppelprojektion des Erdellipsoids auf die Kugel und die Ebene. Königlich Preußisches Geodätisches Institut, Potsdam, Veröffentlichung, new series, no. 60, 43 p.

Krüger L (1922) Zur stereographischen Projektion. Preußisches Geodätisches Institut, Berlin, Veröffentlichung, new series, no. 80, 28 p.

Hermann Heinrich Ernst von Hammer

(Ludwigsburg, 1858–Stuttgart, 1925)

German geodesist and cartographer

He studied (1874–78) at high technical school (Technische Hochschule, now the Vienna University of Technology), where he worked from 1878 to 1884 as a teaching assistant, and from 1884 as a professor of geodesy. In the year 1885, he published his famous textbook of flat and spherical trigonometry with special accent on usage in geodesy and spherical astronomy. This book had several later editions. In 1887, he translated from French to German the famous work of A. Tissot on map projections. He created a modification of the transverse equivalent azimuthal projection in the same way as the Aitoff projection was created by modification of the transverse equidistant azimuthal projection, so it is known as the Hammer-Aitoff projection. The projection is equivalent, the pole is a point, and the relation of Equator length to the length of central meridian is 2:1 (Fig. 12.31).

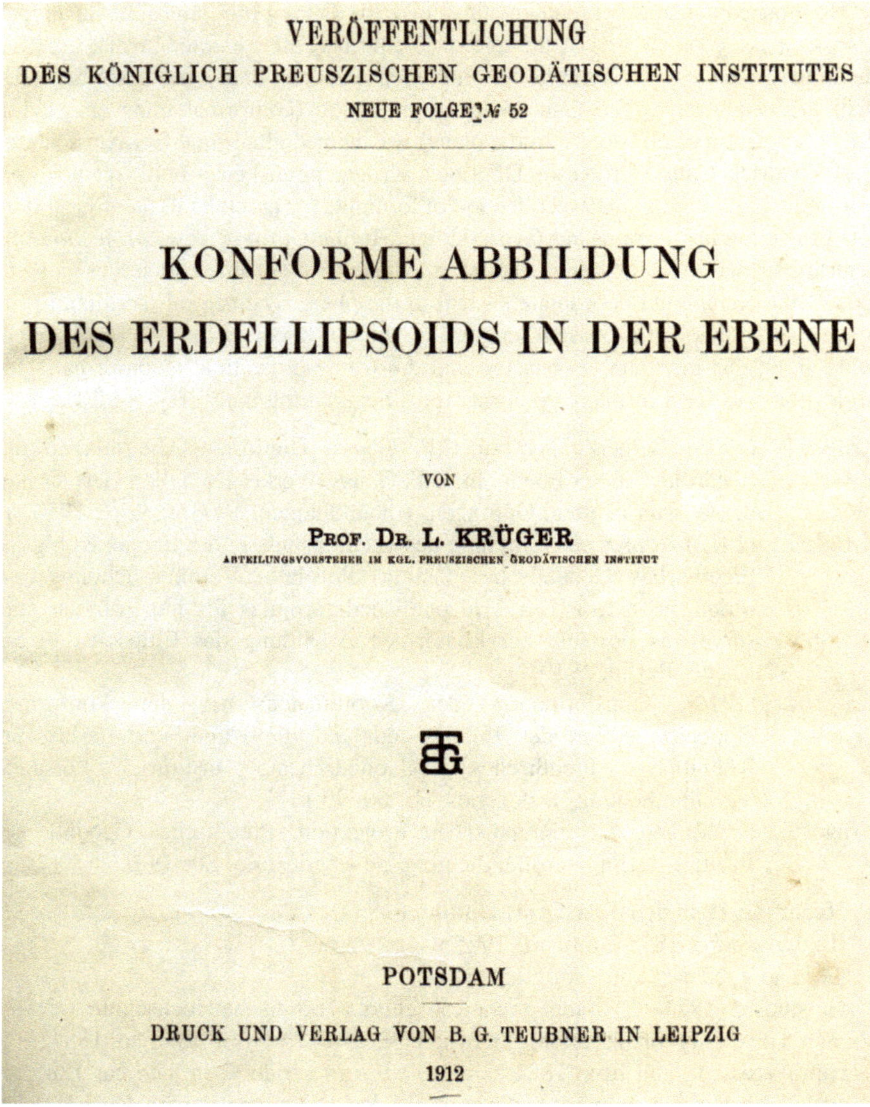

VERÖFFENTLICHUNG
DES KÖNIGLICH PREUSZISCHEN GEODÄTISCHEN INSTITUTES
NEUE FOLGE № 52

KONFORME ABBILDUNG
DES ERDELLIPSOIDS IN DER EBENE

VON

PROF. DR. L. KRÜGER
ABTEILUNGSVORSTEHER IM KGL. PREUSZISCHEN GEODÄTISCHEN INSTITUT

POTSDAM

DRUCK UND VERLAG VON B. G. TEUBNER IN LEIPZIG
1912

Fig. 12.30 Title page of Kruger's work Konforme Abbildung des Erdellipsoids in der Ebene, published in 1912

Hammer E (1885) Lehr- und Handbuch der Ebenen und Sphärischen Trigonometrie, zum Gebrauch beim Selbstunterricht und in Schulen besonders als Vorbereitung auf Geodäsie und Sphärische Astronomie, 2nd ed. 1987, 3rd ed. 1907, 4th ed. 1916, J. B. Metzlersche Buchhandlung, Stuttgart

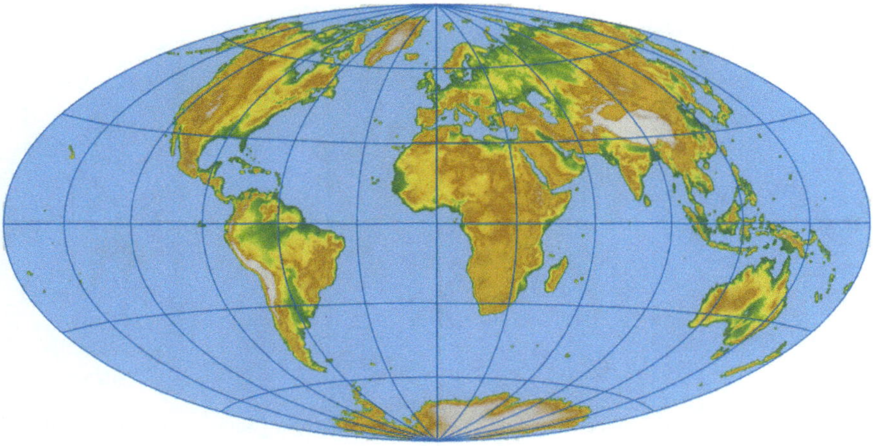

Fig. 12.31 Map of the world in Hammer-Aitoff projection

Tissot A (1887) Die Netzentwürfe geographischer Karten nebst Aufgaben über Abbildung beliebiger Flächen auf Einander, Autorisierte Deutsche Bearbeitung mit einigen Zusätzen bezorgt von E. Hammer, J. B. Metzlersche Buchhandlung, Stuttgart

John Paul Goode

(Stewartville, Minnesota, 1862–Little Point Sable, Michigan, 1932)

American geographer and cartographer

He graduated at the University of Minnesota in 1889. He received his doctorate in economics in 1901 at the University of Pennsylvania, where he taught geography from 1901 to 1917, and from 1917 to 1928 he taught at the University of Chicago as well.

He is known as the inventor of asymmetrical interrupted projections. For about four centuries, interrupted projections were characterized by symmetry of representation. On the contrary, asymmetrical projections prevail today. In the year 1916, Goode suggested a way for reducing distortion in pseudocylindrical projections. By this way, any pseudocylindrical projection can be used for the creation of world maps by certain sections which are joined along the Equator. The central meridian is selected for each section. Its longitude should be selected in a way that distortions on that area are to be as small as possible (Fig. 12.32). Goode is the author of a great number of maps and books in the area of geography. Significant is his atlas Goode's School Atlas (1923; many later editions), which is known today as Goode's World Atlas.

Goode JP (1905) A new method of representing the earth's surface: The Van Der Grinten projection. Journal of Geography, v. 4, no. 9, pp 369–373

Goode JP (1919) Studies in projections: adapting the homalographic projection to the portrayal of the earth's surface entire. Geographical Society of Philadelphia, Bulletin, v. 17, no. 3, pp 103–113 [Pagination for no. 3 alone: pp 21–31]

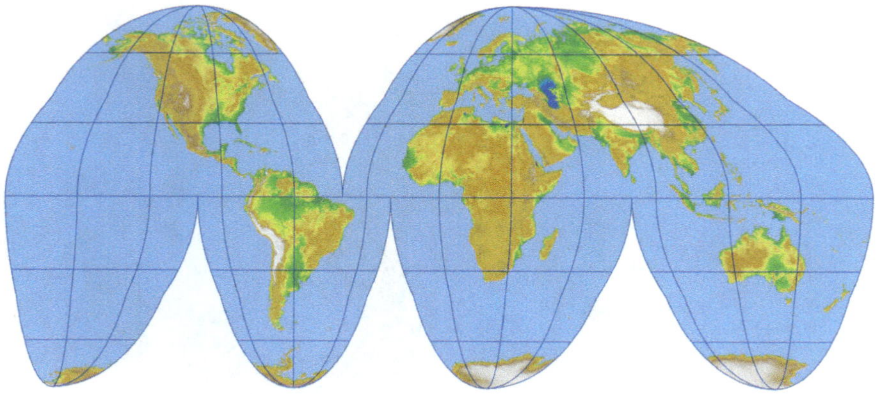

Fig. 12.32 The Goode projection, developed as merging the Sanson and the Mollweide projection

Goode JP (1925) The Homolosine projection: a new device for portraying the Earth's surface entire. Assn. of Amer. Geographers, Annals, v. 15, no. 3, pp 119–125

Goode JP (1929) A new projection for the world map: the Polar Equal Area. Monthly Weather Review, v. 57, no. 4, pp 133–136 [Interrupted Werner projection.]

Goode JP (1929) The Polar Equal Area, a new projection for the world map. Assn. of Amer. Geographers, Annals, v. 19, no. 3, pp 157–161 [Interrupted Werner projection.]

Max Eckert-Greifendorf
(Chemnitz, 1868–Aachen, 1938)
German geographer and cartographer

Eckert studied geography and national economy in Leipzig. From 1907 to 1937, he worked as a professor in vocational university in Aachen, where he taught economic geography and cartography. After World War I, he dedicated himself to cartography. Famous is his work *Die Kartenwissenschat* (published in two volumes in Berlin 1921/25), in which he set the foundations of cartography as a scientific discipline. At the beginning of 20th century, Eckert proposed six new pseudocylindrical projections for the whole world map. The projections are known as Eckert's projections I–VI. In all six

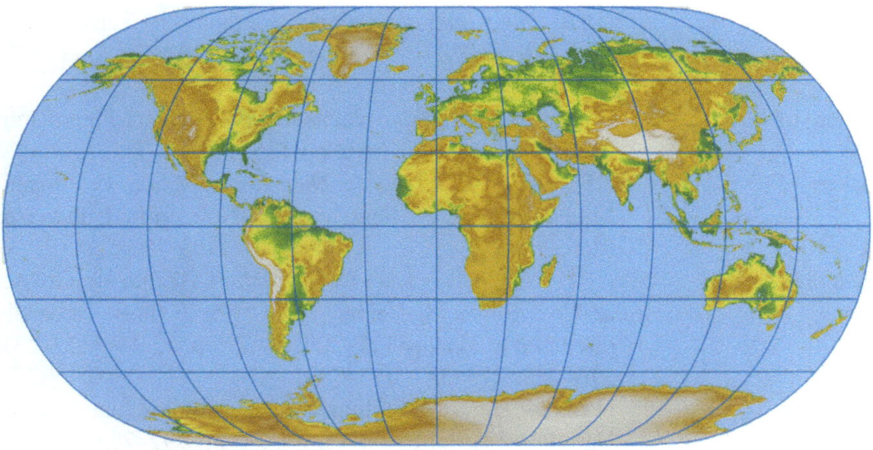

Fig. 12.33 Eckert III projection

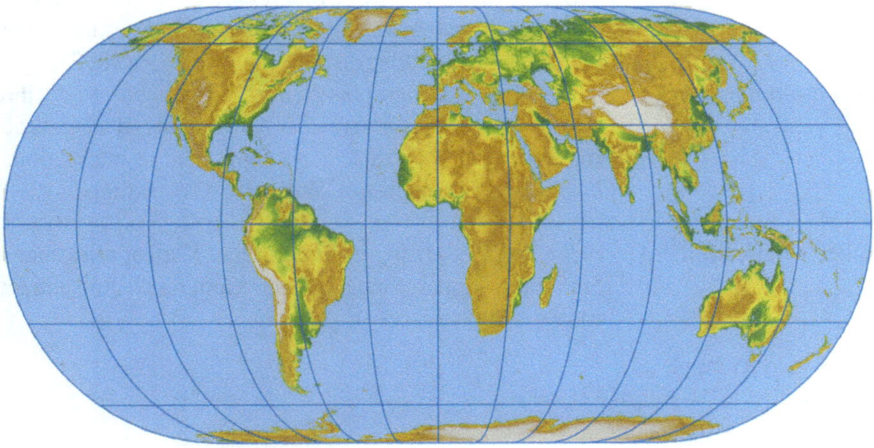

Fig. 12.34 Eckert IV projection

projections the pole is projected as a line half as long as the Equator. In the first two projections, meridians are broken at the Equator. This is their greatest flaw, so they are not used in practice. Eckert's projections III, IV, V and VI (Snyder and Voxland 1989, 1994) can be recommended for the creation of world maps (Figs. 12.33 and 12.34).

Eckert M (1906) Neue Entwürfe für Erdkarten. Petermanns Mitteilungen, v. 52, no. 5, pp 97–109 [New projections for world maps. Six pseudocylindrical projections with lines for poles.]

Eckert M (1909) Eine neue Isochronenkarte der Erde. Petermanns Mitteilungen, v. 55, no. 9, pp 209–216; no. 10, pp 256–263 [Azimuthal Equidistant projection used for isochrone map.]

Eckert M (1910) Die Kartenprojektion. Geographische Zeitschrift, v. 16, no. 6, pp 297–318; no. 7, pp 385–398; no. 8, pp 441–454

Eckert M (1920) Abänderung flächentreuer Netze. Petermanns Mitteilungen, v. 66, June, pp 125–126

Eckert M (1920) Zur Geschichte der Mercatorprojektion. Petermanns Mitteilungen, v. 66, Sept., p. 202. [Note on Etzlaub versus Mercator.]

Eckert M (1921, 1925) Die Kartenwissenschaft. Walter de Gruyter & Co., Berlin & Leipzig, 2 v. [Projections especially discussed v. 1, pp 115–207; v. 2, pp 62–89, 227–231, 276–280, 315–317, 775–787]

Eckert M (1923) Die Projektion der geologischen Karte. Naturwissenschaften, v. 11, no. 38, pp 792–795

Eckert-Greifendorff M (1935) Eine neue flächentreue (azimutaloide) Erdkarte. Petermanns Mitteilungen, v. 81, no. 6, pp 190–192. [Formerly Max Eckert. Compresses equatorial Lambert Azimuthal Equal-Area projection to one-quarter its height, quadrupling longitude; inspired by Hammer (1892)]

Eckert-Greifendorff M (1935) Die kartographischen Projektionen und ihre Anwendung, insbesondere die Anwendung flächentreuer Projektionen in der Geographie. International Geographical Congress, Warsaw, Proceedings, v. 1, pp 145–150

Eckert-Greifendorff M (1938) Die Zielsetzung geographischer Kartenprojektionen. 15th International Geographical Congress, Amsterdam, Proceedings, v. 2, sect. 1, pp 3–13

Oswald Winkel

(Leipzig, 1874–Leipzig, 1953)

German cartographer

He assembled a great number of general and travelling maps for the guide and publisher Karl Beadeker. Oswald Winkel presented in 1921 a projection for the whole world map which was named after him the *Winkel Triple projection*. This arbitrary projection is created as an arithmetical mean between the Aitov and the equidistant cylindrical projection. Since the Aitov projection was created by modification of the transverse azimuthal equidistant projection, the Winkel projection is made of three projections, and this is why it is called triple (Frančula 1971; Fig. 12.35).

Winkel O (1909) Flächentreue, schiefachsige Zylinderprojektion mit längertreuem Grundkreis für eine Karte von Nord-, Mittel- und Südamerika. Petermanns Mitteilungen, v. 55, no. 11, pp 329–330 [Oblique Cylindrical Equal-Area projection. See also table of coordinates at 1:60M in no. 12, pp 379–380]

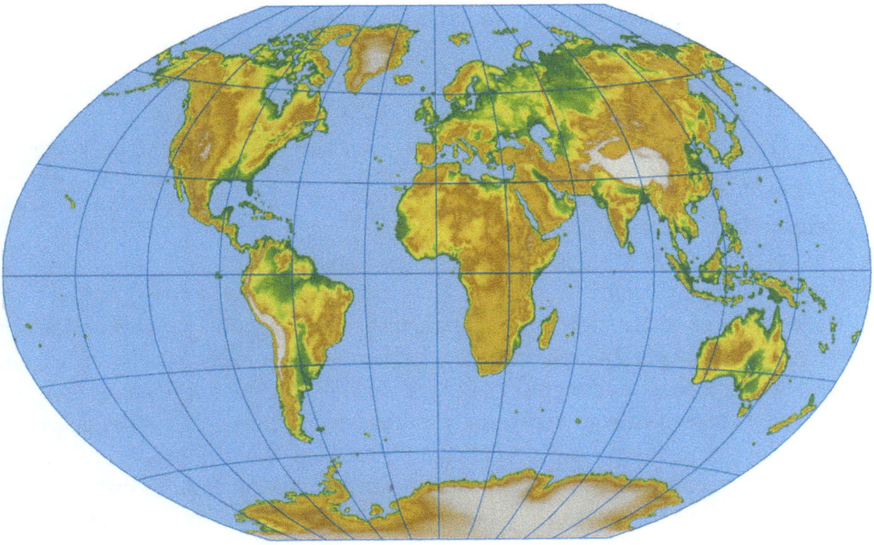

Fig. 12.35 Map of the world in the Winkel projection (standard parallel: 50°28′)

Winkel O (1911) Flächentreue, zwischenständige, azimutale Projektion für eine Karte der Britischen Inseln als praktisches Beispiel eines Kartenentwurfes. Geographischer Anzeiger, v. 12, pp 30–31

Winkel O (1913) Beitrag zur Entwicklung schiefachsigen, speziell zylindrischer Projektionen unter Annahme der Kugelgestalt der Erde. Petermanns Mitteilungen, v. 59–2, Dec., pp 304–306 [Oblique cylindrical projection.]

Winkel O (1921) Neue Gradnetzkombinationen. Petermanns Mitteilungen, v. 67, Dec, pp 248–252 [Proposes three combinations of Equirectangular with Sinusoidal, Mollweide, and Aitoff projections, respectively.]

Winkel O (1922) Die azimutaleschen Erdkartenentwürfe von D. Aitoff und E.v. Hammer. Geographische Zeitschrift, v. 28, pp 112–114 [Aitoff and Hammer projections.]

Winkel O (1922) Allgemeine Betrachtungen über die Abbildung sehr breiter "Zonen". Geographische Zeitschrift, v. 28, no. 5–6, pp 177–181 [General projection considerations in representing latitude zones.]

Winkel O (1927) Die Gewinnung geeigneter Landkartennetze. Geographische Zeitschrift, v. 33, no. 10, pp 599–604

Winkel O (1928) Übersicht der Gradnetzkombinationen. Petermanns Mitteilungen, v. 74, no. 7–8, pp 201–204 [Pseudocylindrical projections.]

Winkel O (1933) Flächentreue oder abstandstreue Projektionen? Geographische Wochenschrift, p. 207ff

Winkel O (1933) Die Projektionswahl bei Erdkarten. Petermanns Mitteilungen, v. 79, no. 5–6, pp 125

Winkel O (1939) 25 Jahre neue Netzkombinationen. Petermanns Geographische Mitteilungen, v. 85, no. 9, pp 278–280

Winkel O (1951) Kurzgefaßte Kartenentwurfslehre. Atlantik-Verlag, Frankfurt-am-Main, 44 p.

Reinholds V. Putniņš

(Bērzpils parish, Balvi, Latvia, 1881–Riga, Latvia, 1934)

Latvian mathematician

He was an associate professor of mathematics and natural sciences at Latvian University. He took part in organization of Latvian national meteorological office and was repeatedly elected as the president of the *Latvian Association of Geographers.*

In 1934 he presented twelve pseudocylindrical projections with simply described relationships resambling those of Eckert in 1906 and Wagner in 1932. On each projection, the central meridian is half the length of equador. Putninš distinguished them with P followed by subscripts 1–6, those without primes having pointed poles, and those with primes having poles half the length of equator (Bugayevskiy and Snyder 1995).

Putniņš P5 projection is pseudocylindrical projection with equally spaced parallels. In this projection scale is constant along any given latitude and the same for the latitude of opposite sign and also constant along the central meridian (Fig. 12.36).

Putniņš RV (1934) Jaunas projekcijas pasaules kartem. Geografiski raksti, Folia Geographica 3 and 4, pp 180–209 [Latvian with extensive French résumé. New projections for world maps. Describes 12 pseudocylindricals, half equal-area and half with equally spaced parallels, half pointed-polar and half with pole-lines, using elliptical, parabolic, or hyperbolic meridians.]

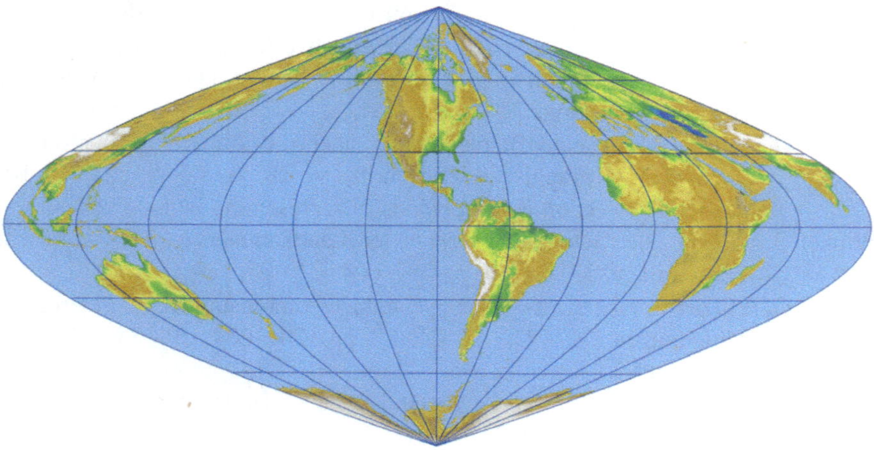

Fig. 12.36 Map of the world in Putniņš P5 projection

Putniņš RV (1934) Application de la projection quadratique équivalente à l'étude des cartes anciennes. International Geographical Congress, Paris, 1931, Comptes Rendus, v. 3, pp 719–729

Putniņš RV (1935) Sur quelques nouvelles projections à méridiens elliptiques. International Geographical Congress, Warsaw, 1934, Comptes Rendus, v. 1, pp 151–157

Vladimir Vladimirovich Kavrayskiy

(Zherebyatnikovo, 1884–Leningrad, 1954)

Russian cartographer, professor, engineer-rear-admiral

Kavrayskiy finished studying at the Kharkov university in 1916. He was employed at the Military-Maritime Academy beginning in 1921. Kavrayskiy was a State Prize winner of the USSR. He was an outstanding representative of the Kharkov scientific school in the field of astrometry and stellar astronomy, which was founded by Ludwig O. Struve (1858–1920). In 1936, Kavrayskiy proposed an equa-area sinusoidal pseudocylindrical projection for the world map (Bugayevskiy and Snyder 1995). The pole in this projection is projected as a line whose length is equal to half of the Equator. The elliptical pseudocylindrical projection of Kavrayskiy is also named after him, and it is classified to the group of arbitrary projections (Fig. 12.37). His monograph in three volumes about the theory of map projections, which was published postmortem, is also well-known (Filatov and Potievskiy 2015).

Kavrayskiy VV (1958) Izbrannye trudy, Tom II; Matematicheskaya kartografiya, Vyp. 1, Obshchaya teorija kartograficheskikh projektsij, Izdanie Upravlenija nachal'nika Gidrograficheskoj sluzhby VMF

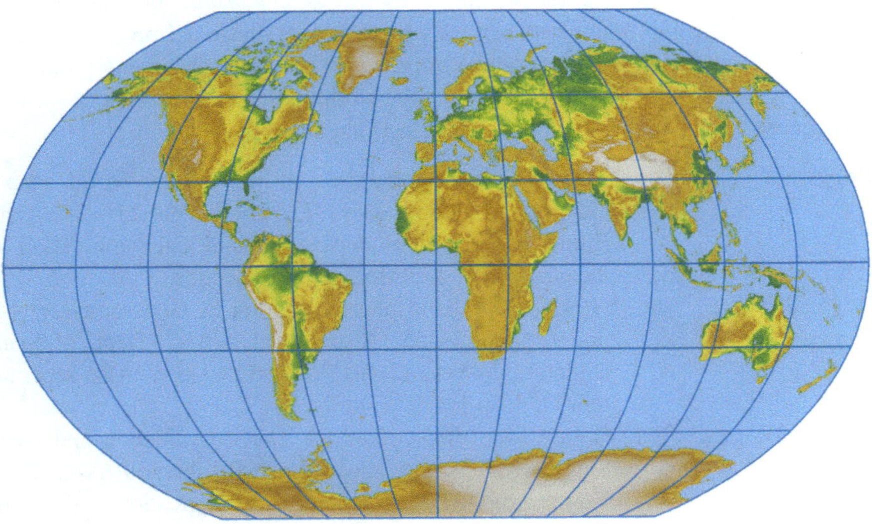

Fig. 12.37 Elliptical pseudocylindrical projection of Kavrayskiy

Kavrayskiy VV (1959) Izbrannye trudy, Tom II; Matematicheskaya kartografiya, Vyp. 2, Konicheskie i tsilindricheskie proyekciji, ih primenenije, Izdanie Upravlenija nachal'nika Gidrograficheskoj sluzhby VMF

Kavrayskiy VV (1960) Izbrannye trudy, Tom II; Matematicheskaya kartografiya, Vyp. 3, Perspektivye, krugovye i drugie vazhnejshie proyekcii. Navigacionnye zadachi, Izdanie Upravlenija nachal'nika Gidrograficheksoj sluzhby VMF

Samuel Whittemore Boggs
(Coolidge, Kansas, 1889–1954)
American geographer

He was a long-time geographer to the *United States Department of State*, recognized as one of the world's leading experts on international boundaries and also as an authority on the subject of map compilation, editing, and cataloguing. He was a constant source of new ideas on cartographic techniques and through his imagination and enthusiasm he was able to contribute in a variety of ways to the development of the field of cartography.

In 1929 he presented a pseudocylindrical equal area projection, which he named the *Eumorphic* projection, today known as *Boggs eumorphic projection*. It is something of a compromise between the parabolic and sinusoidal projections, and was among other things an attempt to improve on the polar shapes of equal-area maps (Fig. 12.38). A political wall map of the world, based on this production, was subsequently marketed by the A.J. Nystrom Company.

Boggs SW (1929) A new equal-area projection for world maps. Geographical Journal, v. 73, no. 3, pp 241–245 [His Eumorphic projection, a pseudocylindrical combination of the Sinusoidal and Mollweide projections.]

Boggs SW (1945) "This hemisphere". Journal of Geography, v. 44, no. 9, pp 345–355 [Also issued as U.S. Dept. of State Bulletin, 1945, v. 12, no. 306, pp 845–850. Also separate reprint, 13 p.]

Boggs SW, Lewis DC (1945) Map projections. In: The classification and cataloging of maps and atlases: Special Libraries Assn., New York, pp 81–91 [Library classification of map projections.]

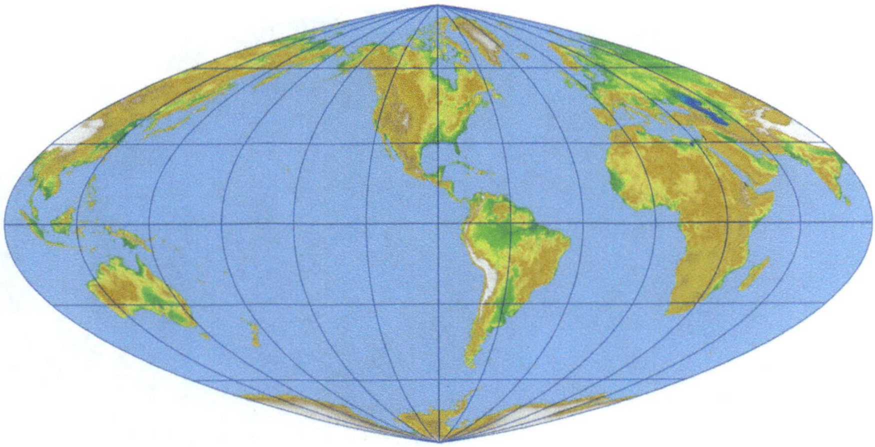

Fig. 12.38 Map of the world in Boggs eumorphic projection

William A. Briesemeister
(1895–1967)
American cartographer

Prominent mapmaker, who retired as chief cartographer of the *American Geographical Society* in 1964, after more than 50 years working there.

Briesemeister Peak in Antarctica was named after him. This peak was photographed from the air by Sir Hubert Wilkins on December 20, 1928, and by the *United States Antarctic Service* in 1940, but Briesemeister recognised this peak on two photographs and established their continuity, an important clue to the identity and correct position of Stefansson Strait.

He supervised the preparation of maps of Antarctica for use during the *International Geophysical Year* (IGY) (1957–1958) and post-IGY programs of the *United States Antarctic Research Program*, including continental maps published at a scale of 1:6,000,000 (1956) and 1:5,000,000 (1962).

His many accomplishments include the *Briesemeister projection*, presented in 1953, which permits a flat map to show land areas in their true relative size (Fig. 12.39).

Briesemeister W (1953) A new oblique equal-area projection. Geographical Review, v. 43, no. 2, pp 260–261 [His oblique aspect of Hammer projection, with axes 1.75 to 1 instead of 2 to 1.]

Briesemeister W (1959) A world equal-area projection for the future. The selection of the most suitable equal-area projection for the purpose of plotting world wide statistics in this present day of super speed, jet planes and intercontinental missiles. Nachrichten aus dem Karten- und Vermessungswesen, series 2, no. 3, pp 60–63. [Published separately, 1958: Chicago, Rand McNally, 4 p. Translated to German as Eine flächentreue

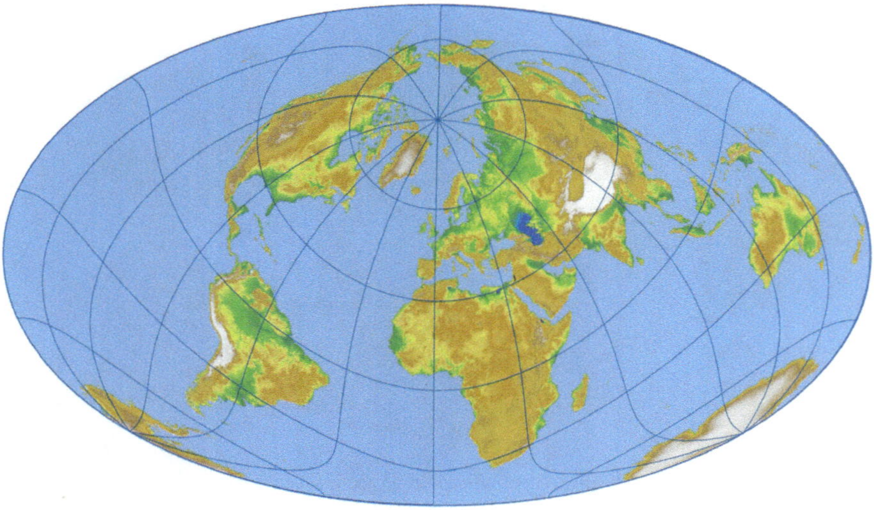

Fig. 12.39 Map of the world in Briesemeister projection

Weltkarten-Projektion für die Zukunft. Nachrichten aus dem Karten- und Vermessungswesen, 1958, series 1, no. 7, pp 72–74]

Briesemeister W (1959) A new equal area projection for the future. Second International Cartographic Conference, Chicago, 1958, Proceedings: Verlag des Instituts für Angewandte Geodäsie, pp 60–63

Osborn Maitland Miller

(Perth, Scotland, 1897–New York, USA, 1979)

American cartographer and geographer

The *Osborn Maitland Miller Cartographic Medal* established in 1968 by the *American Geographical Society Council* was named after him, and it honors "outstanding contributions in the field of cartography or geodesy".

Miller's forty-six year career with the *American Geographical Society* was only one of the many accomplishments of his career. He headed the staff, researched, and taught at the *American Geographical Society's School of Surveying*, specializing in photogrammetry and cartography.

In 1941, in collaboration with William A. Briesemeister, he presented the *bipolar oblique conic conformal projection*, designed specifically for a low-error map of North and South America constructed by the American Geographical Society.

In 1942 he developed the *Miller cylindrical projection*. This projection is used for world maps in numerous American atlases and some other atlases, as a projection resembling the Mercator but having less distortion of area and scale, especially near the poles (Fig. 12.40).

Miller oblated stereographic projection was first applied by Miller to Africa and Europe in 1953 and to other Eastern Hemisphere regions in conjunction with

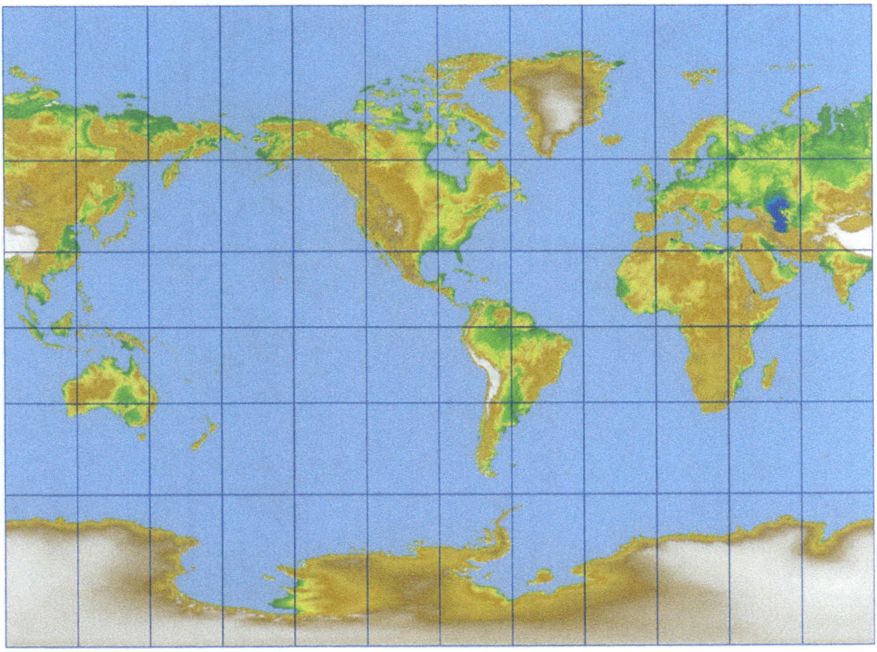

Fig. 12.40 Map of the world in Miller cylindrical projection

several nonconformal fill-in projections in 1955 (Snyder and Steward 1988). This projection is used for an area that can be contained within an oval shape to minimize the scale distortions in the area (Fig. 12.41).

Miller O M (1941) A conformal map projection for the Americas. Geographical Review, v. 31, no. 1, pp 100–104 [Presents Bipolar Oblique Conic Conformal projection.]

Miller O M (1942) Notes on cylindrical world map projections. Geographical Review, v. 32, no. 3, pp 424–430 [Presents his cylindrical projection. See also 1943, v. 33, pp 328–329]

Miller O M (1953) A new conformal projection for Europe and Asia [sic; should read Africa]. Geographical Review, v. 43, no. 3, pp 405–409 [Presents his Oblated Stereographic projection as applied to Europe and Africa only.]

Miller O M (1955) Specifications for a projection system for mapping continuously Africa, Europe, Asia, and Australasia on a scale of 1:5,000,000. American Geographical Society. New York. Contract report to Army Map Service, May 1955. Also corrections in letter by Miller, June 2, 1955, and Supplement, January 31, 1956. [Extends his Oblated Stereographic projection to most landmasses of Eastern Hemisphere. See note in Geographical Review, 1956, v. 46, no. 2, p. 258]

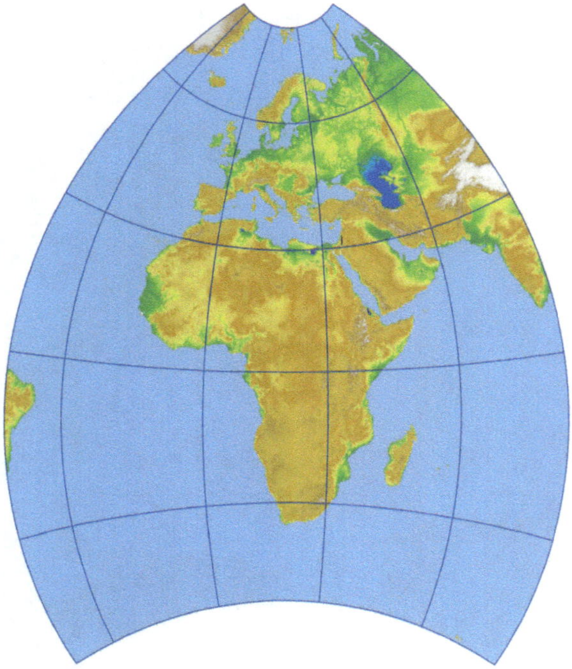

Fig. 12.41 Map of the Europe and Africa in Miller oblated stereographic projection

Miller O M (1962) Map projections. Encyclopædia Britannica, 15th ed., v. 14, pp 841–844 [Reprinted to 1972. Text less than one page by others in later editions.]

Miller O M (1965) Some equivalent map projection transformations in the plane. Survey Review, v. 18, no. 136, pp 73–77

Vladimir Kirilov Khristov (Hristov, Hristow)
(Sofia, 1902–1979)
Bulgarian geodesist, astronomer and cartographer

Khristov studied astronomy, mathematics and physics at Leipzig University, where he graduated in 1925 with his PhD in astronomy.

He worked in the Astronomical and Geographical Department of the State Institute of the Ministry of War from 1925 to 1948. The Department was assigned to organize and carry out necessary astronomical measurements for tasks such as: developing first class triangulation, choosing of a reference ellipsoid, creating working formulas, measuring the ellipsoid, making the orientation of the network, proposing selection of coordinates, and planning and selecting the map projection for topographic maps. Under these conditions, Khristov explored the extensive library of the Leipzig University and works of Helmert dealing with problems of size of the reference ellipsoid and its orientation. Khristov defined and solved the problem of map projection transformations using calculus. Development in series is the basis of his later works, such as the introduction of Gaussian coordinates and calculation of all necessary reductions and transformations in overlapping strips, putting the formulas for transition from one reference geodetic system to another and specific solutions introducing the Bulgarian system in 1950 with the ellipsoid of Krasovsky, such as transformation of the analogue system in 1930 defined on the ellipsoid of Hayford.

Transformation solutions by the method of Khristov were easily applied by using tables and such tables quickly entered the surveying practice.

In 1948, Khristov became Professor of Geodesy at the State Polytechnic (nowadays University of Architecture, Civil Engineering and Geodesy), where he remained until his retirement in 1970. At the University, he laid the foundations, created programs and issued the first Bulgarian textbooks on four main areas—scientific disciplines that formed at that time the students' knowledge of higher geodesy: mathematical cartography, physical geodesy, geodetic astronomy and mathematical geodesy.

Khristov was a corresponding member from 1948, and from 1958 Academician of the Bulgarian Academy of Sciences. He headed the Central Laboratory of Geodesy from 1956 to 1970.

In 1959, academician Khristov managed to restore the Bulgarian National Committee of Geodesy and Geophysics and Bulgaria's membership in the International Union of Geodesy and Geophysics, which dates from 1932, but was suspended during World War II and the years of political isolation. As longtime chairman of this Committee, he developed a particularly fruitful scientific activity and created links with research institutions and communities around the world.

Khristov received recognition in the scientific community and a recognition from the Bulgarian government. Presented with all scientific and honorary titles, he won the highest state awards. He was elected an honorary member of the Hungarian Academy of Sciences in 1969.

Khristov's most important monographs (Snyder and Steward 1988):

Khristov VK (1943) Die Gauß-Krügerschen Koordinaten auf dem Ellipsoid. B.G. Teubner, Leipzig and Berlin [Translated to English by the Army Map Service as The Gauss-Krüger coordinates on the ellipsoid; unpublished.]

Khristov VK (1946) Gausz-Kryugerovite koordinati vyrhu rotacionniya elipszoid. Pecsatnicata na Dyrzsavniya Geografszki Insztitut, Sofia. [Bulgarian. Gauss-Krüger coordinates on the ellipsoid of rotation.]

Khristov VK (1949) Kartii proyektsiy matematicheska kartografie. Sofia, 196 p. [Bulgarian. Map projections in mathematical cartography.]
Khristov VK (1955) Die Gaußschen und geographischen Koordinaten auf dem Ellipsoid von Krassowsky. VEB Verlag Technik, Berlin
Khristov VK (1957) Koordinaty Gaussa-Kryugera na ellipsoide vrashcheniya. Geodezizdat, Moscow 263 p. [Russian. Gauss-Krüger coordinates on the ellipsoid of rotation.]
Khristov VK (1961) Matematicheska kartografiya. Tekhtsmka, Sofia, 155 p. [Bulgarian. Mathematical cartography.]

Karl Heinrich (Karlheinz) Wagner
(Leipzig, 1906–Berlin, 1985)
German cartographer

Karl Heinrich Wagner was son of Eduard Wagner, co-owner of H. Wagner & E. Debes cartographic company in Leipzig, which predestined him for the vocation of geographer/cartographer. He majored in geography, and minored in oceanography, mathematics and geology.

He earned his PhD in 1931 with *Pseudocylindrical projection, their importance and practical application*. In the 1930's, he introduced the Wagner transformation (its original name is *Umbeziffern vor Kartennetzen*, which means renumbering of cartographic grids). This transformation uses a smaller part of a projection's grid, magnified to keep the original area of the projection.

In 1931, he started working as a cartographer for Wagner & Debes, where he worked until 1933. He then went to Barcelona and was employed in the cartographic department of one of the largest Spanish publishers of school books, Editorial Luis Vives. In 1937, he returned to Leipzig and became the manager of Wagner & Debes. The company was completely destroyed in an air strike in 1943, but it was rebuilt in 1945.

From 1944, Wagner also worked on his major book *Kartografische Netzenwürfe* (Map Projections). In his book, Wagner presented three different transformation methods that he applied to various types of map projections and obtained several new graticules.

In 1951, Wagner transferred the company's headquarters to Berlin and renamed it to *Kartografische Anstalt Dr. Karl Heinrich Wagner*. In 1958 and 1959, Wagner worked with the Bibliographische Institut on developing projections for a German world map and a German sea map. Based on these projections, Wagner developed maps for *Atlas zur Ozeanographie* by G. Dietrich. Wagner also reworked *Atlas zur physischen Geographie* and developed relief representations for small scale maps (Ferschke 1966; 1985; Bosse 1976; Mittelstaedt 1976).

He developed several pseudocylindrical and modified azimuthal projections (Bugayevskiy and Snyder 1995), named after him *Wagner projections*.

The *Wagner IV projection* (presented in 1932) is pseudocylindrical equal area projection which has meridians that are only portions of semi ellipses, but parallels are also spaced for equal area.

The *Wagner VII projection* (presented in 1941) is modified azimuthal equal area projection, used for World maps, such as climatic maps prepared by the U.S. Department of Commerce (Figs. 12.42 and 12.43).

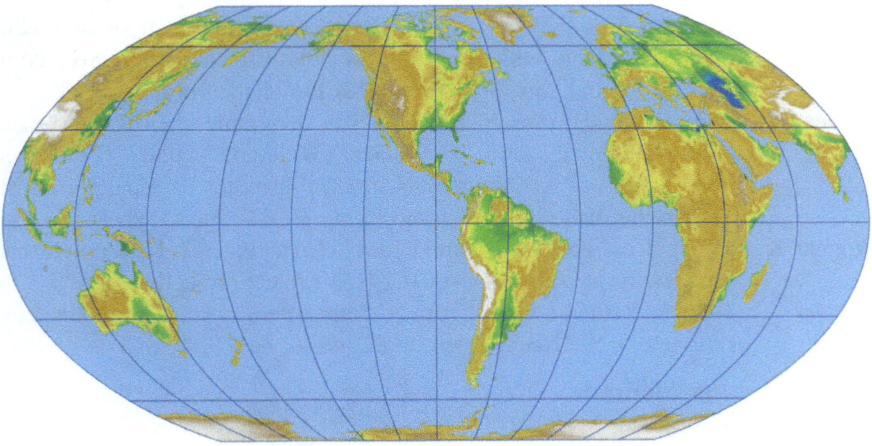

Fig. 12.42 Map of the world in Wagner IV projection

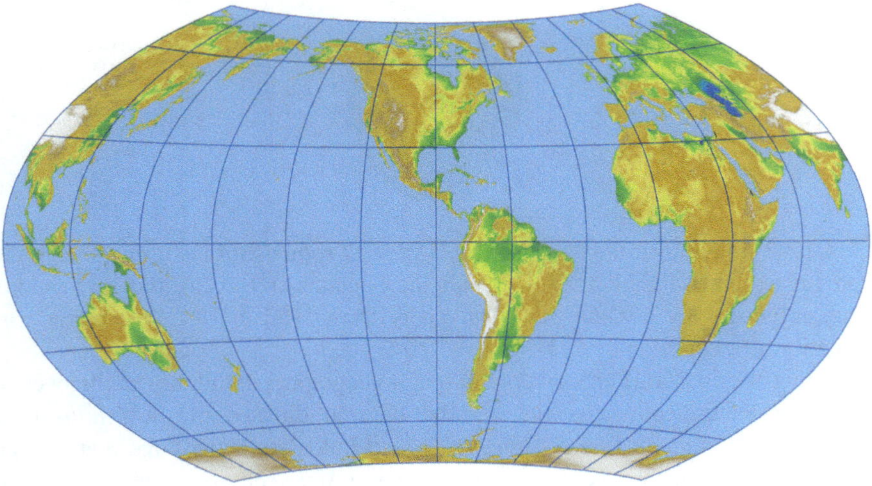

Fig. 12.43 Map of the world in Wagner VII projection

In 1976 he was appointed as Honorary Member of The German Cartographic Society e.V. (Deutsche Gesellschaft für Kartographie e.V.).

Wagner's main publications are:

Wagner K (1932) Die unechten Zylinderprojektionen. Deutsche Seewarte, Archiv, v. 51, no. 4, 68 p. [Presents several pseudocylindrical projections with sinusoidal or elliptical meridians.]

Wagner K (1941) Neue ökumenische Netzentwürfe für die kartographische Praxis. Jahrbuch der Kartographie, pp 176–202 [Presents modifications of Hammer projection with curved lines for poles.]

Wagner K (1944) Umformung von Mercator-Netzen. Petermanns Geographische Mitteilungen, v. 90, no. 11–12, pp 299–306

Wagner K (1949) Kartographische Netzentwürfe. Bibliographisches Institut, Leipzig, 262 p. [2nd ed. 1962. Presents new flat-polar pseudocylindrical projections with sinusoidal or elliptical meridians. Also combines Aitoff and equatorial Azimuthal Equidistant projections.]

Wagner K (1966) Über das Zusammenfügen von geographischen Kartennetzen und die Netze der "Deutschen Weltkarte" und "Deutschen Meereskarte". Bibliographisches Institut, Hauszeitschrift, Wissenschaftliche Redaktion, no. 2, pp 89–117; no. 3, pp 7–55

Wagner K (1973) Das neue Kartenbild des Herrn Peters. Kartographische Nachrichten, v. 23, no. 4, pp 162–163 [Peters projection.]

Wagner K (1982) Bemerkungen zum Umbeziffern von Kartennetzen. Kartographische Nachrichten, v. 32, no. 6, pp 211–218

Felix Webster McBryde

(Lunenburg, Virginia, 1908–1995)

American geographer, cartographic consultant and educator

In 1943 he founded the *American Society for Professional Geographers*. Aside from his research of markets in villages in the Gautemalan Lake Atitlan area and his teaching geography at Ohio State University from 1937 to 1942, his career was primarily as a consultant.

During World War II he worked as a senior geographer in military intelligence in the War Department. After the war he became director of the *Smithsonian Institution's Institute of Social Anthropology* in Lima, Peru. After three years in that position he became chief geographer for the *Latin American program of the United*

States Bureau of the Census that included his establishing the *Ecuadorian Institute of Anthropology and Geography* in Quito.

He also developed new world map projections (Snyder and Voxland 1989, 1994):

- McBryde S3 projection, developed by F. Webster McBryde in 1977 as a merging of the Sinusoidal with the McBryde-Thomas flat-polar sinusoidal projection at the parallels of identical scale on the two projections, latitudes 55° 51′ N and S. U.S. Patent by McBryde
- McBryde-Thomas flat-polar sinusoidal projection, presented by F. Webster McBryde and Paul D. Thomas through the U.S. Coast and Geodetic Survey in 1949
- McBryde-Thomas flat-polar parabolic projection, presented by F. Webster McBryde and Paul D. Thomas through the U.S. Coast and Geodetic Survey in 1949
- McBryde-Thomas flat-polar quartic projection, presented by F. Webster McBryde and Paul D. Thomas through the U.S. Coast and Geodetic Survey in 1949 (Figs. 12.44, 12.45, 12.46 and 12.47).

McBryde FW (1977) New equal-area world map projections and base maps for biological data. Transemantics, Inc., Washington, 19 p. [Merges standard pseudocylindricals with his flat-polar forms.]

McBryde FW (1978) A new series of composite equalarea world map projections. International Cartographic Assn., 9th International Conference on Cartography, College Park, Maryland, Abstracts, pp 76–77

McBryde FW (1981) Pondering projections. Amer. Geophysical Union, Eos Transactions, v. 62, no. 48, p. 1162. [Proposes his interrupted equalarea pseudocylindricals for ocean maps.]

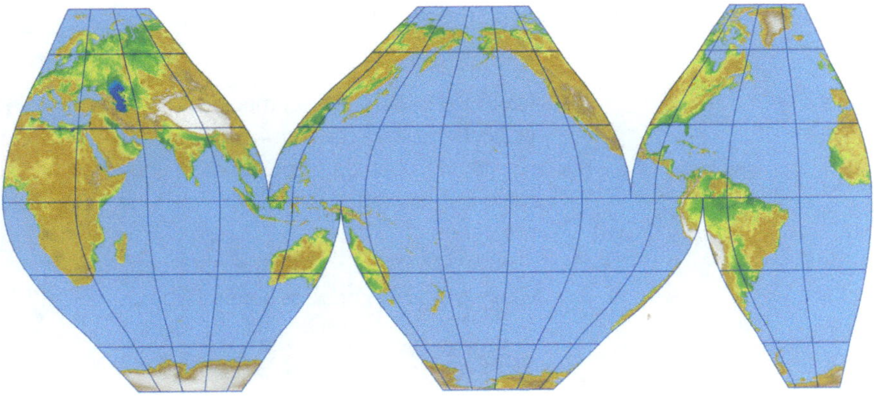

Fig. 12.44 Map of the world in interrupted McBryde S3 projection

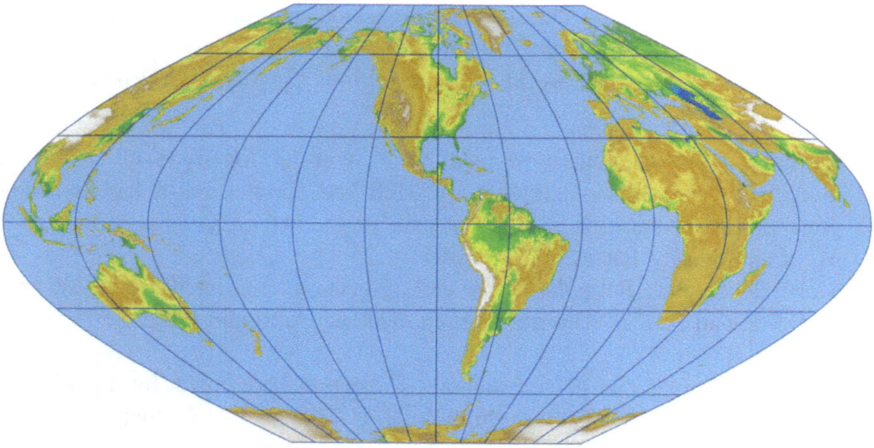

Fig. 12.45 Map of the world in interrupted McBryde-Thomas flat-polar sinusoidal projection

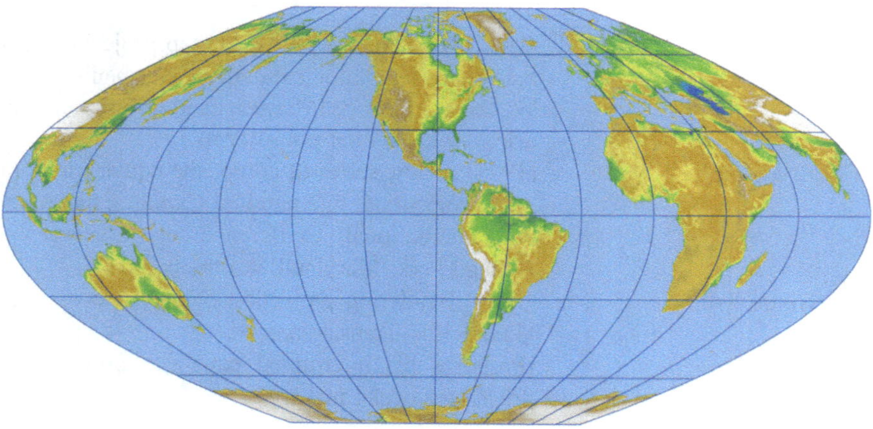

Fig. 12.46 Map of the world in McBryde-Thomas flat-polar parabolic projection

McBryde FW	(1982) Homolinear composite equalarea world projections. U.S. Patent 4,315,747, dated Feb. 16. [Especially his S3 projection, a merging of the regular Sinusoidal with his Flat-Polar Sinusoidal projections.]
McBryde FW	(1982) Mollweide graticule shortcomings. Amer. Geophysical Union, Eos Transactions, v. 63, p. 522

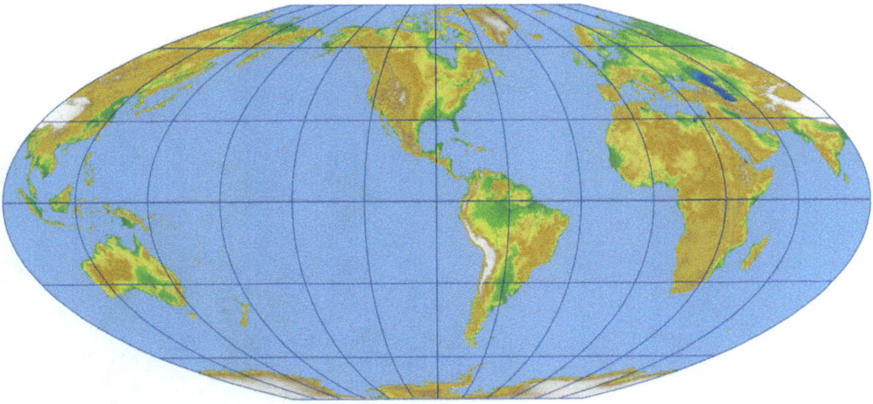

Fig. 12.47 Map of the world in McBryde-Thomas flat-polar quartic projection

McBryde FW, Thomas PD (1949) Equal-area projections for world statistical maps. U.S. Coast and Geodetic Survey Spec. Pub. No. 245, 44 p. [Presents flat-polar forms of sinusoidal, quartic, and parabolic pseudocylindrical projections, and two other pseudocylindricals.]

Laurence Patrick Lee

(England, 1913–1985)

New Zealand surveyor and cartographer

He was educated at Sacred Heart College in Auckland and Auckland University where he undertook study towards a BSc degree. He was tutor in geodesy and astronomy for the Technical Correspondence Institute for many years and wrote the geodesy course for the survey examination in that subject.

His career from 1936 to retirement in 1974 was with *New Zealand Department of Lands and Survey* where he was chief computer during the last ten years and was involved with many major projects. Perhaps the most important were the completion of the First Order Geodetic Triangulation of New Zealand and the establishment of the Geodetic Datum 1949, the change to metric units and the computations for the International Geophysical Year latitude and longitude program during the period 1957–59.

He was deeply interested in map projections and became known worldwide for his work in this field and indeed was an authority on this subject He published many papers and developed many new ideas and methods of computation. He showed the conformal world map in an equilateral triangle and ellipse, a hemisphere in a rectangle, and the world conformally on the faces of all five of the regular Platonic polyhedral (tetrahedron, cube, etc.). Although these projections are truly conformal, they generally have a singular point at each vertex of the bounding polygon and often elsewhere as well, with wide ranges of scale. They principally remain admirable mathematical achievements and artistic curiosities (Fig. 12.48).

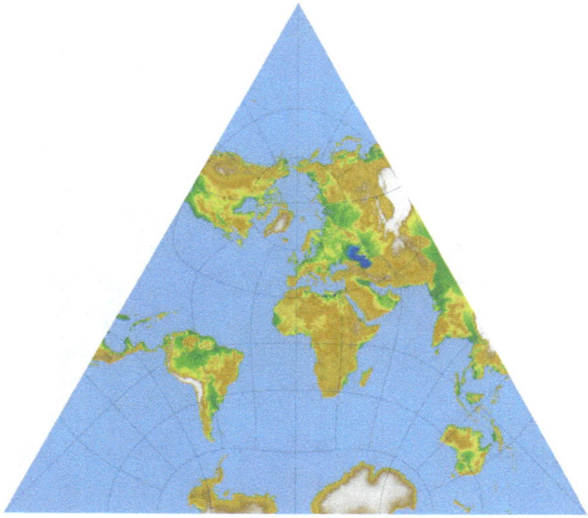

Fig. 12.48 Lee conformal tetrahedric projection of the world

Lee was elected an Honorary Member of the New Zealand Institute of Surveyors in 1971. He was also a foundation member and the first Life Member of the New Zealand Cartographic Society and a member of the Royal Astronomical Society of New Zealand.

New Zealand Cartographic Journal: Obituary: Laurence Patirck Lee, 1985, 15 (1):32

Lee LP (1944) The nomenclature and classification of map projections. Empire Survey Review, v. 7, no. 51, pp 190–200

Lee LP (1945) The transverse Mercator projection of the spheroid. Empire Survey Review, v. 8, no. 58, pp 142–152

Lee LP (1946) The nomenclature of map projections. Empire Survey Review, v. 8, no. 60, pp 217–219

Lee LP (1946) Marginal scales of latitude and longitude on transverse Mercator maps: Empire Survey Review, v. 8, no. 60, pp 220–221

Lee LP (1946) The convergence of meridians. Empire Survey Review, v. 8, no. 61, pp 267–271 [See also 1954, v. 12, no. 91, pp 237–238]

Lee LP (1947) A stereographic device for the solution of spherical triangles. Empire Survey Review, v. 9, no. 65, pp 123–131

Lee LP (1947) The equidistant azimuthal projection of the sphere. New Zealand Geographer, v. 3, no. 1, pp 41–58

Lee LP (1950) The planets 1950. New Zealand Astronomical Handbook, p. 22. [Oblique Mercator projection.]

Lee LP (1953) A transverse Mercator projection of the spheroid alternative to the Gauss-Krüger form. Empire Survey Review, v. 12, no. 87, pp 12–17 [See also 1954, v. 12, no. 91, pp 237–238]

Lee LP (1954) The oblique Mercator projection. New Zealand Geographer, v. 10, no. 2, pp 151–164 [Reprinted in Empire Survey Review, 1956, v. 13, no. 101, pp 321–335]

Lee LP (1962) The transverse Mercator projection of the entire spheroid. Empire Survey Review, v. 16, no. 123, pp 208–217 [See also 1964, v. 17, no. 133, p. 343; comments by Gdowski, Boguslaw, 1966, v. 18, no. 141, pp 339–341]

Lee LP (1963) Scale and convergence on the transverse Mercator projection of the whole spheroid. Empire Survey Review, v. 17, no. 127, pp 49–51

Lee LP (1965) Some conformal projections based on elliptic functions. Geographical Review, v. 55, no. 4, pp 563–580 [Conformal projections of world or hemisphere bounded by a polygon or an ellipse, or on a polyhedron. See also 1968, v. 58, no. 3, pp 490–491]

Lee LP (1973) The search for an improved coordinate system for New Zealand surveys and maps. N.Z., Department of Lands and Survey, Computing Branch, Head Office, Wellington, 20 p.

Lee LP (1973) The conformal tetrahedric projection with some practical applications. Cartographic Journal, v. 10, no. 1, pp 22–28

Lee LP (1974) A conformal projection for the map of the Pacific: New Zealand Geographer, v. 30, no. 1, pp 75–77 [Complex-algebra modification of oblique Stereographic projection.]

Lee LP (1974) The computation of conformal projections. Survey Review, v. 22, no. 172, pp 245–256

Lee LP (1975) Conformal projection with specified scale at selected points. Survey Review, v. 23, no. 178, pp 187–188

Lee LP (1976) Conformal projections based on elliptic functions. Cartographica, Monograph no. 16, supplement no. 1 to Canadian Cartographer, v. 13, 128 p. [Based on his 1965 and other papers.]

Arthur H. Robinson

(Montreal, Canada, 1915–Madison, Wisconsin, 2004)

American geographer and cartographer

He was educated at Miami University in Ohio and received a doctorate at Ohio State in 1947. After his service with military intelligence in the war, he joined the Wisconsin faculty, where he spent the rest of his career.

Robinson was a prolific writer and influential philosopher on cartography. In *The Look of Maps* (1952) he urged cartographers to consider the function of a map as an integral part of the design process. In the publication *The Nature of Maps* (1976), Robinson and co-author Barbara Petchenik created the term map percipient, a map user who interacts with a map in a discerning way and not merely as a casual observer. Robinson also co-authored a widely used textbook, *Elements of Cartography*, the sixth and last edition of which was published in 1995.

He was appointed Professor Emeritus of the University of Wisconsin. He received, among others, the National Geographic society and the British geographical society medals and had the honor of chairing the International Cartographic Association (ICA) as his work turned out to be a great contribution to the development of cartographic science during the 20th century.

One of his most notable accomplishments is the *Robinson projection*. He presented this projection in 1963, at the request of Rand McNally and Company. It uses tabular coordinates rather than mathematical formulas to make the world map "look" right (Fig. 12.49).

Robinson AH (1949) An analytical approach to map projections. Assn. of Amer. Geographers, Annals, v. 39, no. 4, pp 283–290

Robinson AH (1960) Map projections, in McGraw-Hill Encyclopedia of Science and Technology: New York, McGraw-Hill. [Subsequent editions 1966, 1971, 1977 (v. 8, p. 125–127), 1982.]

Robinson AH (1974) A new map projection: Its development and characteristics: International Yearbook of Cartography, v. 14, p. 145–155. [His pseudocylindrical projection. See correction by Richardson, R.T., 1989.]

Robinson AH (1990) Rectangular world maps – no!: Professional Geographer, v. 42, no. 1, p. 101–104. [Opposes general use of cylindrical projections for world maps.]

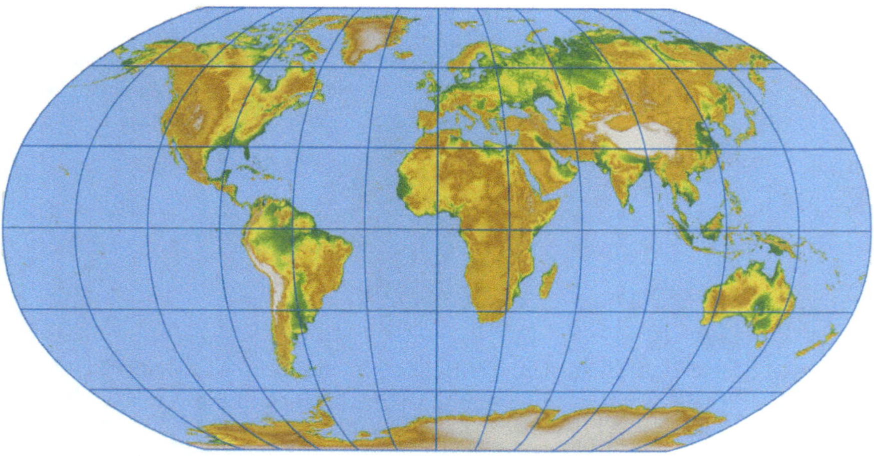

Fig. 12.49 Map of the world in Robinson projection

Robinson AH (1953) Elements of Cartography, 1st ed., published by John Wiley & Sons, New York

Robinson AH, Morrison JL, Muehrcke PC, Kimerling AJ, Guptill SC (1995) Elements of Cartography, 6th ed. John Wiley and Sons, New York

Lev Moiseyevich Bugayevskiy
(Belaya Tserkv' Kievskoy oblasti, 1921–2010)
Russian cartographer professor, colonel

After finishing high school, he enrolled the Leningrad Military-Topographic Higher School in 1939. He participated in defense of Leningrad, as well as in military operations in Leningrad, Belarus and Ukrainian fronts. He was awarded medals for his courage in military operations.

After World War II, he finished studying at the Faculty of Geodesy of the Military-Engineering Academy. He dedicated his entire career to mathematical cartography.

He finished the academy in 1954 and was sent to Irkutsk, where he was an editor and chief of map production department. He was transferred to Moscow in 1967 and continued his military service and scientific research in mathematical cartography. In 1971, he defended his doctoral thesis *Problems of researching and using conformal and related projections in cartography and geodesy*. He was demobilized as a colonel in 1976.

He continued working in the Chair of Map Planning and Composition in the MIIGAiK (Moscow Institute of Engineering Geodesy, Aerial Survey and Cartography), where he stayed until the end of his life. Bugayevskiy's years in MIIGAiK were the most prolific period of his scientific and pedagogical work. He published more than 130 scientific papers (Groshev 2012; Filatov and Potievskiy 2015).

In his monographs, Bugayevskiy was concerned with many theoretical and practical issues, such as:

- General map projection theory
- Distortion theory and mapping one surface onto another
- Theoretical bases of best projections
- Coordinate systems, especially isometric
- Projection transformations, double mapping
- Projection calculations, determining parameters of reference surfaces of celestial bodies

- Using projections to produce topographic, maritime and aerial navigation maps
- Design and nomenclature of map sheets in a particular scale
- Automation in mathematical cartography, etc.

His most famous monographs include:

Bugayevskiy LM, Portnov AM	(1984) Teoriya odinochnykh kosmicheskikh snimkov. Nedra, Moscow [Russian. The theory of single space photographs.]
Bugayevskiy LM, Snyder JP	(1995) Map Projections − A Reference Manual. Taylor & Francis, London, 328 p. [An extensive revision in English of Bugayevskiy and Vakhrameyeva, 1992, Kartograficheskiye proyektsii.]
Bugayevskiy LM	(1998) Matematicheskaya kartografija. Zlatoust, Moscow
Bugayevskiy LM	(1999) Teoriya kartograficheskikh proyektsiy regulyarnykh poverkhnostey. Zlatoust, Moscow
Bugayevskiy LM	(2004) Teoriya kartograficheskikh proyektsiy kvazigeoida. 29 NII MO RF, Moscow
Bugayevskiy LM	(2005) Sferoidicheskaya fotogrammetriya. MIIGAiK, Moscow

Edward Nelson Gilbert

(Woodhaven, New York, 1923–Basking Ridge, New Jersey, 2013)

American mathematician and coding theorist

A longtime researcher at Bell Laboratories whose accomplishments include the Gilbert-Varshamov bound in coding theory, the Gilbert-Elliott model of bursty errors in signal transmission, and the Erdős-Rényi model for random graphs.

He invented and constructed a sphere (1973) that contained a conformal projected sphere on each of its hemispheres. This sphere is named after him *Gilbert's globe*. He had a real globe like this to play tricks on unsuspecting visitors. When people visit Gilbert's office, he liked to ask them what is wrong with his globe. If the visitor cannot see what is wrong, Gilbert gives the globe one slow, complete turn. "Even this hint", he writes, "does not always succeed". Gilbert's globe can be projected into a plane by means of perspective or orthographic projection and accordingly one can obtain the *Gilbert projection*. The Gilbert Two-World Perspective Projection is a useful visual illusion. It resembles the world as people will increasingly see it, from space and in the round (Gardner 1975; DeLucia and Snyder 1986; Lapaine and Frančula 1993; Lapaine et al. 1994). Also, the map communicates its message to the reader in a natural manner, by relying on each viewer's well-developed powers of perspective sight (Fig. 12.50).

Gilbert EN	(ca. 1973) An atlas of oddities. Murray Hill, N.J., Bell Laboratories, unpublished manuscript, 30 p. [Several existing or adapted projections.]
Gilbert EN	(1974) Distortion in maps. SIAM [Society for Industrial and Applied Mathematics] Review, v. 16, no. 1, pp 47–62

Fig. 12.50 Map of the world in Gilbert projection

John Parr Snyder
(1926–1997)
American cartographer

He was president of the *American Cartographic Association* from 1990 to 1991 and also served as a secretary to the *Washington Map Society*. He also taught courses on map projection at George Mason University, but was most known for his work on map projections for the *United States Geological Survey* (USGS). Educated at Purdue University and Massachusetts Institute of Technology as a chemical engineer, he had a lifetime interest in map projections as a hobby, but found the calculations tedious without the benefit of expensive calculators or computers. At a cartography conference in 1976, he learned the need for a map

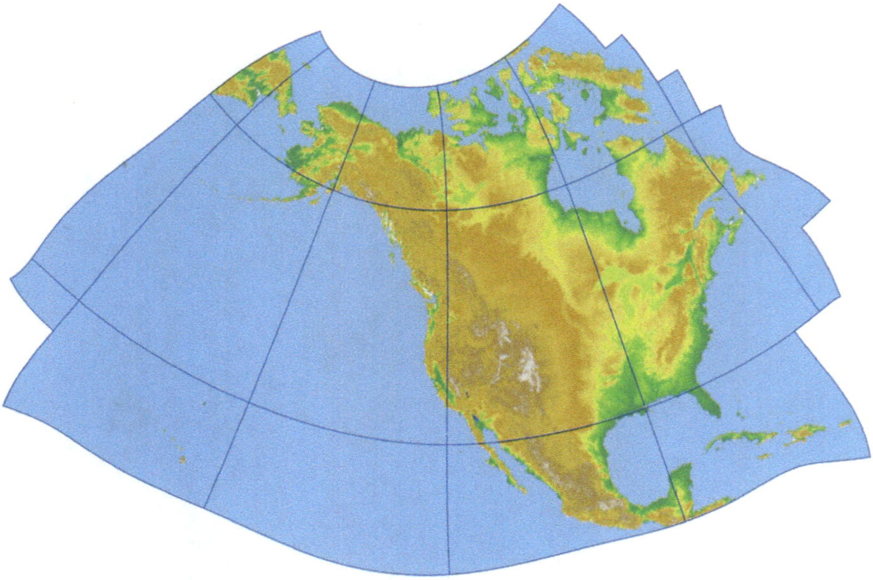

Fig. 12.51 GS50 projection encompassing the regions of the 50 states of the United States

projection that would suit the special needs of Landsat satellite imagery. He subsequently developed a solution with the Space Oblique Mercator projection, which he programmed on a pocket calculator.

He was subsequently offered a job within the USGS within two years, where his work apparently led him to the eventual publication of the definitive technical guide to map projections entitled *Map Projections: A Working Manual among other works*. He also authored *Flattening the Earth: Two Thousand Years of Map Projections* which details the historical development of hundreds of map projections.

Snyder developed conformal projection called *GS50 projection*, which uses a complex polynomial to project the 50 U.S. states with minimal distortion (Fig. 12.51).

Snyder JP (1977) A comparison of pseudocylindrical map projections. Amer. Cartographer, v. 4, no. 1, pp 59–81 [See corrections in 1979, v. 6, no. 1, p. 81.]

Snyder JP (1978) The Space Oblique Mercator projection. Photogrammetric Engineering and Remote Sensing, v. 44, no. 5, pp 585–596 [Two pages reprinted in Explorers Journal, 1978, v. 56, no. 3, pp 112–113]

Snyder JP (1978) Equidistant Conic map projections. Assn. of Amer. Geographers, Annals, v. 68, no. 3, pp 373–383

Snyder JP (1981) Space Oblique Mercator projection – mathematical development. U. S. Geological Survey Bulletin 1518, 108 p. [Also see review by Loon JC, 1982: Photogrammetric Engineering and Remote Sensing, v. 48, no. 10, p. 1581]

Snyder JP	(1982) Map projections used by the U. S. Geological Survey. U. S. Geological Survey Bulletin 1532, 313 p. [2nd ed., 1983, 313 p. See also review by Mugnier CJ, 1983: Surveying and Mapping, v. 43, no. 4, pp 417–420]
Snyder JP	(1984) A low-error conformal map projection for the 50 States. Amer. Cartographer, v. 11, no. 1, pp 27–39 [Uses complex algebra. See correction in Snyder (1987, Map…, p. 207)]
Snyder JP	(1985) Computer-assisted map projection research. U. S. Geological Survey Bulletin 1629, U. S. Government Printing Office, 157 p. [See correction in Snyder (1987, Map…, p. 207)]
Snyder JP	(1987) "Magnifying-glass" azimuthal map projections. Amer. Cartographer, v. 14, no. 1, pp 61–68 [Translated to Chinese in Cartography (Wuhan Technical University, Wuhan, China), 1988, no. 3, pp 32–39]
Snyder JP	(1987) Map projections – a working manual. U.S. Geological Survey Prof. Paper 1395, 383 p. Reprinted 1989 and 1994 with corrections. [General work superseding Snyder, 1982, Map projections used by the U.S. Geological Survey. U.S. Geological Survey Bull. 1532.]
Snyder JP	(1993) Flattening the Earth – two thousand years of map projections. University of Chicago Press, Chicago, 365 p.
Snyder JP, Voxland PM	(1989) An album of map projections. U. S. Geological Survey Prof. Paper 1453, 249 p. Reprinted 1994 with corrections. [Graticules for 90 projections, with brief descriptions and formulas.]

Acknowledgement The authors would like to thank Temenoujka Bandrova, Maxim Nirtsov, Igor Drecki and Geoff Aitken, who provided valuable information on famous cartographers from Bulgaria, Russia and New Zealand.

References

Baily W (1886) A map of the world on Flamsteed's projection: London, Edinburgh, and Dublin Philosophical Magazine, series 5, v. 21, no. 132, pp 415–416 [Same as Sinusoidal projection.]

Bollmann J, Koch WG (2001) Lexikon der Kartographie und Geomatik. Spektrum Akademischer Verlag Heidelberg, Berlin

Bosse H (1976) DGfK-Ehrenmitgliedschaft für Hns Meyerhuber und Karl Heinrich Wagner. Kartographische Nachrichten 6:240

Bugayevskiy LM, Snyder JP (1995) Map projections, a reference manual. Taylor & Francis, London

Curie J (1900) Système de construction des cartes de Babinet et Sanson: Assn. Française pour l'Avancement des Sciences, 28th session, Boulogne-Sur-Mer, 1899, Compte Rendu. Paris, pp 73–86

DeLucia AA, Snyder JP (1986) An innovative world map projection. Am Cartogr 13(2):165–167

Ferschke H (1966) Dr. Karlheinz Wagner 60 Jahre. Kartographische Nachrichten 4:162–163

Ferschke H (1985) Karlheinz Wagner †. Kartographische Nachrichten 3:108–109

Filatov VN, Potievskiy VA (eds) (2015) Otechestvennaya geodeziya, topografiya i kartografiya, Entsiklopediya v litsakh, izdanie vtoroe. BARS, Moscow

Frančula N (1971) Die vorteilhaftesten Abbildungen in der Atlaskartographie. Dissertation. Institut für Kartographie und Topographie der Rheinischen Friedrich-Wilhems-Universität Bonn

Frankich K (1982) Optimization of geographic map projections for Canadian territory. Dissertation. Simon Fraser University, Burnaby BC, Canada

Frischauf J (1905) Die Abbildungslehre und deren Anwendung auf Kartographie und Geodäsie. Druck und Verlag von B. G. Teubner, Leipzig

Gardner M (1975) On mathematical games. Scientific American

Grattan-Guinness I, Pulte H (organisers, 2007) Mini-workshop: the reception of the work of Leonhard Euler (1707–1783), 12–18 Aug 2007, Mathematisches Forschungsinstitut Oberwolfach, Report No. 38/2007 www.mfo.de/programme/schedule/2007/33a/OWR_2007_38.pdf

Groshev VV (2012) Matematicheskaya kartografiya – nauchnoe nasledie L'va Moiseyevicha Bugayevskogo. Geoprofi 3:59–61

Györffy J (2006) Euler map projection for the representation of Europe. In: Zentai L, Györffy J, Török Zs (eds) Térkép – Tudomány, Map – Science, Karte – Wissenschaft, ELTE Térképtudományi és Geoinformatikai Tanszék, Budapest, pp 149–154

Kovačec A (ed) (2002) Hrvatska enciklopedija [Croatian Encyclopaedia, in Croatian], sv. 4, Leksikografski zavod Miroslav Krleža, Zagreb

Lapaine M (2011) Mollweide map projection. KoG 15:7–16

Lapaine M, Frančula N (1993) Gilbert two-world projection. In: Proceedings, volume 1 of the 16th international cartographic conference, Cologne 3–9 May 1993, Bielefeld, pp 66–82

Lapaine M, Sudeta N, Frančula N (1994) Gilbert's globe. In: Proceedings of the 6th international conference on computer graphics and descriptive geometry, Tokyo, vol 1, pp 154–158

Meshcheryakov GA (1958) O postroenii obshchey teorii proyektsiy Eylera: Izvestiya Vysshikh Uchebnykh Zavedeniy. Geodeziya i Aerofotos'emka, no. 6, pp 111–113 [Russian. Construction of the general theory of the Euler projection.]

Meshcheryakov GA (1959) Obshchaya teoriya proyektsiy Eylera: Izvestiya Vysshikh Uchebnykh Zavedeniy. Geodeziya i Aerofotos'emka, no. 4, pp 93–103 [Russian. General theory of the Euler projection. Also in Inz.-Stroitel'n. Fakulteta Chelyabinskogo Politechn. Inst., Trudy, 1959, v. 2, pp 41–53]

Meshcheryakov GA (1963) Obobshcheniye sistemy Eylera-Urmayeva: NIIGAiK, Trudy, v. 17, no. 1, pp 63–73 [Russian. Generalization of the Euler-Urmayev systems.]

Meshcheryakov GA (1964) Vychisliniye na TsAM varianta nailuchshey eylerovoy proyektsii: NIIGAiK, Trudy, v. 17, no. 2, pp 75–94 (Cartography). [Russian. Calculation of the best Eulerian projection with a digital computer.]

Meshcheryakov GA (1968) Teoreticheskie osnovy matematicheskoj kartografii. Nedra, Moscow

Mittelstaedt O (1976) Dr. Karl Heinrich Wagner zum 70. Geburtstag, Kartographische Nachrichten 4:148–149

Monmonier M (2004) Rhumb lines and map wars: a social history of the of the Mercator projection. The University of Chicago Press, Chicago and London

Snyder JP (1993) Flattening of the Earth—two thousand years of map projections. University of Chicago Press, Chicago and London

Snyder JP, Steward H (eds) (1988) Bibliography of map projections. US Geological Survey Bulletin 1856, Denver

Snyder JP, Voxland PhM (1989, 1994) An album of map projections. USGS professional paper 1453. US Geological Survey, Denver

Steffens K (2007) Some aspects of Euler's influence on approximation theory and the St. Petersburg mathematical school. In: Mini-Workshop: the reception of the work of Leonhard Euler (1707–1783), Mathematisches Forschungsinstitut Oberwolfach, Report No. 38/2007, pp 27–30

Urmayev NA (1947) Metody izyskaniya novyh kartograficheskikh proyektsiy, Moscow

Chapter 13
Basics of Geodesy for Map Projections

Miljenko Lapaine

13.1 Practical Geometry (Land Surveying)

Measurements are related to human activity. Hence, the history of measurement has the same beginnings as the history of human kind. There are preserved evidences of high measuring level in the field of the rivers Euphrates and Tigris. The oldest races lead 6000 years back. The excavations have proved that there was a 160 km long irrigation system functioning in that area then (Benčić 1990). The system was maintained successfully for about two thousand years, and today everything is covered with sand. Such an enterprise that can be characterized as grandiose for that time, but also for the present, could not be accomplished without good knowledge of surveying. It becomes obvious when we compare that irrigation system with the Roman water supply system being about 4000 year younger. The Roman water supply system consisting of altogether 19 aqueducts about 400 km long used to be maintained by a full-time crew of 700 people making it possible for the citizens to receive 600–1000 L of water a day per citizen.

Egyptian surveyors were called cord stretchers, after the cord they used to measure distances.

The Greek name for geodesy was first γεωμετρία, and then later γεωδαισία. The first term originates from the word γη, i.e. the Earth, and μετρείν = to measure. Hence, literally translated γεωμετρία = land surveying.

According to the opinion of one of Croatian geodetic doyens, Nikola Neidhardt (1950), the Greek expression γεωμετρία is the best name for the profession. It defines in a linguistically concise way what the profession is dealing with. All later terms are not even close to be so convenient. What has then induced

M. Lapaine (✉)
Faculty of Geodesy, Universtiy of Zagreb, Zagreb, Croatia
e-mail: mlapaine@geof.hr

M. Lapaine and E.L. Usery (eds.), *Choosing a Map Projection*,
Lecture Notes in Geoinformation and Cartography,
DOI 10.1007/978-3-319-51835-0_13

the name to be changed? Why did the old Greek replace a very good term γεωμετρία by much worse expression γεωδαισία? Prof. Neidhardt has explained it in the following way. It is the issue of name abduction or stealing. Geometry, as the science of geometric figures developed from geodesy. Geometry is the daughter of geodesy. It is a custom that a child should get a name once it is born. An inverted procedure took place in this case. The daughter was born, and as it grew up, it took the mother geodesy its name. The mother remained nameless. And it was given the new name geodesy.

Herodotus, the father of history as a science, wrote the following lines: "In Egypt the Nile River floods the land annually and chokes boundaries with mud. After the water withdraws, the boundaries have to be recovered by measurement. It is said, that geometry was developed from such measurements."

It is clear that geometric problems were developed not out of some impulses to go after theoretical speculations and reflections, but out of purely practical, concrete tasks and problems. The notion of triangle or rectangle as geometric figures, the notion of their properties, discovery of formulas for their areas etc. were demanded by purely practical and geodetic tasks that had been solved completely empirically at first. Later the properties were noticed belonging to every triangle, every rectangle etc. The Greek genius was especially inclined to notice general characteristics. Geometry as a science was thus born. One can easily state that it was, and to great extent also mathematics, the daughter of geodesy and geodetic problems.

13.2 Geodesy

Certain entangling or mixing of the name *geometry* on one hand and *geodesy* on the other has been continued even to our very days. A geodetic expert is still called *land surveyor*. Apart from that, some German authors used to call geodesy *Praktische Geometrie*, i.e. practical geometry, unlike theoretical geometry that is taught in high schools, as if geometry were a science, and geodesy only a practical activity. But it is not so. Geodesy is a science on equal footing with other sciences. It is an independent science, with its own problems, very much extended and developed.

Under the influence of Latin, the Roman languages French, Italian, Spanish, and Rumanian were developed. In French and Italian the expression *topography* is used for practical geodesy, and *geodesy* for higher geodesy.

In West European countries the concept of geodesy is referred more and more to 'higher' geodesy. The English and the Americans refer to 'lower' geodesy as *survey* or *surveying* which means to determine the exact form, boundaries, position, extent, etc., of a tract of land, section of a country, etc., by linear and angular *measurements* and the application of the principles of geometry and trigonometry. Analogously, the Germans say *Vermessungskunde*. *Dendrometry*, for example, also deals with surveying, but not of land, but of wood masses in forestry. Electric meters are devices used for measuring power; gasmeters for gas etc., without being geodetic instruments. Hence, many things can be measured. All sciences deal with measuring today without being geodesy. Therefore, the terms survey, surveying or Vermessungskunde are not the best ones. Still, mostly used and internationally well known is *surveyor*. The international association of geodetic professionals is called International Federation of Surveyors (Fédération Internationale des Gèométres—FIG). In English it is called International Federation of Surveyors, in German *International Vereinigung der Vermessungsingenieure*. It seems that there is now some misunderstanding cropping up referring to the distinction between the terms *surveyor* and *geodesist*, because we have seen that these two terms do not have the same meaning everywhere.

Geodesists are very proud to point out their predecessors. About 2330 year ago Eratosthenes measured the length of the meridian arc between Alexandria and Syene and thus determined the radius of the Earth. The Egyptian land surveyors used to recover the plot boundaries after every flood of the Nile River. In the time of great geographic discoveries, Gerhard Mercator gave large contribution to the ocean navigation safety by inventing his conformal cylindrical map projection. The last great contributions in getting better acquainted with our planet were given by geodesy in the second half of the 19th and the first twenty years of the 20th centuries when the last white spots disappeared from the maps of South and North America, Africa and Asia. At that time it was possible to name the highest peak of the Earth after a geodesist, the British surveyor-general of India, Sir George Everest (1790–1866).

There are many definitions of geodesy. One of contemporary definitions reads: "Geodesy is the discipline that deals with the measurement and representation of the earth, including its gravity field, in a three-dimensional time varying space." (Vanícek and Krakiwsky 1987).

13.3 Determination of the Earth's Figure and Size

Earth's physical surface is the surface which divides our planet from its atmosphere. It is very irregular. In order to represent even a small part of Earth's surface in a plane, one requires a simpler surface on which horizontal relations of points are represented, while altitude is written or drawn for certain points.

When the entire Earth is considered, it is advisable the surface be as close to its physical surface as possible. Considering seas and oceans make up $\sim$70% of Earth's total area, the ideal surface of still sea stretched over continents was taken as the reference surface, defining Earth's shape. This shape of Earth was definitely accepted in 1873 and was named the geoid. However, the surface is not regular because of the layout of masses of various densities, making it difficult to use in calculations. A rotational ellipsoid is a geometric surface which approximates the geoid very well. Such an ellipsoid is very close to a sphere (including its interior, i.e. a ball) with radius $\sim$6400 km (minor semi-axis of ellipsoid is $\sim$6400 km, while major semi-axis is only 21 km longer). In solving problems in geodesy, remote sensing, navigation and cartography, we consider Earth a rotational ellipsoid or a sphere.

13.4 Elements of Earth's Ellipsoid

Earth's ellipsoid is made by rotating an ellipse (Fig. 13.1) around its shorter axis, which we assume corresponds with the Earth's axis. In considering properties of a rotational ellipsoid, it is sufficient to know elements of a meridian ellipse, which is rotated to produce an Earth's ellipsoid. We label the ellipse's semi-major axis as a, and its semi-minor axis as b. Dimensions of the ellipsoid are often given with semi-major axis a and flattening f, which is defined as

$$f = \frac{a-b}{a}. \tag{13.1}$$

In addition, second flattening is defined as

$$f' = \frac{a-b}{b}, \tag{13.2}$$

and third flattening is defined as

$$n = \frac{a-b}{a+b}. \tag{13.3}$$

In geodetic and cartographic calculations, one frequently encounters the first and second numerical eccentricities, which are defined as:

$$e = \frac{\sqrt{a^2-b^2}}{a}, \quad e' = \frac{\sqrt{a^2-b^2}}{b}. \tag{13.4}$$

All linear and relative dimensions of the ellipse are named elements of the ellipse. Elements of the ellipse are at the same time elements of a rotational ellipsoid of which the ellipse is a directrice. Therefore, all the above elements of the ellipse and their relations can be transferred to the surface of Earth's ellipsoid.

Numerous geodetic, astronomic, gravimetric and satellite measurements are conducted in order to determine numerical values of Earth's ellipsoid elements.

Fig. 13.1 Ellipse

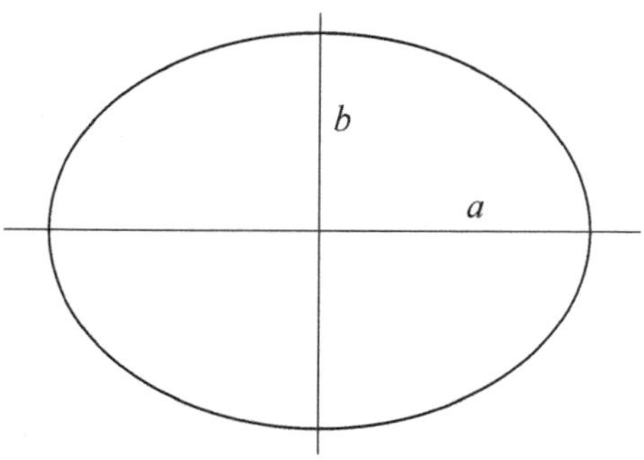

Table 13.1 Dimensions of the Earth's ellipsoid

Name	Year	*a*	*b*	1/*f*
Everest	1830	6,377,276.345	6,356,075.415	300.801726
Bessel	1841	6,377,397.155	6,356,078.963	299.152813
Clarke	1866	6,378,206.400	6,356,583.800	294.978698
Clarke	1880	6,378,249.145	6,356,514.967	293.466308
Hayford International	1909 1924	6,378,388	6,356,911.946	297
Krasovskiy	1940	6,378,245	6,356,863.019	298.3
GRS 1967	1967	6,378,160	6,356,774.516	298.247166
GRS 1980	1980	6,378,137	6,356,752.3141	298.257222
WGS 72		6,378,135	6,356,750.520	298.26
WGS 84	1987	6,378,137	6,356,752.314	298.257223563

Many scholars and institutions were interested in determining dimensions of the Earth's ellipsoid over the centuries. Table 13.1 contains dimensions of the Earth's ellipsoid which have been used the most in geodetic and cartographic practice.

In geodesy, the Earth's ellipsoid is often identified with the Earth's spheroid. A spheroid is a geometrical solid similar in shape to a sphere. Therefore, the Earth's spheroid is considered a rotational ellipsoid with small flattening.

13.5 General Earth's Ellipsoid

The general *Earth's ellipsoid* is the ellipsoid which best represents Earth as a planet. However, it is a mathematical model of Earth which can be defined in several ways. All of them aim for greatest possible congruence between the ellipsoid and Earth. Such an ellipsoid is absolutely oriented in space, i.e. its equatorial plane corresponds to Earth's equatorial plane, and its minor axis corresponds to the central position of Earth's rotational axis (central position for a particular surface).

Available data and mathematical formulas were used to derive numerous parameters of the general ellipsoid, while striving to achieve absolute orientation. This task was not successful until observations with Earth's artificial satellites were made possible.

The ellipsoid to which geodetic measurements are reduced and on which they are analyzed is called a *reference ellipsoid*. Therefore, a general ellipsoid can be a reference ellipsoid if geodetic measurements are reduced to it and analyzed.

The general Earth's ellipsoid was not used as a reference ellipsoid in the past, mostly because it could not be oriented in an adequate way. In addition, it was not possible to calculate the difference between potentials of geoid and a general ellipsoid in at least one point. It was therefore not possible to calculate distances between those two surfaces and reduce measurements to the ellipsoid.

This is why each country or group of countries determined its own reference ellipsoid with dimensions and orientation the most suitable for their area. The consequence was discontinuity—incoherence of geodetic networks and maps of different countries. Economic-political and military integrations of countries which originated after World War II consolidated geodetic work within their areas and transferred them onto a specially selected and oriented ellipsoid.

Contemporary satellite techniques and information technology have enabled accurate determination of Earth's centre of inertia and position of its rotational axis, as well as dimensions and shape. Thus, the general Earth's ellipsoid could be oriented and unique global triangulation could be developed. In addition, new methods could be employed to obtain a great number of high accuracy measurements, which are then processed by computers. Therefore, it is possible to calculate increasingly accurate parameters of Earth's shape, dimensions and gravity field. After they are compared and analyzed, the International Association of Geodesy (IAG) periodically adopts them and recommends them as reference parameters.

For example, in 1979, the 17th Congress of IAG (Moritz 1980) recommended the Geodetic Reference System 1980 (GRS 80). Basic parameters of the corresponding ellipsoid are: $a = 6{,}378{,}137 \pm 2$ m, $1/f = (298\ 257 \pm 1) \times 10^{-3}$.

13.6 WGS 84 Coordinate System

The World Geodetic System 1984 (WGS 84) was developed in USA as a replacement for WGS 72. Development of WGS 84 was initiated by a need to provide more precise geodetic and gravimetric data to navigation and weapon systems of the Department of Defense. The new system represents Earth modeling from geometrical, geodetic and gravimetric perspectives, using data, techniques and technology at the disposal of the Defense Mapping Agency at the beginning of 1984.

The WGS 84 coordinate system is a conventional system obtained by modifying the Navy Navigation Satellite System Doppler Reference Frame (NSWC 9Z-2) in the origin and scale, as well as rotation, which superposes its reference meridian with the prime meridian of Bureau International de l'Heure (BIH).

Therefore, the origin of the WGS 84 coordinate system is in Earth's center of mass. Its Z-axis is parallel in the direction of the Conventional Terrestrial Pole (CTP) for pole movement, as defined by BIH. Its X-axis is the intersection of WGS 84 plane of the reference meridian and the CTP equator plane. The reference meridian is parallel to the prime meridian defined by BIH. The Y-axis completes the right-oriented orthogonal system tied to Earth.

The origin and axes of the WGS 84 coordinate system are also used as the geometric centre and X, Y and Z axes of the WGS 84 ellipsoid. The Z-axis of the WGS 84 coordinate system is the rotational axis of the WGS 84 ellipsoid.

Geodetic applications usually include three different surfaces or Earth shapes. In addition to Earth's natural or physical surface, there are also the geometrical or mathematical reference surface (ellipsoid) and an equipotential surface (geoid). In determining the WGS 84 ellipsoid and its parameters, procedures are accepted of the International Union of Geodesy and Geophysics (IUGG), which established and accepted the Geodetic Reference System 1980 (GRS 80).

The semi-major axis of the WGS 84 ellipsoid equals $a = 6{,}378{,}137 \pm 2$ m. The value is equal to the one of ellipsoid GRS 80 and 2 m longer than the one of ellipsoid WGS 72. The semi-major axis value is based on an estimate of data obtained between 1976 and 1979 from laser, Doppler, radar altimeter and combined measurements.

13.7 Equation of Rotational Ellipsoid

The graticule is used to determine position of points on the Earth's ellipsoid. *Meridians* are semi-ellipses on the surface of a rotational ellipsoid which one obtains by intersecting the ellipsoid with planes which intersect the minor axis. *Parallels* are circles which one obtains by intersecting the ellipsoid with planes perpendicular to the minor axis. The largest parallel is called the *equator*. One of the meridians is called the *prime* or *null meridian*, and the angle between planes of the prime meridian and the meridian through a given point is called the *longitude* of the point and is labeled λ. The longitude is usually measured as an angle in the interval from $-180°$ to $180°$ (180°W, 180°E). The angle between the perpendicular to the ellipsoid in a given point and the equatorial plane is called the *latitude* of the point and is labeled φ. Latitude is measured as an angle from the interval between $-90°$ and $90°$ (90°S, 90°N). The longitude and the latitude are named geographic or geodetic coordinates (Fig. 13.2).

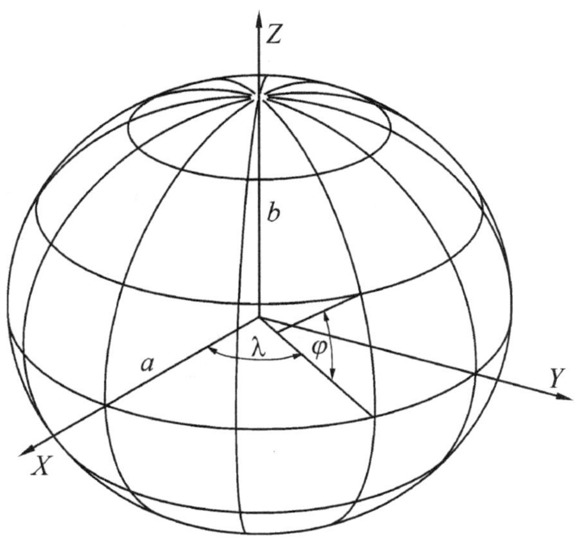

Fig. 13.2 Elements of Earth's ellipsoid

A rotational ellipsoid with semi-axes a and b and centre in the origin of the spatial rectangular coordinate system is the set

$$E = \left\{ (X, Y, Z) : \frac{X^2}{a^2} + \frac{Y^2}{a^2} + \frac{Z^2}{b^2} = 1 \right\}. \tag{13.5}$$

Let $(\varphi, \lambda) \in \Omega = (-90°, 90°) \times (-180°, 180°)$ and mapping $\Omega \rightarrow E$ be given with formulas

$$X = \frac{a}{W} \cos \varphi \cos \lambda, \quad Y = \frac{a}{W} \cos \varphi \sin \lambda, \quad Z = \frac{a(1 - e^2)}{W} \sin \varphi, \tag{13.6}$$

where

$$W = \sqrt{1 - e^2 \sin^2 \varphi}. \tag{13.7}$$

Mapping given with formulas (13.6) is called a *geodetic parameterization* of a rotational ellipsoid E. Equations (13.6) are equations of rotational ellipsoid in parametric form. It is easy to see that parametric φ-curves are meridians, while λ-curves are parallels on a rotational ellipsoid. Points with rectangular coordinates (0, 0, b) and (0, 0, $-b$) are called poles and they are singular points of geodetic parameterization. Longitude is not defined in poles. Singular points are also all points on the meridian diametrally opposite the prime meridian. Namely, longitude at that meridian can be either 180° or −180°.

There are other ellipsoid parameterizations, but the mentioned geodetic parameterization is the one used most often in cartography. In addition to function W, function V is present in many formulas of spheroidal geodesy:

$$V = \sqrt{1 + e'^2 \cos^2 \varphi}. \tag{13.8}$$

V and W are called the basic spheroid functions.

13.8 Curvature Radii of Earth's Ellipsoid

Using expression (13.6) one can calculate partial derivations

$$\begin{bmatrix} X_\varphi & X_\lambda \\ Y_\varphi & Y_\lambda \\ Z_\varphi & Z_\lambda \end{bmatrix} = \begin{bmatrix} -R_1 \sin \varphi \cos \lambda & -R_2 \cos \varphi \sin \lambda \\ -R_1 \sin \varphi \sin \lambda & R_2 \cos \varphi \cos \lambda \\ R_1 \cos \varphi & 0 \end{bmatrix}, \tag{13.9}$$

where

$$R_1 = \frac{a(1-e^2)}{W^3}, \quad R_2 = \frac{a}{W}. \tag{13.10}$$

Using partial derivations (13.9), one can calculate Gauss's fundamental values of the first order

$$E = R_1^2, \quad F = 0, \quad G = R_2^2 \cos^2 \varphi. \tag{13.11}$$

One can now write the first differential form of the rotational ellipsoid

$$ds^2 = Ed\varphi^2 + 2Fd\varphi d\lambda + Gd\lambda^2 = R_1^2 d\varphi^2 + R_2^2 \cos^2 \varphi d\lambda^2. \tag{13.12}$$

Using second partial derivations, one can calculate Gauss's fundamental values of the second order

$$L = R_1, \quad M = 0, \quad N = R_2 \cos^2 \varphi. \tag{13.13}$$

One can now write the second differential form of the rotational ellipsoid

$$Ld\varphi^2 + 2Md\varphi d\lambda + Nd\lambda^2 = R_1 d\varphi^2 + R_2 \cos^2 \varphi d\lambda^2. \tag{13.14}$$

Intersections of the ellipsoid with planes that contain the perpendicular to the given point are called normal intersections. Studying curvature of normal intersections provides the characteristic of ellipsoid curvature in a given point. One can place an infinite number of normal planes through the perpendicular. Each of them intersects the ellipsoid in an ellipse whose curvature in the observed point can be written as a quotient of the second and first differential form and is called the normal curvature of ellipsoid:

$$\kappa_n = \frac{Ld\varphi^2 + 2Md\varphi\ d\lambda + Nd\lambda^2}{Ed\varphi^2 + 2Fd\varphi\ d\lambda + Gd\lambda^2} = \frac{R_1 d\varphi^2 + R_2 \cos^2 \varphi d\lambda^2}{R_1^2 d\varphi^2 + R_2^2 \cos^2 \varphi d\lambda^2}. \tag{13.15}$$

Among the infinite normal intersections in each (regular) point there are two normal intersections with extreme curvatures: the maximum and the minimum. Obtaining the extremes of normal curvature κ_n of a surface parameterized with parameters φ and λ comes down to solving the equation

$$(EM - FL)d\varphi^2 + (EN + 2FM - GL)d\varphi d\lambda + (FN - GM)d\lambda^2 = 0, \tag{13.16}$$

which is simplified in the case of a rotational ellipsoid to

$$(R_1 - R_2)R_1 R_2 \cos^2 \varphi d\varphi d\lambda = 0. \tag{13.17}$$

Equation (13.17) is equivalent to a very simple system of equations

$$d\lambda = 0, \quad d\varphi = 0. \tag{13.18}$$

The solution of the system of differential equations (13.18) is:

$$\lambda = \text{const.}, \quad \varphi = \text{const.} \tag{13.19}$$

which means that curves of extreme or main curvatures are parametric curves, i.e. meridians and parallels. If we substitute values from expression (13.18) in the formula for normal curvature (13.15), we are going to obtain extreme or main curvatures of normal intersections:

$$\kappa_1 = \frac{1}{R_1}, \quad \kappa_2 = \frac{1}{R_2}. \tag{13.20}$$

κ_1 is the curvature in the direction of meridian and κ_2 is the curvature in the direction of parallel. It is easy to see that κ_1 is always greater than κ_2. Based on expression (13.20), we can now geometrically interpret R_1 and R_2: they are the smallest and largest curvature radii among all curvature radii of normal intersections in the observed point. From the expression for the first differential form (13.12), we can read that the curvature radius of meridian is R_1 and the curvature radius of parallel is $R_2 \cos\varphi$. Therefore, we can conclude:

(1) κ_1 is the main curvature of ellipsoid along the meridian on the rotational ellipsoid and it is at the same time the curvature of meridian.

(2) κ_2 is the main curvature of ellipsoid along the parallel on the rotational ellipsoid, but it is not the curvature of the parallel, it is called the curvature of intersection along the first vertical. Namely, a parallel does not generally lie in a normal intersection (see ATA' at the Fig. 13.3). A normal intersection which has the greatest ellipsoid curvature κ_2 in the observed point T is called the *intersection along the first vertical* (see CTC' at Fig. 13.3).

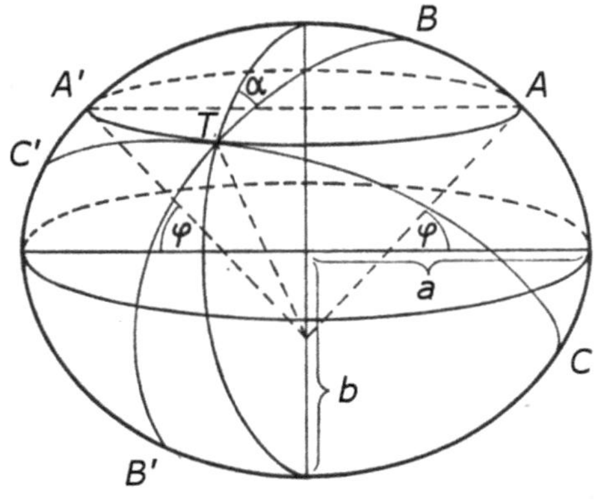

Fig. 13.3 Normal intersections BTB' and CTC' of an ellipsoid through the point T

Poles are singular points of geographic parameterization (13.6) of a rotational ellipsoid. At poles

$$R_1 = R_2 = c = \frac{a^2}{b}, \tag{13.21}$$

i.e. all curvature radii are equal, i.e. curvatures are equal. We now introduce labels which are common in geodetic and cartographic literature: M is the meridian curvature radius, N is the radius of curvature of intersection along the first vertical, and r is the radius of parallel

$$\begin{aligned} M = R_1 &= \frac{a\,(1-e^2)}{\sqrt{(1-e^2\sin^2\varphi)^3}} = \frac{a\,(1-e^2)}{W^3} = \frac{c}{V^3}, \\ N = R_2 &= \frac{a}{\sqrt{1-e^2\sin^2\varphi}} = \frac{a}{W} = \frac{c}{V}, \\ r &= R_2\cos\varphi = N\cos\varphi. \end{aligned} \tag{13.22}$$

Curvature radii M, N and r are indispensable in many geodetic and cartographic calculations, e.g. in calculating meridian arc length, parallel arc length, area on ellipsoid, approximating ellipsoid with a sphere, etc.

13.9 Meridian Arc Length and Parallel Arc Length on Earth's Ellipsoid

The first differential form of rotational ellipsoid according to (13.12) and (13.22) is

$$ds^2 = M^2 d\varphi^2 + N^2 \cos^2\varphi d\lambda^2. \tag{13.23}$$

Along the meridian, $d\lambda = 0$, so we can write the following for the meridian arc length differential:

$$ds_m = M d\varphi. \tag{13.24}$$

Let us note two points on the meridian: T_1 with latitude φ_1 and T_2 with latitude φ_2 (Fig. 13.4). Let us label meridian arc length between points T_1 and T_2 with s_m. By integrating expression (13.24) one obtains:

$$s_m = \int_{\varphi_1}^{\varphi_2} M d\varphi = s_m(\varphi_2) - s_m(\varphi_1), \tag{13.25}$$

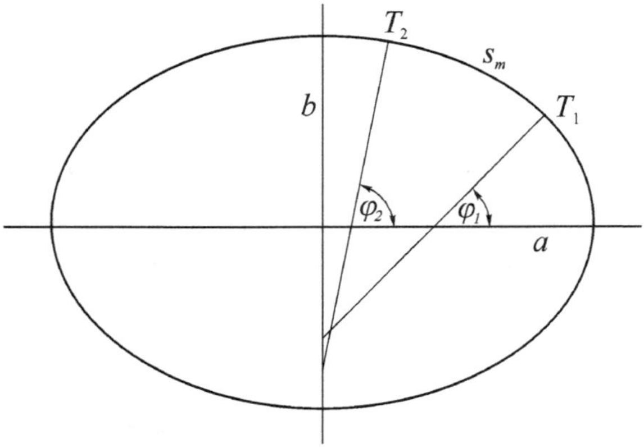

Fig. 13.4 Meridian arc length s_m

where

$$s_m(\varphi) = \int_0^{\varphi} M d\varphi. \tag{13.26}$$

In the last formula (13.26), $s_m(\varphi)$ represents meridian arc length from the equator to the point with latitude φ. Integral (13.26) is an elliptical integral which can not be integrated directly, but it can be integrated by using development in series (Lambert 1772; Helmert 1880; Krüger 1912; König and Weise 1951). Lapaine (1990, 1994) proposed calculating meridian arc length using the following formula:

$$s_m(\varphi) = A[\varphi + \sin 2\varphi(c_1 + (c_2 + (c_3 + (c_4 + c_5 \cos 2\varphi)\cos 2\varphi)\cos 2\varphi)\cos 2\varphi)] + \ldots, \tag{13.27}$$

where

$$A = a\,(1-n)\,(1-n^2)\,(1 + \frac{9}{4}n^2 + \frac{225}{64}n^4 + \ldots) \tag{13.28}$$

$$\begin{array}{llllll} c_1 = & -\frac{3}{2}n & & +\frac{31}{24}n^3 & & -\frac{669}{640}n^5 & +\cdots \\ c_2 = & & \frac{15}{8}n^2 & & -\frac{435}{128}n^4 & & +\cdots \\ c_3 = & & & -\frac{35}{12}n^3 & & +\frac{651}{80}n^5 & +\cdots \\ c_4 = & & & & \frac{315}{64}n^4 & & +\cdots \\ c_5 = & & & & & -\frac{693}{80}n^5 & +\cdots \end{array} \tag{13.29}$$

In previous expressions, n is the third flattening defined by relation (13.3). Since $n^6 \approx 2 \times 10^{-17}$, introducing further terms makes no sense in 16-digit arithmetic.

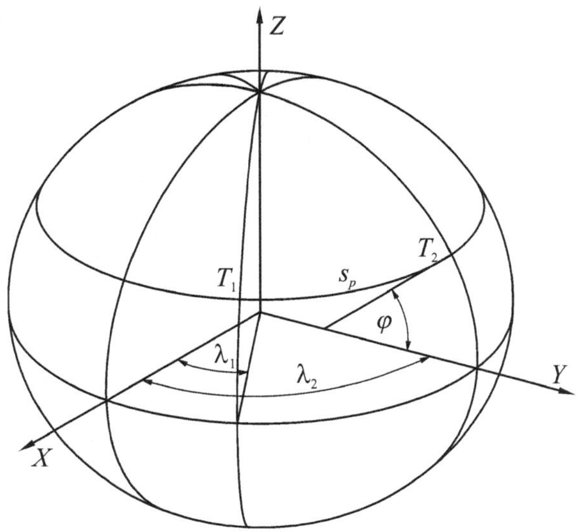

Fig. 13.5 Parallel arc s_p

A parallel is a circle with radius $r = N \cos \varphi$ (13.22). Let us note two points on the parallel: T_1 with longitude λ_1 and T_2 with longitude λ_2 (Fig. 13.5). Parallel arc length between points T_1 and T_2 is equal to

$$s_p = r(\lambda_2 - \lambda_1) = N \cos \varphi(\lambda_2 - \lambda_1). \tag{13.30}$$

13.10 Area of Ellipsoidal Trapezium

Differential of surface dP on a rotational ellipsoid parameterized by geodetic parameterization can be written as

$$dP = \sqrt{EF - G^2} d\varphi d\lambda = MN \cos \varphi d\varphi d\lambda, \tag{13.31}$$

where M and N are main ellipsoid curvature radii (13.22). Area P of curvilinear trapezium bound by parallel φ_1 and φ_2 and meridians λ_1 and λ_2 can be calculated exactly by applying the formula (Lapaine and Lapaine 1991):

$$\begin{aligned} P &= \int\limits_{\lambda_1}^{\lambda_2} \int\limits_{\varphi_1}^{\varphi_2} MN \cos \varphi d\varphi d\lambda \\ &= \frac{b^2}{2}(\lambda_2 - \lambda_1)\left(\frac{\sin \varphi}{1 - e^2 \sin^2 \varphi}\right) + \frac{1}{2e} \ln \frac{1 + e \sin \varphi}{1 - e \sin \varphi}\Bigg|_{\varphi_2}^{\varphi_1} \end{aligned} \tag{13.32}$$

The area of the entire ellipsoid is

$$P = 2b^2\pi\left(\frac{1}{1-e^2} + \frac{1}{2e}\ln\frac{1+e}{1-e}\right) = 2a^2\pi\left(1 + \frac{1-e^2}{2e}\ln\frac{1+e}{1-e}\right). \tag{13.33}$$

13.11 Elements of Earth's Sphere

Earth is often considered a sphere in studying map projections (Fig. 13.6). Radius R of a sphere can be determined in various ways. For example, it can be determined from the condition that the sphere has the same surface as the ellipsoid, i.e.

$$4R^2\pi = P, \tag{13.34}$$

where P is determined by the expression (13.33). From (13.34), it follows that

$$R^2 = \frac{a^2}{2}\left(1 + \frac{1-e^2}{2e}\ln\frac{1+e}{1-e}\right). \tag{13.35}$$

One can then set the condition that meridians on the sphere have the same length as meridians on the ellipsoid, i.e.

$$R\pi = 2s_m\left(\frac{\pi}{2}\right). \tag{13.36}$$

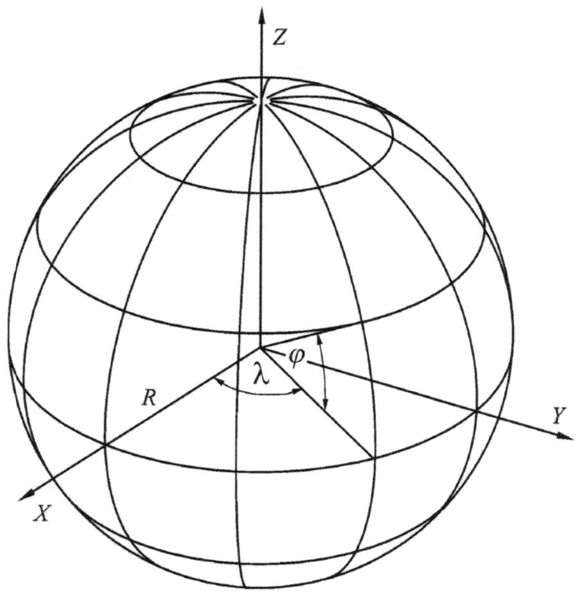

Fig. 13.6 Elements of Earth's sphere

From there, using relation (13.27) for calculating meridian arc length on the ellipsoid, it follows that

$$R = A. \tag{13.37}$$

In the last formula, A is called the *radius of the rectifying sphere*, and it can be calculated for the given ellipsoid according to relation (13.28). For many calculations which do not extend to large areas, we can consider a part of surface a part of a sphere with mean radius like

$$R = \frac{2MN}{M+N} \text{ or } R = \sqrt{MN}. \tag{13.38}$$

However, if one wants to approximate the entire ellipsoid with a sphere, then it can be done using Eqs. (13.35), (13.37), or the arithmetic mean for the mean radius of the Earth's sphere

$$R = \frac{a+a+b}{3} \tag{13.39}$$

or the radius of the sphere which has the same volume as the ellipsoid:

$$R = \sqrt[3]{a^2 b}. \tag{13.40}$$

The radius of the Earth's sphere can also be defined in other ways.

One usually uses the graticule in order to determine the position of a point on the Earth's sphere. In contrast to the graticule on the ellipsoid, meridians are semicircles and all calculations on the sphere are significantly simpler than those on the ellipsoid. *Geographic parameterization* of a sphere with radius R with the centre in the origin of the rectangular coordinate system is mapping determined by formulas

$$X = R \cos\varphi \cos\lambda, \quad Y = R \cos\varphi \sin\lambda, \quad Z = R \sin\varphi. \tag{13.41}$$

Expressions (13.41) are equations of sphere in parametric form. Everything that was said about singular points of geodetic parameterization of a rotational ellipsoid holds true for singular points of geographic parameterization of a sphere.

Let us note two points on the meridian: T_1 with latitude φ_1 and T_2 with latitude φ_2. Considering meridian is a semicircle on a sphere, meridian arc length from point T_1 to T_2 is equal to

$$s_m = R(\varphi_2 - \varphi_1)\,. \tag{13.42}$$

Parallel on a sphere with radius R is a circle with radius $r = R \cos \varphi$. Let us note two points on the parallel: T_1 with longitude λ_1 and T_2 with longitude λ_2. Parallel arc length on the sphere between points T_1 and T_2 is equal to

$$s_p = r(\lambda_2 - \lambda_1) = R \cos \varphi(\lambda_2 - \lambda_1). \tag{13.43}$$

Area P of a curvilinear trapezium on the sphere bound by parallels φ_1 and φ_2 and meridians λ_1 and λ_2 can be calculated by applying the formula

$$P = \int_{\lambda_1}^{\lambda_2} \int_{\varphi_1}^{\varphi_2} R^2 \cos \varphi d\varphi d\lambda = R^2(\lambda_2 - \lambda_1)(\sin \varphi_2 - \sin \varphi_1). \tag{13.44}$$

A special case of the previous formula is the formula according to which the surface of the entire sphere is

$$P = 4R^2\pi. \tag{13.45}$$

All formulas for sphere can be obtained using elementary mathematics or as special cases of corresponding formulas for the rotational ellipsoid if $a = b = R$, i.e. $e = 0$.

References

Benčić D (1990) Geodetski instrumenti (Geodetic Instruments, in Croatian). Školska knjiga, Zagreb, p 548

Helmert FR (1880) Die mathematischen und physikalischen Theorieen der Höheren Geodäsie, Einleitung und I. Teil: Die mathematischen Theorieen. Verlag von B.G. Teubner, Leipzig

Krüger L (1912) Konforme Abbildung des Erdellipsoids in der Ebene. Potsdam: Veröffentlichung des Königlich Preuszischen Geodätischen Institutes, Neue Folge No. 52. BG Teubner, Leipzig

König R, Weise KH (1951) Mathematische Grundlagen der höheren Geodäsie und Kartographie, Erster Band. Springer, Berlin

Lambert JH (1772) Beiträge zum Gebrauche der Mathematik und deren Anwendungen, Dritter Theil, im Verlag der Buchhandlung der Realschule, Berlin. Translated to English with an introduction by WR Tobler: Notes and Comments on the Composition of Terrestrial and Celestial Maps, Michigan Geographical Publication No. 8, Department of Geography, University of Michigan, Ann Arbor, 1972

Lapaine M (1990) Duljina luka meridijana (in Croatian). Geodetski list 4–6:97–108

Lapaine M (1994) Osnovni geodetski zadaci uzduž meridijana i uzduž paralele na rotacijskom elipsoidu (in Croatian). Geodetski list 1:43–52

Lapaine Milj, Lapaine Mir (1991) Površina elipsoidnog trapeza (in Croatian). Geodetski list 1991 (10–12):367–373

Moritz H (1980) Das Geodätische Bezugssystem 1980. Festschrift zur Emeritierung von o.Univ.-Prof. Dipl.-Ing. Dr. techn. Karl Hubeny. Mitt. der geodätischen Institute der Technischen Universität Graz, Folge 35, Graz, pp 153–159
Neidhardt N (1950) Naziv struke (in Croatian). Geodetski list 1–3:72
Vanícek P, Krakiwsky EJ (1987) Geodesy: the concepts, 2nd edn. Elsevier Science, Amsterdam

Appendix
Glossary of Map Projections

Miljenko Lapaine, Nedjeljko Frančula and E. Lynn Usery

All items are listed in the Glossary by alphabetical order. If an item consists of two or more words, the first is always a noun. For example: *Azimuthal projection* is listed as *projection, Azimuthal.* Comma means that the usual order of the word is inverted. The synonym in English is marked with *also*. The advantage is given to the first stated item. If there are synonyms in French and German, they are separated by a semicolon (;). See refers to the terms that were used in the definition of a certain item or are connected with them.

Almucantar
Also: parallel of altitude
Small circle on the surface of the Earth's globe along which all points are equally distant from a point on the globe that we consider a pole of a pseudo-geographic coordinate system.
Note: In astronomy, the circles on the celestial sphere parallel with the horizon.
Fr. parallèle de hauteur
Ger. Netzbreite; Höhenkreis

Aspect of the projection
The position of the projection axis in relation to the geographic sphere parameterization axis.
See: axis of the projections

Aspect, normal
The aspect of a projection in which the projection axis coincides with the geographic sphere parameterization axis.
See: projection, map; aspect of the projection, axis of the projection
Fr. projection directe
Ger. Abbildung, normalachsige

M. Lapaine and E.L. Usery (eds.), *Choosing a Map Projection*,
Lecture Notes in Geoinformation and Cartography,
DOI 10.1007/978-3-319-51835-0

Aspect, oblique

The aspect of a projection that is neither normal nor transverse.

Remark: In the group of perspective projections, these are the projections in which the developable surface axis or perpendicular to the projection plane falls onto the Earth's surface in any point between the geographic pole and the equator.

See: projection, map; surface, developable

Fr. projection oblique

Ger. Abbildung, schiefachsige

Aspect, transverse

The aspect of the projection in which the projection axis is perpendicular to the geographic sphere parameterization axis.

Remark: In the group of perspective projections, these are the projections in which the developable surface axis or the perpendicular to the projection plane is placed in the equator plane.

See: projection, map

Fr. projection transverse

Ger. Abbildung, querachsige

Axis of rotation

The straight line around which a sphere is created by the rotation of a semicircle, or a rotational ellipsoid is created by the rotation of a semiellipse.

Remark: A sphere and a rotational ellipsoid are surfaces by means of which the Earth's form is usualy approximated. The axis of rotation runs through the poles.

Fr. axe de rotation

Ger. Rotationsachse

Axis of the geographic parameterization of a sphere

The straight line intersecting the North and South Poles and the coordinate system origin and centre of the sphere; axis Z in the geographic parameterization of a sphere.

Axis of the projection

The axis of pseudogeographic parameterization of a sphere, based on which the basic equations of map projection are defined. If the basic equations of a map projection are given using geographic coordinates, then the projection axis is identical to the axis of geographic sphere parameterization.

Axis of the pseudogeographic parameterization of a sphere

The straight line intersecting the North and South Pseudopoles and the coordinate system origin and centre of the sphere; axis Z' in the pseudogeographic parameterization of a sphere.

Directions, principal

Also: directions, base

Two mutually perpendicular straight lines in a point on the ellipsoid or sphere and the appropriate mutually perpendicular straight lines in the plane of projection along which the linear scale has extreme values—maximum and minimum.

See: scale, linear
Fr. directions principales
Ger. Hauptverzerrungsrichtungen

Equations of map projection, basic
Map projection equations which define a map projection in a pseudogeographic system.
Note: The selection of basic equations for a map projection is a question of agreement and/or custom. By selecting the basic equations of a map projection, one of its aspects is implicitly or explicitly defined. For cylindrical projections, meridians are represented as parallel straight lines, while parallels are represented as parallel straight lines perpendicular to meridian images. For Robinson's or Winkel Tripel projections, they take the form conceived by their authors. In fact, if we have projection equations in a geographic coordinate system, then we obtain equations in the pseudogeographic system by formally replacing geographic coordinates with pseudogeographic coordinates. However, this still does not guarantee the basic equations of a map projection, because the equations of any projection in any aspect can be written in the geographic coordinate system.

Geodesic
Also: line, geodesic; line, geodetic
Geometrically interpreted, it is the shortest line connecting two points of a not too large area on a surface.
See: orthodrome
Fr. ligne géodésique
Ger. Linie, geodätische

Graticule
Image of coordinate lines in a plane of projection.
Note: The graticule presented by the lines of meridians and parallels is called the basic graticule.
Ger. Kartennetz

Latitude, geographic
(1) Generic (general) term for geodetic and astronomical latitude. (2) The angle between the equatorial plane and the direction of the normal to the Earth's sphere through the given point; regarded as positive Northwards. (3) The parameter φ in the geographic parameterization.
Remark: The geographic latitude for an ellipsoid can be defined analogously.
Fr. latitude géographique
Ger. Breite, geographische

Line, rhumb
Also: line of constant bearing; loxodrome
A line on the rotational surface intersecting all meridians at the same angle.
Remark: Ships sail along the rhumb line when sailing continuously in the same course on their way between two positions.

See: meridian
Fr. loxodromie
Ger. Loxodrome

Longitude, geographic

(1) Generic (general) term for geodetic or astronomic longitude. (2) The angle between the plane of the prime meridian and the plane of the meridian through the given point; it is considered positive Eastward. (3) The parameter λ in the geographic parameterization.
Fr. longitude géographique
Ger. Länge, geographische

Loxodrome

See: line, rhumb

Meridian

(1) Generic (general) term for astronomic and geodetic meridian. (2) A line on the Earth's sphere obtained by the intersection of the sphere with the half-plane with the boundary straightline coinciding with the sphere rotational axis. (3) The parametric curves (semicircles) $\lambda = const.$ connecting the North and South Poles in the geographic parameterization of a sphere.
Remark: The meridian on an ellipsoid can be defined analogously.
See: longitude, geographic
Fr. méridien
Ger. Meridian

Orthodrome

Geodesic on a sphere.
Note: On a sphere, the orthodromes are the arcs of great circles.
See: geodesic
Fr. orthodrome
Ger. Orthodrome

Parallel, geographic

(1) Generic (general) term for astronomic and geodetic parallel. (2) The line on the Earth's sphere obtained by the intersection of the sphere with the plane perpendicular to the rotational axis of the sphere. (3) The parametric curve (circle) $\varphi = const.$ in a plane perpendicular to the Z axis of the geographic parameterization of a sphere.
See: latitude, geographic
Fr. parallèle de latitude
Ger. Breitenkreis; Parallelkreis

Parameterization of a sphere, geographic

Mapping $(\varphi, \lambda) \in \left[-\frac{\pi}{2}, \frac{\pi}{2}\right] \times [-\pi, \pi] \to (X, Y, Z)$ given with the formulae $X = R\cos\varphi\cos\lambda$, $Y = R\cos\varphi\sin\lambda$, $Z = R\sin\varphi$, where R is the given radius.

Parameterization of a sphere, pseudogeographic

The generalization of geographic parameterization of the same sphere obtained by rotation around the origin. The geographic latitude and longitude are transformed into the pseudogeographic latitude and longitude. The two corresponding 3D rectangular coordinate systems X, Y, Z and X', Y', Z' have a mutual origin (0, 0, 0), their coordinate axes are generally going to be placed at certain angles, and the relation between the two systems can be described using a rotation matrix:

$$\begin{bmatrix} X' \\ Y' \\ Z' \end{bmatrix} = \begin{bmatrix} l_1 & m_1 & n_1 \\ l_2 & m_2 & n_2 \\ l_3 & m_3 & n_3 \end{bmatrix} \begin{bmatrix} X \\ Y \\ Z \end{bmatrix},$$

where $l_i, m_i, n_i,\ i = 1, 2, 3$, are cosines of directions of the new axes in relation to the old axes. The connection between pseudogeographic and geographic parameterizations is given by three independent parameters. If the rotation is identity, then the pseudogeographic parameterization and geographic parameterization are identical.

Plane, equatorial

The plane containing the equator.

Fr. plan d'équateur

Ger. Äquatorebene

Plane of projection

The plane into which the surface of the Earth or a celestial body, assumed to be an ellipsoid or sphere, is mapped (projected).

Fr. plan de projection

Ger. Abbildungsebene

Pole, North

The point with coordinates (0, 0, R) in the geographic parametrization of a sphere.

Pole, South

The point with coordinates (0, 0, $-R$) in the geographic parametrization of a sphere.

Projection, Arbitrary

Map projection that is neither equivalent, nor conformal, nor equidistant. In this projection the surface of the Earth's ellipsoid or sphere is mapped into the plane under some special conditions.

See: projection, map; projection, conformal; projection, equivalent; projection, equidistant

Fr. projection aphylactique

Ger. Abbildung, vermittelnde

Projection, Azimuthal

Also: projection, zenithal

Projection in which pseudomeridians are represented as straight lines intersecting at a certain point and forming an angle equal to the difference of corresponding pseudogeographic latitudes, while parallels are concentric circles with a common centre at the point of intersecting pseudomeridians. Azimuthal projection equations in a polar coordinate system in the plane by using the pseudogeographic system are: $\rho = \rho(\varphi'), \delta = \lambda'$.

Remark: Perspective azimuthal projection—special type of azimuthal projection in which the Earth is considered to be a sphere, and the points from the sphere are projected following the laws of linear perspective from the point of view to a projection plane. The projection plane is perpendicular to the line connecting the point of view with the globe centre.

See: projection, map; pseudomeridian; pseudoparallel; system, pseudogeographic coordinate

Fr. projection azimutale

Ger. Azimutalabbildung

Projection, Conformal

Also: projection, orthomorphic

Map projection preserving angles.

Remark: Map projection in which there are no angular distortions. In conformal projection the linear scale in every point is equal in all directions, so in these projection the similarity of infinitesimal parts of the representation is preserved.

See. projection, map; scale, linear

Fr. projection conforme

Ger. Abbildung, konforme; Abbildung, winkeltreue

Projection, Conic

Also: projection, conical

Projections in which pseudomeridians are represented as straight lines intersecting at a certain point, while pseudoparallels are represented as concentric circle arcs, with the angle between any two pseudomeridians being lesser than the corresponding difference of the corresponding pseudogeographic latitudes. Conical projection equations in a polar coordinate system in the plane by using the pseudogeographic system are: $\rho = \rho(\varphi')$, $\delta = k\lambda'$, $0 < k < 1$.

Remark: Perspective conic projection—perspective projection in which the cone is used as a developable surface.

See: projection, map; projection, perspective; pseudomeridian; pseudoparallel; system, pseudogeographic coordinate

Fr. projection conique

Ger. Kegelabbildung

Projection, Cylindrical

Projection in which pseudomeridians are represented as parallel straight lines and pseudoparallels are represented as parallel straight lines perpendicular to meridian images. Cylindrical projection equations in a Cartesian coordinate system in the plane by using the pseudogeographic system are: $y = y(\varphi')$, $x = k\lambda'$, $0 < k \leq 1$.

Remark: Perspective cylindrical projections—perspective projection in which the cylinder is used as a developable surface.

See: projection, perspective; pseudomeridian; pseudoparallel; system, pseudogeographic coordinate

Fr. projection cylindrique

Ger. Zylinderabbildung

Projection, Equidistant

Map projection preserving distances in a particular direction.

Remark: Map projection on which the linear scale along one principal direction is equal to the unit, i.e. in any point there exists a direction with no linear distortion along it.

See: scale, linear; directions, principal

Fr. projection équidistante

Ger. Abbildung, abstandstreue

Projection, Equivalent

Also: projection, equal-area; projection, authalic

Map projection preserving areas.

Remark: An equivalent map projection has the property that in any point the area scale is equal to 1, i.e. there are no area distortions in any point.

See: projection, map

En. projection, equivalent; projection, equal-area

Fr. projection équivalente

Ger. Abbildung, flächentreue

Projection, Gauss-Krüger

Also: projection, Transverse Mercator

Conformal transverse cylindrical projection with the property that the central meridian of the given area is mapped as a straight line and serves as the x axis of the rectangular coordinate system in the plane; the central meridian is mapped without linear distortions or the linear scale along this meridian is constant.

Remark: In English speaking area, the projection is known as the Transverse Mercator projection.

See: projection, conformal; aspect, transverse; projection, cylindrical

En. projection, Gauss-Krüger

Fr. projection de Gauss-Krüger

Ger. Gauss-Krüger-Abbildung

Projection, Geodetic

A map projection of an ellipsoid into a plane which is applied in state survey, numerical processing of geodetic networks, solving various practical geodetic problems and producing topographic maps and plans in larger scales.

Remark: Geodetic projections are usually conformal and most commonly used are the Gauss-Krüger or transverse Mercator projection, the Lambert conic conformal projection and the stereographic projection.

En. projection, geodetic

Fr. projection géodésiqu

Ger. Abbildung, geodätische

Projection, Gnomonic

Also: projection, central

Perspective azimuthal projection in which the point of view is in the globe centre.

Remark: In this projection, the orthodromes are represented as straight lines.

See: projection, azimuthal; orthodrome

Fr. projection gnomonique

Ger. Zentralprojektion; Abbildung, gnomonische

Projection, map

The method of representing the Earth or a celestial body, assumed to be an ellipsoid or sphere, in a plane. It is mostly defined by map projection equations $x = f_1(\varphi, \lambda), y = f_2(\varphi, \lambda)$, where φ, λ are geographic coordinates on the ellipsoid or sphere, and x, y the coordinates in the projection plane. It can also be defined with the table of coordinates or the description of map graticule construction. According to the distortion characteristics, they are classified into conformal, equivalent, equidistant and arbitrary projections. Depending on the orientation of the axis of projection map projections can be divided into normal (direct), transverse and oblique aspects of the projections. According to the shape of the graticule, they are classified into conic, cylindrical, azimuthal, pseudoconic, pseudocylindrical, polyconic, and other projections. They are often named after their authors, e.g. Mercator, Sanson, Robinson. As a special group of map projections we separate geodetic projections, i.e. projections needed in state surveys.

See: graticule, normal; aspect, normal

Fr. projection cartographique

Ger. Abbildung, kartographische

Projection, Mercator

Conformal cylindrical projection.

Remark: Normal aspect has special importance in navigation, because the rhumb lines are represented as straight lines in this projection. Transverse aspect is used in many countries for official cartography. Universal Transverse Mercator (UTM) is used in military (NATO).

See: projection, cylindrical; projection, conformal; aspect, normal; aspect, transverse; line, rhumb; UTM

Fr. projection de Mercator

Ger. Mercatorabbildung

Projection, Orthographic

Perspective azimuthal projection in which the point of view is placed in infinity, so the projection rays are mutually parallel.

See: projection, azimuthal

Fr. projection orthographique

Ger. Abbildung, orthographische; Parallelprojektion

Projection, Perspective

Map projection in which the points from the ellipsoid or sphere are projected following the laws of linear perspective from the point of view into the projection plane or developable surface.

Remark: Of all perspective projections, the azimuthal projections are most often applied in practice, so the term perspective projection often denotes only this group of projections.

See: projection, map; plane, projection; surface, developable

Fr. projection perspective

Ger. Projection

Projection, Polyconic

Map projection on which the pseudomeridians are mapped as curves symmetrical about the straight central meridian, and pseudoparallels as nonconcentric circular arcs with centres on the central meridian.

See: projection, map; aspect, normal; pseudomeridian; pseudoparallel

Fr. projection policonique

Ger. Abbildung, polykonische

Projection, Polyhedric

Map projection in which the Earth's surface is divided by meridians and parallels into ellipsoidal trapeziums; each trapezium is mapped into the plane separately, providing that its sides are mapped as the parts of the straight lines. The lengths of trapezium bases are equal to the lengths of the arcs of the corresponding parallels, while the trapezium altitude or the trapezium legs are equal to the length of the meridian arc between the two parallels.

See: projection, map; meridian; parallel, geographic

Fr. projection polyedrique

Ger. Polyederabbildung

Projection, Pseudoconic

Projections in which pseudomeridians are represented as curves symmetrical to the central pseudomeridian, which is mapped as a straight line, while pseudoparallels are mapped as arcs of concentric circles. Pseudoconic projection equations in a polar coordinate system in the plane by using pseudogeographic system are: $\rho = \rho(\varphi')$, $\delta = \delta(\varphi', \lambda')$

See: pseudomeridian; pseudoparallel

Fr. projection méríconique

Ger. Abbildung, unechtkonische

Projection, Pseudocylindrical

Projection in which pseudomeridians are represented as curves symmetrical to the central pseudomeridian, which is mapped as a straight line, while pseudoparallels are represented as parallel straight lines perpendicular to the central pseudomeridian image. Pseudocylindrical projection equations in the Cartesian coordinate system in the plane by using pseudogeographic system are: $y = y(\varphi')$, $x = x(\varphi', \lambda')$.

See: pseudomeridian; pseudoparallel
Fr. projection méricylindrique
Ger. Abbildung, unechtzylindrische

Projection, Transverse Mercator

(1) Map projection that is a Mercator projection and in transverse aspect. (2) In the English speaking area it is the name for the Gauss-Krüger projection.

See: projection, Mercator; aspect, transverse; projection, Gauss-Krüger
Fr. projection transverse de Mercator
Ger. Mercatorabbildung, transversale

Pseudomeridian

The parametric curves (semicircles) $\lambda' = const.$ connecting the North and South Pseudopoles in the pseudogeographic parameterization of a sphere.

Pseudoparallel

The parametric curve (circle) $\varphi' = const.$ in a plane perpendicular to the Z' axis of the pseudogeographic parameterization of a sphere.

Scale, linear

Also: scale factor

The ratio of the differential of the arc length in the plane of projection and the corresponding differential on the surface of the Earth or a celestial body, assumed to be an ellipsoid or sphere.

Note: Linear scale varies from point to point on a map, and is different in every direction in any given point, which is why we differentiate: linear scale along the meridian, linear scale along the parallel, linear scale along principal directions. If at some point in a certain direction there are no linear distortions, the linear scale is equal to the unit.

See: directions, principal
Fr. échelle des longueurs
Ger. Längenmaßstab

Surface, developable

The surface that can be developed into the plane (cone and cylinder) into which the points are projected from the globe or ellipsoid surface in perspective conic and cylindrical projections.

See: projection, perspective
Fr. surface auxiliaire de projection
Ger. Hilfsabbildungsfläche

System, geographic coordinate
The coordinate system on a sphere defined by the geographic parameterization of a sphere.

System, pseudogeographic coordinate
The coordinate system on a sphere defined by the pseudogeographic parameterization of a sphere.

System, UTM (Universal Transverse Mercator) coordinate
Sixty systems of the transverse Mercator projection with each of them covering the area of six degrees of longitude. The point of origin in each system is in the intersection of the central meridian with the longitude 3°, 9°, 15° etc. and the equator. Linear scale along the central meridian is 0.9996.
See: projection, transverse Mercator
Fr. projection UTM
Ger. UTM-Abbildung

Vertical
Every great circle on the sphere passing through the pole of the adopted pseudogeographic coordinate system.
Remark: In astronomy, great circles on the sky sphere passing through zenith.
Fr. cercle vertical
Ger. Netzmeridian; Vertikalkreis

References
Bugayevskiy LM, Snyder JP (1995) Map projections—a reference manual. Taylor & Francis, London

Canters F (2002) Small-scale map projection design. Taylor & Francis, London

Frančula N, Lapaine M (2003) Dictionary of geodesy and geoinformatics (Geodetsko geoinformatički rječnik). State Geodetic Administration, Zagreb

Frančula N, Lapaine M (2016) Map projections aspect. Int J Cartogr 2(1):38–58. doi: 10.1080/23729333.2016.1184554

Index

M. Lapaine and E.L. Usery (eds.), *Choosing a Map Projection*,
Lecture Notes in Geoinformation and Cartography,
DOI 10.1007/978-3-319-51835-0

Kolibri 360 GmbH
Ferdinand-Jühlke-Straße 7
99095 Erfurt, Deutschland
produktsicherheit@kolibri360.de